Abiotic Stress and Plant Responses

Abiotic Stress and Plant Responses

Editor: Zeke Adams

MURPHY & MOORE
www.murphy-moorepublishing.com

www.murphy-moorepublishing.com

ⓂMURPHY & MOORE

Cataloging-in-Publication Data

Abiotic stress and plant responses / edited by Zeke Adams.
   p. cm.
Includes bibliographical references and index.
ISBN 978-1-63987-755-3
1. Plants--Effect of stress on. 2. Plant physiology. 3. Plant genomes. I. Adams, Zeke.
QK754 .A25 2023
581.7--dc23

Murphy & Moore Publishing
1 Rockefeller Plaza,
New York City,
NY 10020, USA

ISBN 978-1-63987-755-3

Contents

Preface

This book has been an outcome of determined endeavour from a group of educationists in the field. The primary objective was to involve a broad spectrum of professionals from diverse cultural background involved in the field for developing new researches. The book not only targets students but also scholars pursuing higher research for further enhancement of the theoretical and practical applications of the subject.

Crops are exposed to frequent episodes of abiotic stresses such as soil salinity, draughts, extreme temperatures and nutrient deficiencies. All these stresses inhibit plant growth and development. A number of researches are being undertaken to find mitigation strategies for abiotic stresses and increase yield productivity under unfavorable conditions. Genetic engineering is a major tool that is being used to develop crop plants, which are resistant to these stresses. Plants respond to different types of stresses through complex and diverse responses on cellular, molecular and physiological levels. Some of the methods which have been developed to increase crop resistance to abiotic stresses include genetically engineered organisms (GMO's) and new breeding technologies such as clustered regularly interspaced short palindromic repeats (CRISPR/Cas) edition strategy. The application of biostimulants is also an effective method to improve abiotic stress tolerance in plant species. This book provides a comprehensive understanding of plant responses to abiotic stresses. A number of latest researches have been included to keep the readers updated with the latest concepts in this area of study. Students, researchers, experts and all associated with agriculture and plant science will benefit alike from this book.

It was an honour to edit such a profound book and also a challenging task to compile and examine all the relevant data for accuracy and originality. I wish to acknowledge the efforts of the contributors for submitting such brilliant and diverse chapters in the field and for endlessly working for the completion of the book. Last, but not the least; I thank my family for being a constant source of support in all my research endeavours.

Editor

Function and Mechanism of WRKY Transcription Factors in Abiotic Stress Responses of Plants

Weixing Li [†][ID], Siyu Pang [†], Zhaogeng Lu and Biao Jin *[ID]

College of Horticulture and Plant Protection, Yangzhou University, Yangzhou 225009, China;
liwx@yzu.edu.cn (W.L.); pangsiyu_0212@163.com (S.P.); luzhaogeng@163.com (Z.L.)
* Correspondence: bjin@yzu.edu.cn
† Contributed equally to this work.

Abstract: The WRKY gene family is a plant-specific transcription factor (TF) group, playing important roles in many different response pathways of diverse abiotic stresses (drought, saline, alkali, temperature, and ultraviolet radiation, and so forth). In recent years, many studies have explored the role and mechanism of WRKY family members from model plants to agricultural crops and other species. Abiotic stress adversely affects the growth and development of plants. Thus, a review of WRKY with stress responses is important to increase our understanding of abiotic stress responses in plants. Here, we summarize the structural characteristics and regulatory mechanism of WRKY transcription factors and their responses to abiotic stress. We also discuss current issues and future perspectives of WRKY transcription factor research.

Keywords: WRKY transcription factor; abiotic stress; gene structural characteristics; regulatory mechanism; drought; salinity; heat; cold; ultraviolet radiation

1. Introduction

As a fixed-growth organism, plants are exposed to a variety of environmental conditions and may encounter many abiotic stresses, for example, drought, waterlogging, heat, cold, salinity, and Ultraviolet-B (UV-B) radiation. To adapt and counteract the effects of such abiotic stresses, plants have evolved several molecular mechanisms involving signal transduction and gene expression [1,2]. Transcription factors (TFs) are important regulators involved in the process of signal transduction and gene expression regulation under environmental stresses. TFs can be combined with *cis*-acting elements to regulate the transcriptional efficiency of target genes by inhibiting or enhancing their transcription [3,4]. Accordingly, plants may show corresponding responses to external stresses via TFs regulating target genes. Although some TF families (MYB, bZIP, AP2/EREBP, NAC) are associated with adversity [2,5], WRKY is the most extensively studied TF family in plant stress responses.

The WRKY family is a unique TF superfamily of higher plants and algae, which play important roles in many life processes, particularly in response against biotic and abiotic stress [6,7]. In 1994, the SWEET POTATO FACTOR1 (*SPF1*) gene of the WRKY family was first found in *Impoea batatas* [8]. Later, *ABF1* and *ABF2* were found in wild *Avena sativa*, and showed regulatory roles in seed germination [9]. A previous study successively cloned *WRKY1*, *WRKY2*, and *WRKY3* from *Petroselinum crispum*, named the WRKY TF, and proved for the first time that WRKY protein can regulate plant responses to pathogens [10]. With an increase in available published genomes, many members of the WRKY TF family have been identified in various species, including 104 from *Populus* [11], 37 from *Physcomitrella patens* [12], 45 from *Hordeum vulgare* [13], 55 from *Cucumis sativus* [14], 74 from *Arabidopsis thaliana* [15], 83 from *Pinus monticola* [16], 81 from *Solanum lycopersicum* [17], and 102 from *Oryza sativa* [18]. WRKY TFs exist as gene families in plants, and the number of WRKY TFs varies among species. In plants exposed to abiotic stresses (salt, drought, temperature, and so forth), WRKY family members play important roles in diverse stress responses. In addition, these TFs affect the growth and development of plants [19,20]. Therefore, WRKY TFs have attracted broad attention. Although some reviews on WRKY TFs are available, in this review we focus on the structural characteristics and regulatory mechanisms of WRKY TFs and summarize recent progress in understanding the roles of WRKY TFs during exposure to abiotic stresses such as drought, temperature, salt, and UV radiation.

2. Structural Characteristics of WRKY TFs

The WRKY structure consists of two parts: the N-terminal DNA binding domain and the C-terminal zinc-finger structure [21]. The DNA binding domain sequence of WRKY is based on the heptapeptide WRKYGQK (Figure 1), but there are some differences, such as WRKYGQK, WRKYGKK, WRKYGMK, WSKYGQK, WKRYGQK, WVKYGQK, and WKKYGQK [17,22]. Zinc-finger structures mainly include C_2H_2 type and C_2HC type [23], whereas some exist in the form of $CX_{29}HXH$ and $CX_7CX_{24}HXC$ [17] (Figure 1). According to the number of WRKY domains and the structure of their zinc-finger motifs, WRKY can be divided into groups I, II, and III [23] (Figure 1). Group I usually contains two WRKY domains and one C_2H_2 zinc-finger structure. Those in group II and group III contain only one WRKY domain. The difference is that the zinc-finger structure in group II is C_2H_2 and that in group III is C_2HC [19,21,23] (Figure 1). According to the phylogenetic relationship of the amino acid sequence of the primary structure, group II can be further divided into subgroups a–e [7,23,24]. Evolutionary analyses have shown that the WRKY of group II is not generally a single source, mainly including five categories I, IIa + IIb, IIc, IId + IIe, and III [7,24]. In addition, some WRKY proteins contain a glutamate enrichment domain, a proline enrichment domain, and a leucine zipper structure [25].

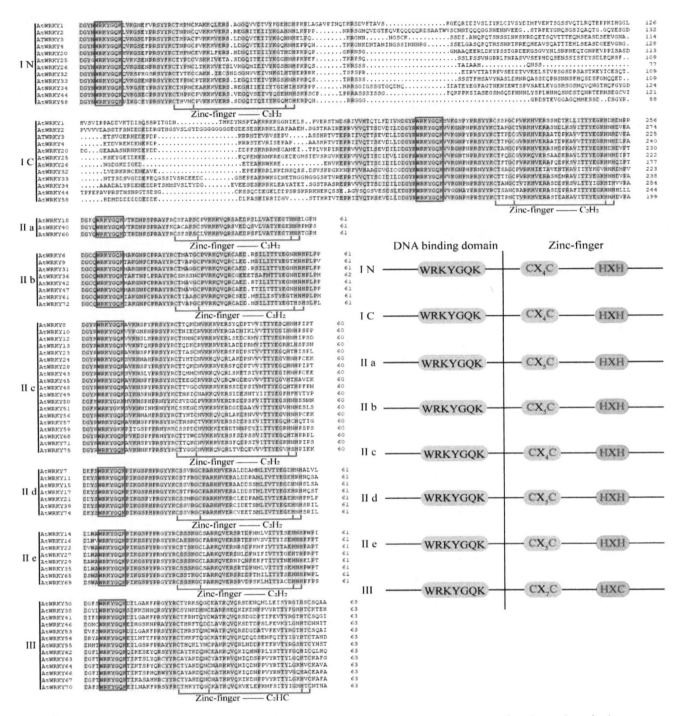

Figure 1. The domain of WRKY genes in *Arabidopsis thaliana*. The WRKY gene family is classified into the **I** (**I N** and **I C**), **IIa**, **IIb**, **IIc**, **IId**, **IIe**, and **III** subfamilies. The aligned conserved domains (DNA binding and zinc-finger structures) are highlighted (left panel) and simplified (right panel).

3. Regulatory Mechanism of WRKY TFs

WRKY family members have diverse regulatory mechanisms. Briefly, WRKY protein can be effectively combined with W-box elements to activate or inhibit the transcription of downstream target genes. Moreover, it can also bind other acting elements to form protein complexes, which enhances the activity of transcription binding [21].

WRKY TFs can effectively activate the expression of downstream genes by binding conserved W-box *cis*-acting elements in the downstream gene promoter region [21,26]. There are abundant W-box elements in the self-promoter of most WRKY TFs. Therefore, these WRKY TFs can bind with their own promoters to achieve self-regulation or cross-regulation networks by combining with other WRKY TFs [27]. For example, *CaWRKY6* of *Capsicum frutescens* can activate *CaWRKY40* and make the plant more tolerant to high temperature and humidity. *Glycine max GmWRKY27* not only inhibits the activity of downstream *GmNAC29* promoter by independent inhibition, but also cooperatively interacts with *GmMYB174* to inhibit the expression of *GmNAC29*, thereby increasing drought and salt stress resistances [28]. Moreover, chromatin immunoprecipitation (ChIP) studies have shown that when *Petroselinum crispum* is infected by pathogenic bacteria, *PcWRKY1* promoter can effectively bind to itself and the W-box of *PcWRKY3* promoter, and transcriptional activation can be achieved through self-negative feedback regulation and cross-regulation with other WRKY proteins [29]. In addition, WRKY TFs can interact with non-W-box elements. For example, *Oryza sativa OsWRKY13* can interact with PRE4 (TGCGCTT) elements [30]. *Hordeum vulgare HvWRKY46* and *Nicotiana tabacum NtWRKY12* can effectively combine with the sucrose response element SURE [31,32]. These results indicate that there are multiple binding modes between WRKY TFs and structural genes. Different binding patterns and preferences of binding sites allow for the regulation of downstream target genes, providing WRKY TFs with versatile functions in the plant transcriptional regulation network.

4. WRKY TF Involved in Abiotic Stress Responses

When plants sense stress, the corresponding signaling is activated and transferred to the cell interior. Reactive oxygen species (ROS) and Ca^{2+} ions are usually exchanged as the signal transduction in the cell. Protein kinases such as MPKs are subsequently activated to regulate the activities of related TFs. Consequently, the plant presents a stress response [31,32]. In response to abiotic stresses, some WRKY TFs can be rapidly differentially expressed, promoting signal transduction and regulating the expression of related genes [33]. The expression patterns and functional identifications of WRKYs in most studies are generally based on transcriptome analyses, real-time fluorescence quantitative PCR, gene chip analyses, and genetic transformation. Hence, WRKY genes can function effectively in most abiotic stress responses or tolerances in various plants (Table 1, Figure 2).

Table 1. WRKY transcription factors (TFs) involved in abiotic stress responses in plants.

No.	Gene	Species	Induced by Factors	Function	References
1	*AtWRKY25/26*	*Arabidopsis*	Heat	Tolerance to heat	[34]
2	*AtWRKY33*	*Arabidopsis*	NaCl, mannitol, H_2O_2	Tolerance to heat and NaCl, negative regulator in oxidative stress and abscisic acid (ABA)	[33]
3	*AtWRKY34*	*Arabidopsis*	Cold	Negative regulator in cold stress	[35]
4	*AtWRKY39*	*Arabidopsis*	Heat	Tolerance to heat	[36]
5	*AtWRKY53*	*Arabidopsis*	Drought, salt	Reduced drought resistance and H_2O_2, sensitive to salt	[37,38]
6	*AtWRKY57*	*Arabidopsis*	Drought	Tolerance to drought	[39]
7	*AtWRKY63*	*Arabidopsis*	ABA	Tolerance to drought, regulated ABA signaling	[40]
8	*AtWRKY54*	*Arabidopsis*	Heat	Response to heat stress	[41]
9	*POWRKY13*	*Populus tomentosa*	Heat	Response to heat stress	[42]
10	*GhWRKY21*	*Gossypium hirsutum*	Drought	Tolerance to drought	[43]
11	*GhWRKY25*	*Gossypium hirsutum*	Drought	Tolerance to salt, reduced drought resistance	[44]
12	*GhWRKY68*	*Gossypium hirsutum*	Salt, drought	Reduced salt tolerance and drought resistance, positive regulator in ABA signaling	[45]
13	*VvWRKY24*	*Vitis vinifera*	Cold	Upregulated expression at all stages of hypothermia	[46]

Table 1. *Cont.*

No.	Gene	Species	Induced by Factors	Function	References
14	CaWRKY40	Capsicum annuum	Heat	Tolerance to heat	[47]
15	BdWRKY36	Brachypodium distachyon	Drought	Tolerance to drought	[48]
16	FcWRKY70	Fortunella crassifolia	Salt	Tolerance to salt	[49]
17	OsWRKY11	Oryza sativa	Heat, drought	Tolerance to drought and heat	[50]
18	OsWRKY72	Oryza sativa	Drought, NaCl, ABA	Sensitive to salt, drought, sucrose, and ABA	[51]
19	OsWRKY74	Oryza sativa	Pi deprivation, cold	Tolerance to cold and Pi deprivation	[52]
20	OsWRKY76	Oryza sativa	Cold	Tolerance to cold	[53]
21	OsWRKY89	Oryza sativa	ABA, UV-B	Tolerance to UV	[54]
22	GmWRKY13	Soybean	Salt, drought	Sensitive to salt and mannitol, negative regulator in ABA signaling	[55]
23	GmWRKY17	Soybean	Salt	Reduced salt tolerance	[56]
24	GmWRKY54	Soybean	Salt, drought	Tolerance to salt and drought	[55]
25	GmWRKY21	Glycine max	NaCl, drought, cold	Tolerance to cold	[55]
26	ZmWRKY17	Zea mays	ABA, salt	Reduced salt tolerance	[57]
27	TaWRKY2/19	Triticum aestivum	NaCl, drought, ABA	Tolerance to salt and drought	[58]
28	BcWRKY46	Brassica campestris	NaCl, drought, cold	Tolerance to salt and drought	[59]
29	BhWRKY1	Boea hygrometrica	Dehydration, ABA	Tolerance to drought	[60]
30	VpWRKY1	Vitis pseudoreticulata	NaCl, ABA	Tolerance to salt	[61]
31	VpWRKY2	Vitis pseudoreticulata	Cold, NaCl, ABA	Tolerance to salt and cold	[61]
32	VpWRKY3	Vitis pseudoreticulata	Drought, ABA, salicylic acid (SA)	Tolerance to salt	[62]
33	TcWRKY53	Thlaspi caerulescens	Cold, PEG, NaCl	Negative regulator in osmotic stress	[63]
34	NaWRKY3	Nicotiana attenuate	Mechanical damage	Sensitive to mechanical damage	[64]
35	JrWRKY2/7	Juglans regia	Drought, cold	Tolerance to drought and cold	[65]
36	SbWRKY30	Sorghum bicolor	Salt, drought	Tolerance to salt and drought	[66]
37	SbWRKY50	Sorghum bicolor	Salt	Tolerance to salt	[67]
38	IbWRKY47	Ipomoea batatas	Salt	Tolerance to salt	[68]
39	IbWRKY2	Ipomoea batatas	Salt, drought	Tolerance to salt and drought	[69]
40	MdWRKY30	Malus domestica	Salt, osmotic stress	Tolerance to salt and osmotic stress	[70]
41	MdWRKY100	Malus domestica	Salt	Sensitive to salt	[71]
42	SlWRKY81	Solanum lycopersicum	Drought	Reduced drought tolerance	[72]
43	GbWRKY1	Gossypium barbadense	Salt	Tolerance to salt	[73]
44	VbWRKY32	Verbena bonariensis	Cold	Tolerance to cold	[74]
45	PgWRKY33/62	Pennisetum glaucum	Salt, drought	Tolerance to salt and drought	[75]
46	PagWRKY75	Populus alba	Drought	Negative regulator in salt and osmotic stress	[76]

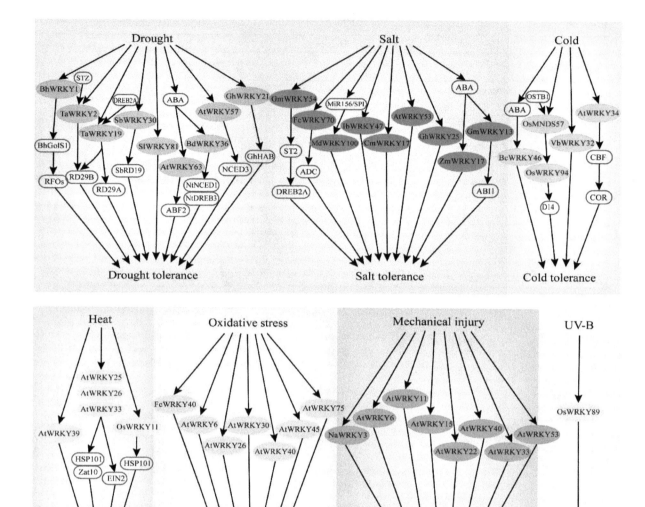

Figure 2. Some WRKY genes involved in the response pathways of major abiotic stresses (drought, salt, cold, heat, oxidative stress, mechanical injury, UV-B).

4.1. WRKY TFs and Drought Stress

Drought has a major impact on plant growth and development, resulting in a significant decrease in grain and other types of crop yield [77]. Under drought stress, drought-tolerant plants can accumulate oligosaccharides through sucrose metabolism to improve drought resistance. For example, when *Arabidopsis* is subjected to drought stress, the expression of *AtWRKY53* combined with the Qua-Quine Starch (QQS) promoter sequence is rapidly induced, hydrogen peroxide content is reduced, and the glucose metabolism pathway is significantly enhanced, thereby regulating stomatal opening and ultimately affecting drought tolerance [37]. In *Boea hygrometrica*, *BhWRKY1* effectively regulates the expression of the *BhGolS1* gene, and the overexpression of *BhGolS1* and *BhWRKY1* induces the accumulation of raffinose family oligosaccharides (RFOs) in transgenic *Nicotiana tabacum*, thus improving the ability of seedlings to resist drought [60].

WRKY protein can directly regulate the expression of drought-resistant genes. For example, in sorghum, *SbWRKY30* regulates the drought stress response gene *SbRD19* by binding with W-box elements of the *SbRD19* promoter, and protects plant cells from the damage of reactive oxygen species by improving ROS scavenging capability, enhancing drought tolerance [66]. *TaWRKY2* of wheat can bind to *STZ* and downstream drought-resistant gene *RD29B* promoter, with a positive regulatory effect on the expression of *RD29B* [58]. *DREB2A* regulates the expression of dehydration stress-related genes [78], while *TaWRKY19* can bind to *DREB2A* promoter, ultimately activating the expression of *RD29A*, *RD29B*, and *Cor6.6* in transgenic *Arabidopsis* plants [58]. Similarly, *Arabidopsis AtWRKY57* positively regulates the expression of *RD29A* and *NCED3* genes by binding their W-box elements in the promoter regions [39]. In addition, WRKY protein can act on other TFs to play regulatory roles in drought tolerance. For example, *TcWRKY53* of *Thlaspi arvense* significantly inhibits the expression of *NtERF5* and *NterEBp-1* of the AP2/ERF TF family, thus improving plant resistance to drought stress [63].

WRKY TFs also regulate plant tolerance through abscisic acid (ABA) and ROS-related signaling pathways. During drought stress, higher ABA levels were accumulated in plants, and leaf stomata were closed to reduce transpiration rate, thus regulating water balance in plants. ABA accumulation in cells, integrated with a variety of stress signals, regulates the expression of downstream genes, consequently sensing and responding to the adverse environment [40]. *Arabidopsis AtWRKY63* has a specific effect on ABA-mediated stomatal closure and other signal transduction pathways, thus affecting the drought response [40]. *GhWRKY21* regulates ABA-mediated cotton drought tolerance by promoting the expression of *GhHAB* [43]. Overexpression of *BdWRKY36* in tobacco reduces the accumulation of ROS, activated *NtLEA5*, *NtNCED1*, and *NtDREB3* in the ABA biosynthetic pathway, and significantly enhances the drought resistance of plants [48]. In *Solanum lycopersicum*, *SlWRKY81* increases the drought tolerance of plants by inhibiting the accumulation of H_2O_2, playing a negative regulation role of stomatal closure [72].

4.2. WRKY TFs and Salt Stress

Salt stress is an important abiotic stress affecting crop productivity, particularly in arid and semiarid regions. WRKY TFs play essential roles in regulating the response to salt stress. To date, a total of 47 WRKY genes have been found to be expressed under salt stress in the wheat genome [79]. *STZ* is a protein related to ZPT2, which acts as a transcriptional inhibitor to downregulate the deactivation of other transcription factors. *GmWRKY54* of *Glycine max* inhibits *STZ* expression and responds to salt stress by positively regulating the DREB2A-mediated pathway [55]. *FcWRKY70* promotes the upregulation of arginine decarboxylase (ADC) expression, which is heterologously expressed in tobacco, and the content of lemon putrescine is significantly increased, thus enhancing the salt tolerance of plants [49]. The *IbWRKY47* gene positively regulates stress resistance-related genes and significantly improves the salt tolerance of *Ipomoea batatas* [68]. MiR156/SPL modulates salt tolerance by upregulation of *Malus domestica* salt tolerance gene *MdWRKY100* [71]. In *Sorghum bicolor*, *SbWRKY50* could directly bind to the upstream promoter of *SOS1* and *HKT1* and participate in plant salt response by controlling ion homeostasis [67]. In addition, some *WRKY* genes function as negative regulation factors involved in salt stress resistance. *Arabidopsis* RPD3-like histone deacetylase HDA9 inhibits salt stress tolerance by regulating the DNA binding and transcriptional activity of *WRKY53* [38]. *Chrysanthemum CmWRKY17* overexpressed in *Arabidopsis* allows the plants to be more sensitive to salt stress. The expression level of stress resistance-related genes in transgenic *Arabidopsis* is lower than that in wild-type plants, indicating that *CmWRKY17* may be involved in negatively regulating the salt stress response in *Chrysanthemum* [80]. The expression of *GhWRKY68* is strongly induced in upland cotton and decreases salt tolerance [45]. In contrast, a high expression level of *GhWRKY25* enhances the salt tolerance of upland cotton, while transgenic tobacco shows a relatively weaker tolerance to drought stress [44], indicating that the regulatory effects of different WRKY TFs involved in drought response are different.

Plants can also respond to saline–alkali stress through ABA, H_2O_2, and other signal pathways. In *Glycine max*, the negative regulatory factor *ABI1* in the ABA pathway may be the downstream target gene of *GmWRKY13*. Genetic transformation experiments in *Arabidopsis* have shown that overexpression of *GmWRKY13* significantly increases the expression of *ABI1*, but plants show a low tolerance to salt stress [55]. Overexpression of *ZmWRKY17* has an inhibitory effect on the sensitivity of exogenous ABA treatment, resulting in a relatively lower tolerance to high levels of salinity [57]. Under salt-induced H_2O_2 and cytosolic Ca^{2+} stimulation, the activity of antioxidant enzymes increases, thus improving the tolerance to high-salinity environments [81]. ABA-induced WRKY gene expression is largely related to salt stress. Exogenous application of ABA and NaCl also induce *AtWRKY25* and *AtWRKY33* in *Arabidopsis* [33], *OsWRKY72* in rice [51], *GbWRKY1* in *Verbena bonariensis* [73], and *VpWRKY1/2* [61] and *VpWRKY3* [62] in grape.

4.3. WRKY TFs and Temperature Stress

Both low- and high-temperature stress can reduce crop yield and quality in plants. WRKY TFs play a role in the stress response through different signal transduction pathways. For example, in *Verbena bonariensis*, *VbWRKY32* as a positive regulator, upregulates the transcriptional level of cold response genes, which increases the antioxidant activity, maintains membrane stability, and enhances osmotic regulation ability, thereby improving the survival ability under cold stress [74]. The *BcWRKY46* gene of *Brassica campestris* is strongly induced by low temperature and ABA, activating related genes in the ABA signaling pathway to improve the low-temperature tolerance of plants [59]. *CBF* TFs regulate the expression of *COR*, and the overexpressed transgenic lines of *CBF1*, *CBF2*, and *CBF3* show stronger cold resistance [82]. *AtWRKY34* has a negative regulatory effect on the CBF-mediated cold response pathway; it is specifically expressed in mature pollen grains after exposure to low temperatures, resulting in resistance to low temperatures [35]. In addition, plants respond to temperature changes by coordinating organ development in an adverse environment. At low temperatures, rice MADS-Box TF *OsMADS57* and its interacting protein *OsTB1* synergistically activate the transcriptional regulation of *OsWRKY94*, preventing tillering by inhibiting transcription of the organ development gene *D14* [83].

Due to global climate change, high-temperature stress has attracted significant attention. There is evidence that, to a certain extent, high temperatures will lead to biochemical changes in plants [84]. Thermal stimulation can activate Ca^{2+} channels to maintain a higher intracellular Ca^{2+} concentration, thereby activating calmodulin protein expression and inducing thermal-shock protein transcriptional expression [85]. In *Arabidopsis*, *AtWRKY54* significantly responds to heat shock whereas basic leucine zipper factors (bZIPs) respond to prolonged warming [41]. Overexpression of *AtWRKY39* can make plants more heat-sensitive. *AtWRKY39* is highly homologous to *AtWRKY7*, and both of them can effectively bind calmodulin in plants, indicating a similar function [36]. In addition, *AtWRKY25*, *AtWRKY26*, and *AtWRKY33* can improve tolerance to high-temperature stress in transgenic *Arabidopsis* by regulating the *Hsp101* and *Zat10* genes [34]. Plants subjected to heat stress can also activate the oxidative stress response through ethylene [86]. Under high-temperature stress, the expressions of *AtWRKY25*, *AtWRKY26*, and *AtWRKY33* in *Arabidopsis* are induced by ethylene, the feedback factor *EIN2* is transcriptionally regulated, and the effective activation of ethylene signal transduction contribute to relatively stronger heat resistance. In *Oryza sativa*, *HSP101* promoter can activate the expression of the *OsWRKY11* gene. Under heat treatment, the leaves wilted more slowly and the green part of the plant was less damaged, which makes it more heat-resistant [50]. In addition, some noncoding RNAs, such as miR396, play a role in the response to heat stress by regulating its target *WRKY6* [87].

4.4. WRKY TFs and Other Abiotic Stresses

WRKY TFs are also involved in oxidative stress, mechanical damage, UV radiation, and other abiotic stresses (Figure 3). *FcWRKY40* overexpression can significantly enhance the resistance of transgenic tobacco to oxidative stress [88]. When *Arabidopsis* is treated with ROS, the expressions of *AtWRKY30*, *AtWRKY40*, *AtWRKY75*, *AtWRKY6*, *AtWRKY26*, and *AtWRKY45* are significantly upregulated [89]. After mechanical injury, the expression levels of *AtWRKY11*, *AtWRKY15*, *AtWRKY22*, *AtWRKY33*, *AtWRKY40*, *AtWRKY53* [90] and *AtWRKY6* [64] are upregulated. Similarly, *NaWRKY3* is strongly expressed in tobacco. By contrast, the sensitivity of transgenic plants is increased when *NaWRKY3* is knocked out [64]. In two previous studies, UV-B radiation treatment induced three WRKY genes in *Arabidopsis* and the *OsWRKY89* gene in rice, resulting in a thick waxy substance on the leaf surface and improved tolerance to heat [54,91].

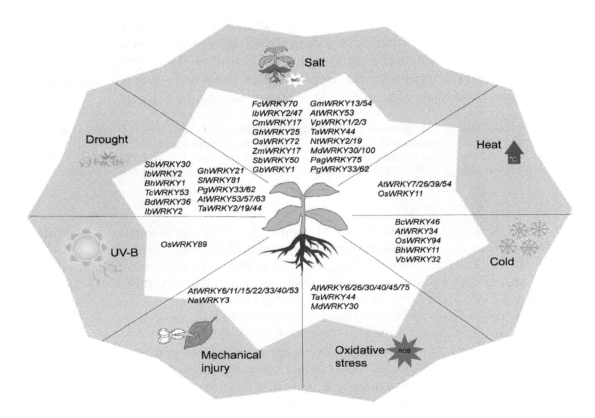

Figure 3. WRKY transcription factors in response to abiotic stresses.

In addition, a single WRKY TF can play multiple roles in different stress responses via various signal pathways and regulatory networks. For example, *TaWRKY44* expression in tobacco can improve resistance to drought, salt stress, and osmotic stress [92], while *PgWRKY62* and *PgWRKY33* in *Pennisetum glaucum* respond to salt and drought simultaneously [75]. *BhWRKY1* protein in *Boea hygrometrica* binds to the promoter of *BhGolS1* and is associated with both low-temperature resistance and drought tolerance [60]. *IbWRKY2* can interact with *IbVQ4*, and drought and salt treatment can induce the expression of *IbVQ4*, thus improving the tolerance of plants to drought and salt stress [69]. *MdWRKY30* overexpression enhances tolerance to salt and osmotic stress in transgenic apple callus through transcriptional regulation of stress-related genes [70]. *PagWRKY75* negatively regulates the tolerance of 84 K poplar (*Populus alba* × *P. glandulosa*) to salt and osmotic stress by reducing the scavenging capacity of ROS and the accumulation of proline, thus actively regulates the rate of leaf water loss [76].

5. Conclusions and Perspectives

As one of the largest TF families, WRKY plays an important and indispensable role in normal life activities of plants. Over the years, it has been shown that WRKY TFs not only participate in plant growth and development, but also show complex regulatory mechanisms and networks involved in external abiotic stresses. A large number of WRKYs have been functionally characterized in model plants, providing abundant functional references for other plants. Given that crops usually face various stresses and WRKYs play important roles in stress responses, further in-depth studies on WRKY genes in more crops are required. As increasing plant genomes have been sequenced, particularly of economically important crops, the genome-wide identification of WRKY genes will facilitate screening for stress resistance-related functional genes in plants. Moreover, previous studies of WRKY gene functions were largely dependent on transcriptomics and functional predictions, whereas more applications of genetic verification combined with new technologies are accelerating the research progress of WRKY's novel functions. In addition, characterization of the downstream genes regulated by WRKY TFs or WRKY TF self-regulation will help clarify the regulatory network of abiotic stress responses. Furthermore, noncoding RNAs and epigenetic modifications involved in the regulation of WRKY TFs should be explored in future studies. Ultimately, using WRKY TFs to screen for stress-resistant plant cultivars and improve plant stress resistance will significantly benefit agricultural crop yield and quality in the context of aggravated climate change.

Author Contributions: Conceptualization, W.L. and B.J.; methodology, B.J.; software, S.P.; formal analysis, S.P. and Z.L.; investigation, S.P. and Z.L.; resources, S.P.; writing—original draft preparation, S.P. and W.L.; writing—review and editing, B.J., S.P., Z.L. and W.L.; visualization, B.J. and S.P.; supervision, B.J. and W.L.; project administration, W.L.; funding acquisition, W.L. All authors have read and agreed to the published version of the manuscript.

References

1.   Yoon, Y.; Seo, D.H.; Shin, H.; Kim, H.J.; Kim, C.M.; Jang, G. The role of stress-responsive transcription factors in modulating abiotic stress tolerance in plants. *Agronomy* **2020**, *10*, 788. [CrossRef]

2.   Ma, Q.; Xia, Z.; Cai, Z.; Li, L.; Cheng, Y.; Liu, J.; Nian, H. *GmWRKY16* enhances drought and salt tolerance through an ABA-mediated pathway in *Arabidopsis thaliana*. *Front. Plant Sci.* **2019**, *9*, 9. [CrossRef]

3.   Liu, Y.; Yang, T.; Lin, Z.; Guo, B.; Xing, C.; Zhao, L.; Dong, H.; Gao, J.; Xie, Z.; Zhang, S.-L.; et al. A WRKY transcription factor *PbrWRKY53* from *Pyrus betulaefolia* is involved in drought tolerance and AsA accumulation. *Plant Biotechnol. J.* **2019**, *17*, 1770–1787. [CrossRef] [PubMed]

4.   Shrestha, A.; Khan, A.; Dey, N. cis-trans Engineering: Advances and perspectives on customized transcriptional regulation in plants. *Mol. Plant* **2018**, *11*, 886–898. [CrossRef] [PubMed]

5.   Yamasaki, K.; Kigawa, T.; Inoue, M.; Watanabe, S.; Tateno, M.; Seki, M.; Shinozaki, K.; Yokoyama, S. Structures and evolutionary origins of plant-specific transcription factor DNA-binding domains. *Plant Physiol. Biochem.* **2008**, *46*, 394–401. [CrossRef] [PubMed]

6.   Zhang, Y.; Wang, L. The WRKY transcription factor superfamily: Its origin in eukaryotes and expansion in plants. *BMC Evol. Biol.* **2005**, *5*, 1. [CrossRef]

7.   Pandey, S.P.; Somssich, I.E. The role of WRKY transcription factors in plant immunity. *Plant Physiol.* **2009**, *150*, 1648–1655. [CrossRef] [PubMed]

8.   Ishiguro, S.; Nakamura, K. Characterization of a cDNA encoding a novel DNA-binding protein, *SPF1*, that recognizes SP8 sequences in the 5′ upstream regions of genes coding for sporamin and beta-amylase from sweet potato. *Mol Gen Genet* **1994**, *244*, 563–571. [CrossRef]

9.   Rushton, P.J.; Macdonald, H.; Huttly, A.K.; Lazarus, C.M.; Hooley, R. Members of a new family of DNA-binding proteins bind to a conserved cis-element in the promoters of Amy2 genes. *Plant Mol. Biol.* **1995**, *29*, 691–702. [CrossRef]

10.   Rushton, P.J.; Torres, J.T.; Parniske, M.; Wernert, P.; Hahlbrock, K.; Somssich, I.E. Interaction of elicitor-induced DNA-binding proteins with elicitor response elements in the promoters of parsley PR1 genes. *EMBO J.* **1996**, *15*, 5690–5700. [CrossRef] [PubMed]

11.   He, H.; Dong, Q.; Shao, Y.; Jiang, H.; Zhu, S.; Cheng, B.; Xiang, Y. Genome-wide survey and characterization of the WRKY gene family in *Populus trichocarpa*. *Plant Cell Rep.* **2012**, *31*, 1199–1217. [CrossRef] [PubMed]

12.   Rensing, S.A.; Lang, D.; Zimmer, A.D.; Terry, A.; Salamov, A.A.; Shapiro, H.; Nishiyama, T.; Perroud, P.-F.; Lindquist, E.; Kamisugi, Y.; et al. The *Physcomitrella* genome reveals evolutionary insights into the conquest of land by plants. *Science* **2007**, *319*, 64–69. [CrossRef] [PubMed]

13.   Mangelsen, E.; Kilian, J.; Berendzen, K.W.; Kolukisaoglu, Ü.; Harter, K.; Jansson, C.; Wanke, D. Phylogenetic and comparative gene expression analysis of barley (*Hordeum vulgare*) WRKY transcription factor family reveals putatively retained functions between monocots and dicots. *BMC Genom.* **2008**, *9*, 194. [CrossRef]

14.   Ling, J.; Jiang, W.; Zhang, Y.; Yu, H.; Mao, Z.; Gu, X.; Huang, S.; Xie, B. Genome-wide analysis of WRKY gene family in *Cucumis sativus*. *BMC Genom.* **2011**, *12*, 1–20. [CrossRef]

15.   Berri, S.; Abbruscato, P.; Faivre-Rampant, O.; Brasileiro, A.C.M.; Fumasoni, I.; Satoh, K.; Kikuchi, S.; Mizzi, L.; Morandini, P.; Pè, M.E.; et al. Characterization of WRKY co-regulatory networks in rice and *Arabidopsis*. *BMC Plant Biol.* **2009**, *9*, 1–22. [CrossRef]

16. Liu, J.-J.; Ekramoddoullah, A.K. Identification and characterization of the WRKY transcription factor family in *Pinus monticola*. *Genome* **2009**, *52*, 77–88. [CrossRef] [PubMed]

17. Huang, S.; Gao, Y.; Liu, J.; Peng, X.; Niu, X.; Fei, Z.; Cao, S.; Liu, Y. Genome-wide analysis of WRKY transcription factors in *Solanum lycopersicum*. *Mol. Genet. Genom.* **2012**, *287*, 495–513. [CrossRef]

18. Ross, C.A.; Liu, Y.; Shen, Q.J. The WRKY gene family in rice (*Oryza sativa*). *J. Integr. Plant Biol.* **2007**, *49*, 827–842. [CrossRef]

19. Rushton, P.J.; Somssich, I.E.; Ringler, P.; Shen, Q.J. WRKY transcription factors. *Trends Plant Sci.* **2010**, *15*, 247–258. [CrossRef]

20. Ülker, B.; Somssich, E.I. WRKY transcription factors: From DNA binding towards biological function. *Curr. Opin. Plant Biol.* **2004**, *7*, 491–498. [CrossRef]

21. Phukan, U.J.; Jeena, G.S.; Shukla, R.K. WRKY transcription factors: Molecular regulation and stress responses in plants. *Front. Plant Sci.* **2016**, *7*, 760. [CrossRef] [PubMed]

22. Xie, Z.; Zhang, Z.-L.; Zou, X.; Huang, J.; Ruas, P.; Thompson, D.; Shen, Q.J. Annotations and functional analyses of the rice WRKY gene superfamily reveal positive and negative regulators of abscisic acid signaling in aleurone cells. *Plant Physiol.* **2004**, *137*, 176–189. [CrossRef]

23. Eulgem, T.; Rushton, P.J.; Robatzek, S.; Somssich, I.E. The WRKY superfamily of plant transcription factors. *Trends Plant Sci.* **2000**, *5*, 199–206. [CrossRef]

24. Rushton, P.J.; Bokowiec, M.T.; Han, S.; Zhang, H.; Brannock, J.F.; Chen, X.; Laudeman, T.W.; Timko, M.P. Tobacco transcription factors: Novel insights into transcriptional regulation in the *Solanaceae*. *Plant Physiol.* **2008**, *147*, 280–295. [CrossRef]

25. Chen, L.; Song, Y.; Li, S.; Zhang, L.; Zou, C.; Yu, D. The role of WRKY transcription factors in plant abiotic stresses. *Biochim. Biophys. Acta BBA Bioenerg.* **2012**, *1819*, 120–128. [CrossRef]

26. Ciolkowski, I.; Wanke, D.; Birkenbihl, R.P.; Somssich, I.E. Studies on DNA-binding selectivity of WRKY transcription factors lend structural clues into WRKY-domain function. *Plant Mol. Biol.* **2008**, *68*, 81–92. [CrossRef]

27. Ezentgraf, U.; Laun, T.; Miao, Y. The complex regulation of *WRKY53* during leaf senescence of *Arabidopsis thaliana*. *Eur. J. Cell Biol.* **2010**, *89*, 133–137. [CrossRef]

28. Cai, H.; Yang, S.; Yan, Y.; Xiao, Z.; Cheng, J.; Wu, J.; Qiu, A.; Lai, Y.; Mou, S.; Guan, D.; et al. *CaWRKY6* transcriptionally activates *CaWRKY40*, regulates Ralstonia solanacearum resistance, and confers high-temperature and high-humidity tolerance in pepper. *J. Exp. Bot.* **2015**, *66*, 3163–3174. [CrossRef]

29. Turck, F.; Zhou, A.; Somssich, I.E. Stimulus-dependent, promoter-specific binding of transcription factor WRKY1 to Its native promoter and the defense-related gene *PcPR1-1* in *Parsley*. *Plant Cell* **2004**, *16*, 2573–2585. [CrossRef]

30. Cai, T.; Flanagan, L.B.; Jassal, R.S.; Black, T.A. Modelling environmental controls on ecosystem photosynthesis and the carbon isotope composition of ecosystem-respired CO_2 in a coastal Douglas-fir forest. *Plant Cell Environ.* **2008**, *31*, 435–453. [CrossRef]

31. Grierson, C.; Du, J.-S.; Zabala, M.D.T.; Beggs, K.; Smith, C.; Holdsworth, M.; Bevan, M.W. Separate cis sequences and trans factors direct metabolic and developmental regulation of a potato tuber storage protein gene. *Plant J.* **1994**, *5*, 815–826. [CrossRef] [PubMed]

32. Sun, C.; Palmqvist, S.; Olsson, H.; Borén, M.; Ahlandsberg, S.; Jansson, C. A novel WRKY transcription factor, *SUSIBA2*, participates in sugar signaling in barley by binding to the sugar-responsive elements of the iso1 promoter. *Plant Cell* **2003**, *15*, 2076–2092. [CrossRef] [PubMed]

33. Jiang, Y.; Deyholos, M.K. Functional characterization of *Arabidopsis* NaCl-inducible *WRKY25* and *WRKY33* transcription factors in abiotic stresses. *Plant Mol. Biol.* **2008**, *69*, 91–105. [CrossRef]

34. Li, S.; Fu, Q.; Chen, L.; Huang, W.-D.; Yu, D. *Arabidopsis thaliana* WRKY25, WRKY26, and WRKY33 coordinate induction of plant thermotolerance. *Planta* **2011**, *233*, 1237–1252. [CrossRef]

35. Zou, C.; Jiang, W.; Yu, D. Male gametophyte-specific WRKY34 transcription factor mediates cold sensitivity of mature pollen in *Arabidopsis*. *J. Exp. Bot.* **2010**, *61*, 3901–3914. [CrossRef]

36. Park, C.Y.; Lee, J.H.; Yoo, J.H.; Moon, B.C.; Choi, M.S.; Kang, Y.H.; Lee, S.M.; Kim, H.S.; Kang, K.Y.; Chung, W.S.; et al. WRKY group IId transcription factors interact with calmodulin. *FEBS Lett.* **2005**, *579*, 1545–1550. [CrossRef] [PubMed]

37. Sun, Y.; Yu, D. Activated expression of *AtWRKY53* negatively regulates drought tolerance by mediating stomatal movement. *Plant Cell Rep.* **2015**, *34*, 1295–1306. [CrossRef] [PubMed]

38. Zheng, Y.; Ge, J.; Bao, C.; Chang, W.; Liu, J.; Shao, J.; Liu, X.; Su, L.; Pan, L.; Zhou, D.-X. Histone deacetylase *HDA9* and WRKY53 transcription factor are mutual antagonists in regulation of plant stress response. *Mol. Plant* **2020**, *13*, 598–611. [CrossRef]

39. Jiang, Y.; Liang, G.; Yu, D. Activated expression of WRKY57 confers drought tolerance in *Arabidopsis*. *Mol. Plant* **2012**, *5*, 1375–1388. [CrossRef]

40. Ren, X.; Chen, Z.; Liu, Y.; Zhang, H.; Zhang, M.; Liu, Q.; Hong, X.; Zhu, J.-K.; Gong, Z. ABO3, a WRKY transcription factor, mediates plant responses to abscisic acid and drought tolerance in *Arabidopsis*. *Plant J.* **2010**, *63*, 417–429. [CrossRef]

41. Wang, L.; Ma, K.-B.; Lu, Z.-G.; Ren, S.-X.; Jiang, H.-R.; Cui, J.-W.; Chen, G.; Teng, N.-J.; Lam, H.-M.; Jin, B. Differential physiological, transcriptomic and metabolomic responses of *Arabidopsis* leaves under prolonged warming and heat shock. *BMC Plant Biol.* **2020**, *20*, 1–15. [CrossRef]

42. Ren, S.; Ma, K.; Lu, Z.; Chen, G.; Cui, J.; Tong, P.; Wang, L.; Teng, N.; Jin, B. Transcriptomic and metabolomic analysis of the heat-stress response of *Populus tomentosa*. *Carr. For.* **2019**, *10*, 383. [CrossRef]

43. Wang, J.; Wang, L.; Yan, Y.; Zhang, S.; Li, H.; Gao, Z.; Wang, C.; Guo, X. *GhWRKY21* regulates ABA-mediated drought tolerance by fine-tuning the expression of *GhHAB* in cotton. *Plant Cell Rep.* **2020**, 39. [CrossRef]

44. Liu, X.; Song, Y.; Xing, F.; Wang, N.; Wen, F.; Zhu, C. *GhWRKY25*, a group I WRKY gene from cotton, confers differential tolerance to abiotic and biotic stresses in transgenic *Nicotiana benthamiana*. *Protoplasma* **2015**, *253*, 1265–1281. [CrossRef]

45. Jia, H.; Wang, C.; Wang, F.; Liu, S.; Li, G.; Guo, X. *GhWRKY68* reduces resistance to salt and drought in transgenic *Nicotiana benthamiana*. *PLoS ONE* **2015**, *10*, e0120646. [CrossRef]

46. Wang, M.; Vannozzi, A.; Wang, G.; Liang, Y.-H.; Tornielli, G.B.; Zenoni, S.; Cavallini, E.; Pezzotti, M.; Cheng, Z.-M. Genome and transcriptome analysis of the grapevine (*Vitis vinifera* L.) WRKY gene family. *Hortic. Res.* **2014**, *1*, 14016. [CrossRef] [PubMed]

47. Dang, F.F.; Wang, Y.N.; Yu, L.; Eulgem, T.; Lai, Y.; Liu, Z.Q.; Wang, X.; Qiu, A.L.; Zhang, T.X.; Lin, J.; et al. *CaWRKY40*, a WRKY protein of pepper, plays an important role in the regulation of tolerance to heat stress and resistance to Ralstonia solanacearum infection. *Plant Cell Environ.* **2013**, *36*, 757–774. [CrossRef]

48. Sun, J.; Hu, W.; Zhou, R.; Wang, L.; Wang, X.; Wang, Q.; Feng, Z.-J.; Yu, H.; Qiu, D.; He, G.; et al. The *Brachypodium distachyon BdWRKY36* gene confers tolerance to drought stress in transgenic tobacco plants. *Plant Cell Rep.* **2014**, *34*, 23–35. [CrossRef]

49. Gong, X.; Zhang, J.; Hu, J.; Wang, W.; Wu, H.; Zhang, Q.; Liu, J.-H. *FcWRKY70*, a WRKY protein of *Fortunella crassifolia*, functions in drought tolerance and modulates putrescine synthesis by regulating arginine decarboxylase gene. *Plant Cell Environ.* **2015**, *38*, 2248–2262. [CrossRef]

50. Wu, X.; Shiroto, Y.; Kishitani, S.; Ito, Y.; Toriyama, K. Enhanced heat and drought tolerance in transgenic rice seedlings overexpressing *OsWRKY11* under the control of *HSP101* promoter. *Plant Cell Rep.* **2009**, *28*, 21–30. [CrossRef] [PubMed]

51. Song, Y.; Chen, L.; Zhang, L.; Yu, D. Overexpression of *OsWRKY72* gene interferes in the abscisic acid signal and auxin transport pathway of *Arabidopsis*. *J. Biosci.* **2010**, *35*, 459–471. [CrossRef]

52. Dai, X.; Wang, Y.; Zhang, W.-H. *OsWRKY74*, a WRKY transcription factor, modulates tolerance to phosphate starvation in rice. *J. Exp. Bot.* **2016**, *67*, 947–960. [CrossRef]

53. Yokotani, N.; Sato, Y.; Tanabe, S.; Chujo, T.; Shimizu, T.; Okada, K.; Yamane, H.; Shimono, M.; Sugano, S.; Takatsuji, H.; et al. WRKY76 is a rice transcriptional repressor playing opposite roles in blast disease resistance and cold stress tolerance. *J. Exp. Bot.* **2013**, *64*, 5085–5097. [CrossRef]

54. Wang, H.; Hao, J.; Chen, X.; Hao, Z.; Wang, X.; Lou, Y.; Peng, Y.; Guo, Z. Overexpression of rice WRKY89 enhances ultraviolet B tolerance and disease resistance in rice plants. *Plant Mol. Biol.* **2007**, *65*, 799–815. [CrossRef]

55. Zhou, Q.-Y.; Tian, A.-G.; Zou, H.-F.; Xie, Z.-M.; Lei, G.; Huang, J.; Wang, C.-M.; Wang, H.-W.; Zhang, J.-S.; Chen, S.-Y. Soybean WRKY-type transcription factor genes, *GmWRKY13*, *GmWRKY21*, and *GmWRKY54*, confer differential tolerance to abiotic stresses in transgenic *Arabidopsis* plants. *Plant Biotechnol. J.* **2008**, *6*, 786–503. [CrossRef]

56. Yan, H.; Jia, H.; Chen, X.; Hao, L.; An, H.; Guo, X. The cotton WRKY transcription factor *GhWRKY17* functions in drought and salt stress in transgenic *Nicotiana benthamiana* through ABA signaling and the modulation of reactive oxygen species production. *Plant Cell Physiol.* **2014**, *55*, 2060–2076. [CrossRef]

57. Cai, R.; Dai, W.; Zhang, C.; Wang, Y.; Wu, M.; Zhao, Y.; Ma, Q.; Xiang, Y.; Cheng, B. The maize WRKY transcription factor *ZmWRKY17* negatively regulates salt stress tolerance in transgenic *Arabidopsis* plants. *Planta* **2017**, *246*, 1215–1231. [CrossRef]

58. Niu, C.-F.; Wei, W.; Zhou, Q.-Y.; Tian, A.-G.; Hao, Y.-J.; Zhang, W.; Ma, B.; Lin, Q.; Zhang, Z.-B.; Zhang, J.-S.; et al. Wheat WRKY genes *TaWRKY2* and *TaWRKY19* regulate abiotic stress tolerance in transgenic Arabidopsis plants. *Plant Cell Environ.* **2012**, *35*, 1156–1170. [CrossRef]

59. Wang, F.; Hou, X.; Tang, J.; Wang, Z.; Wang, S.; Jiang, F.; Li, Y. A novel cold-inducible gene from Pak-choi (*Brassica campestris* ssp. chinensis), *BcWRKY46*, enhances the cold, salt and dehydration stress tolerance in transgenic tobacco. *Mol. Biol. Rep.* **2011**, *39*, 4553–4564. [CrossRef]

60. Wang, Z.; Zhu, Y.; Wang, L.; Liu, X.; Liu, Y.; Phillips, J.; Deng, X. A WRKY transcription factor participates in dehydration tolerance in *Boea hygrometrica* by binding to the W-box elements of the galactinol synthase (*BhGolS1*) promoter. *Planta* **2009**, *230*, 1155–1166. [CrossRef]

61. Li, H.; Xu, Y.; Xiao, Y.; Zhu, Z.; Xie, X.; Zhao, H.; Wang, Y. Expression and functional analysis of two genes encoding transcription factors, *VpWRKY1* and *VpWRKY2*, isolated from Chinese wild *Vitis pseudoreticulata*. *Planta* **2010**, *232*, 1325–1337. [CrossRef]

62. Zhu, Z.; Shi, J.; Cao, J.; He, M.; Wang, Y. *VpWRKY3*, a biotic and abiotic stress-related transcription factor from the Chinese wild *Vitis pseudoreticulata*. *Plant Cell Rep.* **2012**, *31*, 2109–2120. [CrossRef] [PubMed]

63. Wei, W.; Zhang, Y.; Han, L.; Guan, Z.; Chai, T. A novel WRKY transcriptional factor from *Thlaspi caerulescens* negatively regulates the osmotic stress tolerance of transgenic tobacco. *Plant Cell Rep.* **2008**, *27*, 795–803. [CrossRef]

64. Skibbe, M.; Qu, N.; Galis, I.; Baldwin, I.T. Induced plant defenses in the natural environment: *Nicotiana attenuata* WRKY3 and WRKY6 coordinate responses to herbivory. *Plant Cell* **2008**, *20*, 1984–2000. [CrossRef]

65. Yang, G.; Zhang, W.; Liu, Z.; Yi-Maer, A.Y.; Zhai, M.; Xu, Z. Both *JrWRKY2* and *JrWRKY7* of *Juglans regia* mediate responses to abiotic stresses and abscisic acid through formation of homodimers and interaction. *Plant Biol.* **2017**, *19*, 268–278. [CrossRef]

66. Yang, Z.; Chi, X.; Guo, F.; Jin, X.; Luo, H.; Hawar, A.; Chen, Y.; Feng, K.; Wang, B.; Qi, J.; et al. *SbWRKY30* enhances the drought tolerance of plants and regulates a drought stress-responsive gene, *SbRD19*, in sorghum. *J. Plant Physiol.* **2020**, *246*, 153142. [CrossRef]

67. Song, Y.; Li, J.; Sui, Y.; Han, G.; Zhang, Y.; Guo, S.; Sui, N. The sweet sorghum *SbWRKY50* is negatively involved in salt response by regulating ion homeostasis. *Plant Mol. Biol.* **2020**, *102*, 603–614. [CrossRef]

68. Qin, Z.; Hou, F.; Li, A.; Dong, S.; Wang, Q.; Zhang, L. Transcriptome-wide identification of WRKY transcription factor and their expression profiles under salt stress in sweetpotato (*Ipomoea batatas* L.). *Plant Biotechnol. Rep.* **2020**, *14*, 599–611. [CrossRef]

69. Zhu, H.; Zhou, Y.; Zhai, H.; He, S.; Zhao, N.; Liu, Q. A novel sweetpotato WRKY transcription factor, *IbWRKY2*, positively regulates drought and salt tolerance in transgenic. *Arabidopsis* **2020**, *10*, 506. [CrossRef]

70. Dong, Q.; Zheng, W.; Duan, D.; Huang, D.; Wang, Q.; Liu, C.; Li, C.; Gong, X.; Li, C.; Mao, K.; et al. *MdWRKY30*, a group IIa WRKY gene from apple, confers tolerance to salinity and osmotic stresses in transgenic apple callus and *Arabidopsis* seedlings. *Plant Sci.* **2020**, *299*, 110611. [CrossRef] [PubMed]

71. Ma, Y.; Xue, H.; Zhang, F.; Jiang, Q.; Yang, S.; Yue, P.; Wang, F.; Zhang, Y.; Li, L.; He, P.; et al. The miR156/SPL module regulates apple salt stress tolerance by activating *MdWRKY100* expression. *Plant Biotechnol. J.* **2020**, *18*. [CrossRef]

72. Ahammed, G.J.; Li, X.; Yang, Y.; Liu, C.; Zhou, G.; Wan, H.; Cheng, Y. Tomato WRKY81 acts as a negative regulator for drought tolerance by modulating guard cell H_2O_2–mediated stomatal closure. *Environ. Exp. Bot.* **2020**, *171*, 103960. [CrossRef]

73. Luo, X.; Li, C.; He, X.; Zhang, X.; Zhu, L.-F. ABA signaling is negatively regulated by GbWRKY1 through JAZ1 and ABI1 to affect salt and drought tolerance. *Plant Cell Rep.* **2019**, *39*, 181–194. [CrossRef]

74. Wang, M.-Q.; Huang, Q.-X.; Lin, P.; Zeng, Q.-H.; Li, Y.; Liu, Q.-L.; Zhang, L.; Pan, Y.-Z.; Jiang, B.-B.; Zhang, F. The overexpression of a transcription factor gene *VbWRKY32* enhances the cold tolerance in *Verbena bonariensis*. *Front. Plant Sci.* **2020**, *10*, 1746. [CrossRef]

75. Chanwala, J.; Satpati, S.; Dixit, A.; Parida, A.; Giri, M.K.; Dey, N. Genome-wide identification and expression analysis of WRKY transcription factors in pearl millet (*Pennisetum glaucum*) under dehydration and salinity stress. *BMC Genom.* **2020**, *21*, 1–16. [CrossRef]

76. Zhao, K.; Zhang, D.; Lv, K.; Zhang, X.; Cheng, Z.; Li, R.; Zhou, B.; Jiang, T. Functional characterization of poplar WRKY75 in salt and osmotic tolerance. *Plant Sci.* **2019**, *289*, 110259. [CrossRef]

77. Anjum, S.A.; Xie, X.-Y.; Wang, L.-C.; Saleem, M.F.; Man, C.; Lei, W. Morphological, physiological and biochemical responses of plants to drought stress. *Afr. J. Agr. Res.* **2011**, *6*, 2026–2032. [CrossRef]

78. Sakuma, Y.; Maruyama, K.; Osakabe, Y.; Qin, F.; Seki, M.; Shinozaki, K.; Yamaguchi-Shinozaki, K. Functional analysis of an *Arabidopsis* transcription factor, *DREB2A*, involved in drought-responsive gene expression. *Plant Cell* **2006**, *18*, 1292–1309. [CrossRef] [PubMed]

79. Hassan, S.; Lethin, J.; Blomberg, R.; Mousavi, H.; Aronsson, H. In silico based screening of WRKY genes for identifying functional genes regulated by WRKY under salt stress. *Comput. Biol. Chem.* **2019**, *83*, 107131. [CrossRef]

80. Linxiao, W.; Song, A.; Gao, C.; Wang, L.; Wang, Y.; Sun, J.; Jiang, J.; Chen, F.; Chen, S. *Chrysanthemum* WRKY gene *CmWRKY17* negatively regulates salt stress tolerance in transgenic chrysanthemum and *Arabidopsis* plants. *Plant Cell Rep.* **2015**, *34*, 1365–1378. [CrossRef]

81. Shen, Z.; Yao, J.; Sun, J.; Chang, L.; Wang, S.; Ding, M.; Qian, Z.; Zhang, H.; Zhao, N.; Sa, G.; et al. *Populus euphratica* HSF binds the promoter of *WRKY1* to enhance salt tolerance. *Plant Sci.* **2015**, *235*, 89–100. [CrossRef] [PubMed]

82. Jaglo-Ottosen, K.R. Arabidopsis CBF1 overexpression induces COR genes and enhances freezing tolerance. *Science* **1998**, *280*, 104–106. [CrossRef]

83. Chen, L.; Zhao, Y.; Xu, S.; Zhang, Z.; Xu, Y.; Zhang, J.; Chong, K. *OsMADS57* together with *OsTB1* coordinates transcription of its target *OsWRKY94* and *D14* to switch its organogenesis to defense for cold adaptation in rice. *New Phytol.* **2018**, *218*, 219–231. [CrossRef]

84. Jin, B.; Wang, L.; Wang, J.; Jiang, K.-Z.; Wang, Y.; Jiang, X.X.; Ni, C.-Y.; Wang, Y.; Teng, N.-J. The effect of experimental warming on leaf functional traits, leaf structure and leaf biochemistry in *Arabidopsis thaliana*. *BMC Plant Biol.* **2011**, *11*, 35. [CrossRef] [PubMed]

85. Liu, H.-T.; Li, G.-L.; Chang, H.; Sun, D.-Y.; Zhou, R.-G.; Li, B. Calmodulin-binding protein phosphatase PP7 is involved in thermotolerance in *Arabidopsis*. *Plant Cell Environ.* **2007**, *30*, 156–164. [CrossRef] [PubMed]

86. Larkindale, J.; Huang, B. Effects of abscisic acid, salicylic acid, ethylene and hydrogen peroxide in thermotolerance and recovery for creeping bentgrass. *Plant Growth Regul.* **2005**, *47*, 17–28. [CrossRef]

87. Zhao, J.; He, Q.; Chen, G.; Wang, L.; Jin, B. Regulation of non-coding RNAs in heat stress responses of plants. *Front. Plant Sci.* **2016**, *7*, 1213. [CrossRef]

88. Gong, X.-Q.; Hu, J.-B.; Liu, J.-H. Cloning and characterization of *FcWRKY40*, A WRKY transcription factor from *Fortunella crassifolia* linked to oxidative stress tolerance. *Plant Cell Tissue Organ Cult. PCTOC* **2014**, *119*, 197–210. [CrossRef]

89. Cheong, Y.H.; Chang, H.-S.; Gupta, R.; Wang, X.; Zhu, T.; Luan, S. Transcriptional profiling reveals novel interactions between wounding, pathogen, abiotic stress, and hormonal responses in *Arabidopsis*. *Plant Physiol.* **2002**, *129*, 661–677. [CrossRef]

90. Robatzek, S.; Somssich, I.E. A new member of the *Arabidopsis* WRKY transcription factor family, *AtWRKY6*, is associated with both senescence- and defence-related processes. *Plant J.* **2001**, *28*, 123–133. [CrossRef]

91. Kilian, J.; Whitehead, D.; Horak, J.; Wanke, D.; Weinl, S.; Batistic, O.; D'Angelo, C.; Bornberg-Bauer, E.; Kudla, J.; Harter, K. The AtGenExpress global stress expression data set: Protocols, evaluation and model data analysis of UV-B light, drought and cold stress responses. *Plant J.* **2007**, *50*, 347–363. [CrossRef]

92. Han, Y.; Zhang, X.; Wang, Y.; Ming, F. The suppression of WRKY44 by GIGANTEA-miR172 pathway is involved in drought response of *Arabidopsis thaliana*. *PLoS ONE* **2013**, *8*, e73541. [CrossRef] [PubMed]

2

In Silico Analyses of Autophagy-Related Genes in Rapeseed (*Brassica napus* L.) under Different Abiotic Stresses and in Various Tissues

Elham Mehri Eshkiki [1], **Zahra Hajiahmadi** [2], **Amin Abedi** [2], **Mojtaba Kordrostami** [3](ID)
and **Cédric Jacquard** [4],*(ID)

[1]   Department of Agricultural Biotechnology, Payame Noor University (PNU),
      Tehran P.O. Box 19395-4697, Iran; elham.mehri66@gmail.com
[2]   Department of Biotechnology, Faculty of Agricultural Sciences, University of Guilan,
      Rasht P.O. Box 41635-1314, Iran; z.hajiahmadi1366@gmail.com (Z.H.); abedi.amin@yahoo.com (A.A.)
[3]   Nuclear Agriculture Research School, Nuclear Science and Technology Research Institute (NSTRI),
      Karaj P.O. Box 31485498, Iran; kordrostami009@gmail.com
[4]   Resistance Induction and Bioprotection of Plants Unit (RIBP)—EA4707, SFR Condorcet FR CNRS 3417,
      University of Reims Champagne-Ardenne, Moulin de la Housse, CEDEX 2, BP 1039, 51687 Reims, France
*   Correspondence: cedric.jacquard@univ-reims.fr.

Abstract: The autophagy-related genes (ATGs) play important roles in plant growth and response to environmental stresses. *Brassica napus* (*B. napus*) is among the most important oilseed crops, but *ATGs* are largely unknown in this species. Therefore, a genome-wide analysis of the *B. napus ATG* gene family (*BnATGs*) was performed. One hundred and twenty-seven ATGs were determined due to the *B. napus* genome, which belongs to 20 main groups. Segmental duplication occurred more than the tandem duplication in *BnATGs*. Ka/Ks for the most duplicated pair genes were less than one, which indicated that the negative selection occurred to maintain their function during the evolution of *B. napus* plants. Based on the results, *BnATGs* are involved in various developmental processes and respond to biotic and abiotic stresses. One hundred and seven miRNA molecules are involved in the post-transcriptional regulation of 41 *BnATGs*. In general, 127 simple sequence repeat marker (SSR) loci were also detected in *BnATGs*. Based on the RNA-seq data, the highest expression in root and silique was related to *BnVTI12e*, while in shoot and seed, it was *BnATG8p*. The expression patterns of the most *BnATGs* were significantly up-regulated or down-regulated responding to dehydration, salinity, abscisic acid, and cold. This research provides information that can detect candidate genes for genetic manipulation in *B. napus*.

Keywords: codon usage bias; duplication; expression pattern; gene family; gene ontology

1. Introduction

Autophagy consists of transferring the cargo into vacuole in plants, lysosomes, and animals and subsequently, decomposition [1]. In general, there are two types of autophagic processes, including macroautophagy and microautophagy [2]. Macroautophagy is a basic route for the decomposition of substrates using special double-membrane vesicles, called autophagosomes, to trap the materials [3]. Microphagy is a non-selective decomposition process in which the substances are directly decomposed by the lysosomes in animals or vacuoles in plants [4]. More than 36 types of autophagy-related genes (ATG) were already characterized, more than half of them encode core autophagy proteins and are conserved in the most studying organisms, including plants [5]. The most *ATGs* are involved in decomposition processes, but they also contributed to biosynthesis processes. In plants, there research about the role of these genes in growth, development, and responding to environmental stresses [6–9]. The *ATG* gene family is composed of several subfamilies in which *ATG4* and *ATG8* subfamilies have generally more members than other subfamilies in yeast, animals, and plants [10,11]. The plants are constantly exposed to environmental changes, which can be considered as biotic and abiotic stressors when they are too strong. Many molecular and physiological mechanisms are involved in plants' tolerance to environmental stresses. The abiotic stresses include cold, heat, drought, salinity, nutrient deficiencies, heavy metal toxicity, and oxidative stress. Many parts of the world encounter the drought and salinity conditions limiting the crop yields. Abiotic stresses can increase the level of reactive oxidative species (ROS) in plants, which can be reduced by autophagy [12]. It is known that *ATG18a*, *ATG10b*, and *ATG8* are involved in responding to the salinity and drought stresses in *Arabidopsis*, rice, and wheat, respectively [13–15]. Likewise, the autophagosome formation was immediately induced under salinity conditions in *atg2* and *atg7 Arabidopsis* mutants, which displayed a hypersensitive phenotype [9]. The overexpression of many-core *ATGs* improves plant growth under abiotic stresses [16,17]. It has been reported that the destruction of *ATG* in *Arabidopsis* resulted in the accelerated leaf senescence [18]. Under normal conditions, the basal level of autophagy requires to maintain cellular homeostasis [19]. Although *ATG* mutants in the model plants such as *Arabidopsis thaliana*, *Oryza sativa*, and *Zea mays* can complete their life cycle, they have impaired in growth or stress response compared to wild-type plants [20]. It has been reported that *ATG* genes, in addition to playing a role in plant stress response, are involved in other plant biological processes such as nutrient metabolism, nutrient recycling, lipid metabolism, root development, aging, reproductive development, and crop yield [21,22]. Surveying the expression pattern of *Arabidopsis ATGs* in tissues and developmental stages indicated the specific and different functions of these genes in multiple tissues and cells [23]. For instance, *ATG8* is involved in developing the endosperm in maize and its expression increases from 18 to 30 days after pollination [24]. During the leaf senescence, the expression of 15 *Arabidopsis ATGs* is induced [25]. Likewise, during the seed development of *Arabidopsis*, all *ATGs* are already expressed in siliques [26]. *ATG* genes also play an important role in productivity and yield. For instance, rice *ATG7* knockout line caused the abnormal development of anther, spikelet sterility, decreased nitrogen efficiency, premature senescence leaf, and finally reduced the yield [27,28]. Autophagy plays an important role in the plant's response to nutrient starvation, and the expression of *ATGs* is induced under carbon or nitrogen starvation. Maize plants with mutations in *ATG12* (*atg12*) have normal growth similar to wild-type plants under control conditions but, in nitrogen starvation conditions, stopped seedling growth, premature senescence leaf, and reduced seed production mutant plants have been observed [29].

Rapeseed (*Brassica napus* L.) is one of the most important oilseed crops in the world and is an allopolyploid due to the hybridization of *B. rapa* (2n = 2x = 20) and *B. oleracea* (2n = 2x = 18). This plant is used as one of the healthiest edible oils due to its high content of unsaturated fatty acids and proteins. *B. napus* oil contains omega-3, essential vitamins, and minerals [30]. *B. napus* are used as an excellent species for the genetic and molecular studies of development and adaptation to stresses such as heavy metals stress due to its outstanding features, including rapid growth, high biomass, accumulation of heavy metals in the stem without showing any symptoms. *B. napus* is also a good candidate for phytoremediation [31,32]. This plant is relatively cold and salinity tolerant. However, spring frost can cause serious damage and kill *B. napus* seedlings, followed by a severe reduction in yield [33]. Identifying the abiotic stress-responsive genes in *B. napus* can increase our knowledge to improve commercial *B. napus* cultivars and increase its tolerance to abiotic stresses such as cold and salinity. Although the genome of many plants is known, *ATGs* were studied in a few plants, including *Arabidopsis* [34], *O. sativa* [35], *Nicotiana tabacum* [36], *Z. mays* [29], *Capsicum annuum* [37], *Setaria italica* [38], *Musa acuminate* [39], *Vitis vinifera* [12], *Citrus sinensis* [40], and *Ricinus communis* [41]. Further investigations are necessary to characterize these genes in the various plant species to increase our knowledge about *ATGs* in plants. Therefore, in the present study, identification, evolutionary relationship, duplication, and selection pressure, exon–intron structure, promoter analysis, transcript-targeted miRNA, and simple sequence repeat markers (SSRs) prediction, RNA-seq data analysis, codon usage, and gene ontology of the *BnATGs* genes were investigated.

2. Results

2.1. The Identification of B. napus ATG Genes

To identify the *ATG* genes in the *B. napus* genome, the BLASTP program was applied using the query sequence of *Arabidopsis* [42], *C. Sinensis* [40], *V. vinifera* [12], and *O. sativa* [35] ATG proteins against the *B. napus* genome database [43]. One hundred and twenty-seven genes were determined after omitting the sequences without ATG domains (Supplementary Table S1). All *BnATGs* were grouped due to their specific domains and numbered based on the chromosomal locus except the *ATG2* subfamily, which was not identified. All identified *BnATGs* were encoded by gene families, including 26 members for *BnATG8*, 24 members for *BnATG18*, 11 members for *BnATG1*, eight members for each of the *BnATG20* and *BnVTI12*, six members for the *BnATG11*, five members for each *BnATG4*, *BnATG14*, and *BnATG101*, four members for each of the *BnATG3*, *BnATG5*, *BnATG6*, and *BnATG13*, three members for each of the *BnATG9*, *BnATG12*, and *BnTOR*, and two members for each *BnATG7*, *BnATG10*, *BnATG16*, *BnVPS15*, and *BnVPS34*. The physicochemical analysis of *BnATGs* was carried out using the ProtParam tool. The length of protein sequences encoded by these 127 *BnATGs* ranged from 93 amino acids of BnATG12b to 2478 amino acids of BnTORc. The molecular weight (MW) of BnATG proteins ranged from 10.29 to 278.5 kDa with the isoelectric points (pI) varying from 4.63 to 9.74 (Supplementary Table S1). Due to the prediction of subcellular localization, BnATGs have activity in nucleus, cytoplasm, chloroplast, mitochondria, plasma membrane, and extracellular space (Supplementary Table S1). The diversity in length and location of BnATG proteins indicates their diverse functions in various cellular processes.

2.2. The Phylogenetic Analysis of B. napus ATG Gene Family

A neighbor-joining phylogenetic tree was constructed to assess the relationship of *B. napus* ATG proteins. The results indicated that each ATGs had high similarity to their counterparts in *Arabidopsis thaliana* (At), *Citrus sinensis* (Cs), *Oryza sativa* (Os), *Vitis vinifera* (Vv), and *Nicotiana tabacum* (Nt) (Figures 1 and 2). Almost in all ATGs, OsATGs were clustered separating from the dicot clades. All BnATGs were clustered in different branches due to the existence of multiple isoforms except BnATG6–7, BnATG10, BnATG16, and BnTOR. Therefore, it can be suggested that they may have different functions in the same subfamily.

2.3. Gene Location, Duplication, and Selection Pressure of BnATGs

Based on Figure 3, the chromosomal distribution of 127 *BnATGs* was uneven. Chromosome (Chr) A03, A09, C05, and C09 contained the highest number of *BnATGs*, whereas on ChA06, A6_random, A7_random, A10_random, C01_random, C06_random, C08_random, and C09_random, only one gene has been determined related to *BnATG13b*, *BnTORb*, *BnATG8j*, *BnATG20d*, *BnATG13c*, *BnATG6d*, *BnATG8y*, and *BnATG18x*, respectively. Based on gene duplication analysis, segmental duplication has occurred more than tandem duplication in the *B. napus* ATG gene family (Supplementary Table S2). Only five pairs of *BnATG* genes (*BnATG6b/BnATG6c* (ChC06), *BnATG8n/BnATG8m* (ChC01), *BnATG14c/BnATG14d* (ChC06), *BnATG16a/BnATG16b* (ChA01), and *BnATG18u/BnATG18t* (ChC08)) revealed tandem repeats; thus, the role of segmental duplication in the expansion of the *BnATG* family is more important than tandem duplication. The Ka/Ks ratio of 165 paired genes was measured to show the selection pressure among duplicated genes (Supplementary Table S2). Ka/Ks was <1 for the most of paired genes, suggesting a negative selection to maintain their function during the evolution of *B. napus* plants. For three duplicated pairs (*BnATG16a/16d*, *BnATG8m/BnATG8a*, and *BnATG18g/BnATG18r*), the Ka/Ks ratio was >1, indicating a positive selection, leading to their different functions due to these mutations during the evolutionary process.

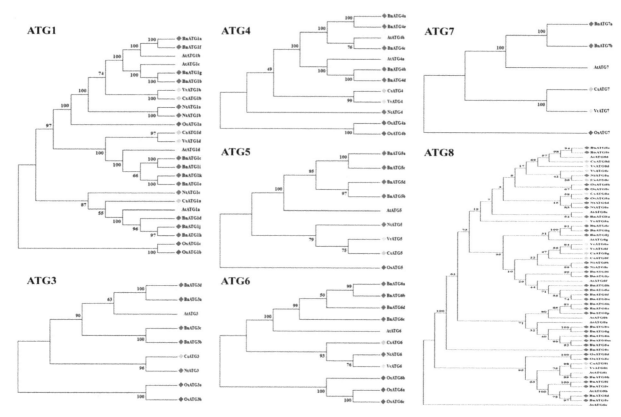

Figure 1. Phylogenetic relationships of BnATG1-BnATG8 proteins from *Brassica napus* (Bn), *Arabidopsis thaliana* (At), *Citrus sinensis* (Cs), *Oryza sativa* (Os), *Vitis vinifera* (Vv), and *Nicotiana tabacum* (Nt). The tree was constructed using MEGA 7 by the neighbor-joining (NJ) method with 1000 bootstraps. The names with pink rhombus are *B. napus* ATGs.

2.4. The Exon–Intron Structure and Conserved Motifs of B. napus ATGs

Considering the exon–intron structure analysis of *BnATG*, two to 56 introns exhibited a high structural diversity among the *ATG* subfamilies (Figures 4 and 5). *BnTORb* exhibited the longest intron. For the *B. napus ATG* gene family, three intron splicing phases were observed, including zero, one, and two, which resulted from splicing after the codon's third, first, and second nucleotide, respectively [44]. Most of *BnATGs* had all three splicing phases except *BnATG10, BnATG12, BnVTI12*, and *BnATG16*. *BnATGs* had three to 17 exons, whereas three genes named *BnTORa–c* had 55 and 56 exons. All of *BnATGs* contained introns, but some had no untranslated region, including *BnATG6c, BnATG8b, BnATG9b, BnATG11f, BnATG16a, BnATG18d, BnATG18n, BnATG101c-d*, and *BnTORb*. The number of exons was similar in each ATG group. For instance, the *BnATG5, BnATG10, BnATG12*, and *BnATG20* revealed 8, 6, 5, and 10 exons, respectively. The diversity in the number of exons in some subfamilies indicated selective pressure to achieve different functions during plant evolution [45]. The similar exon–intron structure among A and C homologous copies showed that the homologous genes maintained their function during *B. napus* evolution. The intron splicing phases for each group were similar. *BnATG1, BnATG3, BnATG4* (except *BnATG4b* and *BnATG4d* with splicing phases zero and two), *BnATG5, BnATG6* (except *BnATG6d* with splicing phases zero and two), *BnATG7, BnATG8, BnATG9, BnATG14, BnATG15, BnATG34, BnTOR*, and *BnATG101* subfamilies revealed all three intron splicing phases. Likewise, in the *BnATG18* subfamily, all three splicing phases were observed except *BnATG18a, BnATG18b, BnATG18k, and BnATG18w* genes with splicing phases zero and two, *BnATG18b* and *BnATG18x* with splicing phases zero and one, and *BnATG18a* with only splicing phase zero. *BnATG11* (except *BnATG11b* with all three intron splicing phases) and *BnATG16* subfamilies demonstrated splicing phases zero and two whereas *BnATG10, BnATG20* (except *BnATG20b* with all three intron splicing phases), and *BnVTI12* subfamilies showed splicing phases zero and one. To detect conserving motifs in BnATG proteins, Multiple Em for Motif Elicitation (MEME) was used (Figures 6 and 7). Twenty motifs were nearly identified in almost all subfamilies, but the lowest number of motifs was observed in the BnVIT12 subfamily with four motifs, followed by ATG20 subfamily with six motifs. The main BnATGs conserved domains were Pkinase_Tyr, Autophagy_act_C, ATG, APG, WD40, BCAS3, Vps, PX, Peptidase_C54, PI3_PI4_kinase, and V-SNARE (Supplementary Table S3).

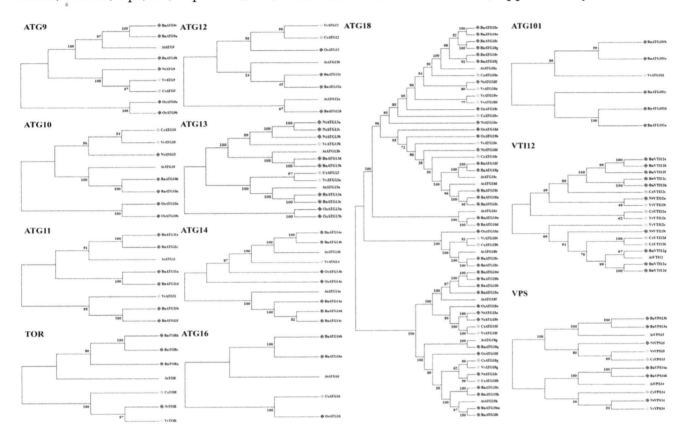

Figure 2. Phylogenetic relationships of BnATG9-BnATG114, BnATG16, BnATG18, BnVTI12, BnTOR, BnVPs15, BnVPS334, and BnVTI12 from *Brassica napus* (Bn), *Arabidopsis thaliana* (At), *Citrus sinensis* (Cs), *Oryza sativa* (Os), *Vitis vinifera* (Vv), and *Nicotiana tabacum* (Nt). The tree was constructed using MEGA 7 by the neighbor-joining (NJ) method with 1000 bootstraps. The names with pink rhombus are *B. napus* ATGs.

2.5. The Gene Ontology Annotations and Cis-Regulatory Elements of BnATGs

The gene ontology (GO) annotations of 127 BnATGs proteins were determined using TBtools [46] due to GO terms (Figure 8, Supplementary Table S4). Considering Figure 8, BnATGs are involved in the various biological processes, molecular functions, and cellular components. As excepted, most of BnATGs are predicted to the function in autophagy-related processes including autophagosome organization (13 BnATGs), autophagosome assembly (13), vacuole organization (13), organelle assembly (13), catabolic process (52), autophagy (52), a process utilizing autophagic mechanism (52), macroautophagy (13), and cellular catabolic process (52). Moreover, the cellular component prediction indicated that rapeseed ATGs were primarily localized in the intracellular (45), which was followed by cytoplasm (43). BnATGs in the molecular function category were mainly involved in protein binding (39), soluble N-ethylmaleimide-sensitive factor attachment protein (SNAP) receptor activity (8), protein-macromolecule adaptor activity (8), and macromolecule adaptor activity (8).

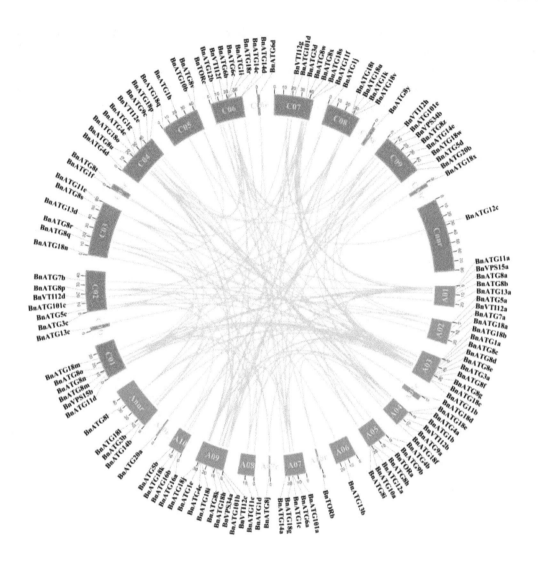

Figure 3. Chromosomal locations of *B. napus* autophagy-related genes. The location of genes on chromosomes and the duplication relationship between them were presented using TBtools. Chromosomes are represented by colored boxes. Pink curves connecting the genes indicate duplications.

To better evaluate the biological function of *BnATGs*, 1.5 Kb upstream of their transcriptional start codon (ATG) was examined using the PlantCare database to find cis-acting elements (Supplementary Table S5). A total of 91 cis-elements were found in the *B. napus ATG* gene family, which can regulate gene expression responding to five groups, including regulatory elements, light, developmental stages, phytohormones, and environmental stresses. The highest frequency of cis-elements was related to Myb (87.40%), followed by MYC (82.67%), ARE (76.37%), and G-box (70.07%). The lowest frequency of cis-acting regulatory elements was also related to re2f-1 (only in *BnVPS15b*), NON (only in *BnVPS15a*), L-box (only in *BnATG5d*), HD-Zip3 (only in *BnATG1h*), GATT-motif (only in *BnATG14a*), CAG-motif (only in *BnATG1h*), Box II-like sequence (only in *BnATG5d*), AUXRE (only in *BnATG18p*), and 4cl-CMA1b (only in *BnATG18m*). Eight hundred and sixty-three stress-responsive elements were found in *BnATGs*, indicating that they might have a potential function to regulate the *B. napus* response to various stresses. Likewise, 41, 22, 5, and 3 cis-regulatory elements were found in *BnATGs* related to meristem, endosperm, shoot, and seed, respectively (Supplementary Table S5). Due to the promoter analysis results, the *BnATGs* could have a potential function in different processes, including *B. napus* growth, development, and response to various environmental stresses.

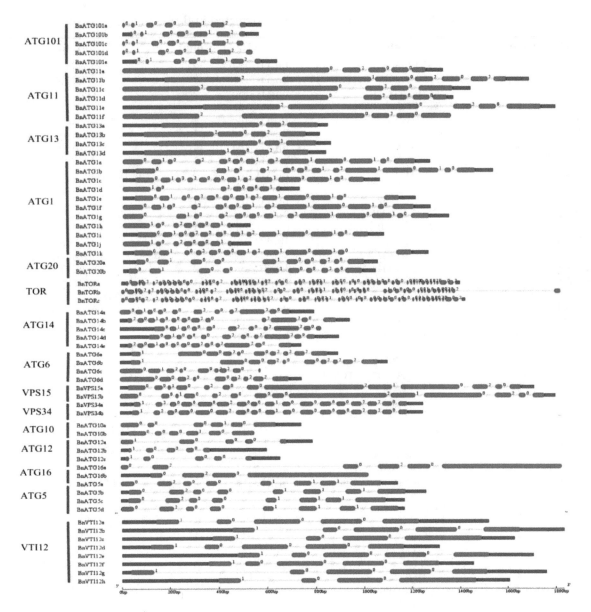

Figure 4. The exon–intron structure of *ATG1*, *ATG5–6*, *ATG10–14*, *ATG16*, and *ATG18* in *B. napus*. Exons and introns were represented by red boxes and green lines, respectively. The exon–intron structure of the *BnATGs* was determined using a gene structure display server (GSDS).

2.6. The Prediction of Simple Sequence Repeats (SSRs) in BnATGs

137 SSRs were detected in 73 out of 127 *BnATG* genes (Table 1). Most genes had a single SSR except *BnATG9a* (7 SSRs), *BnATG9c* (6 SSRs), *BnATG101c* (5 SSRs), *BnTORa* (5 SSRs), *BnATG1b* (4 SSRs), *BnATG6a* (3 SSRs), *BnATG11b* (3 SSRs), *BnATG18a* (3 SSRs), *BnTORb* (3 SSRs), *BnATG18l* (4 SSRs), *BnATG8l* (3 SSRs), *BnATG8q* (3 SSRs), *BnATG18x* (3 SSRs), *BnATG20b* (3 SSRs), *BnATG101d* (3 SSRs), *BnVTI12b* (3 SSRs), *BnATG1a* (2 SSRs), *BnATG1f* (2 SSRs), *BnATG1i* (2 SSRs), *BnATG5a* (2 SSRs), *BnATG8c* (2 SSRs), *BnATG8d* (2 SSRs), *BnATG8h* (2 SSRs), *BnATG8j* (2 SSRs), *BnATG8y* (2 SSRs), *BnATG11a* (2 SSRs), *BnATG13a* (2 SSRs), *BnATG13d* (2 SSRs), *BnATG18m* (2 SSRs), *BnATG20a* (2 SSRs), *BnATG101e* (2 SSRs), *BnTORc* (2 SSRs), *BnVPS15a* (2 SSRs), *BnVPS15b* (2 SSRs), and *BnVTI12f* (2 SSRs). The highest frequency was related to tri-nucleotide repeats (62 SSRs), which were followed by tetra-nucleotide repeats (40 SSRs), di-nucleotide repeats (21 SSRs), penta-nucleotide repeats (11 SSRs), and hexa-nucleotide repeats (3 SSRs).

Table 1. Simple sequence repeats detected in *BnATGs*.

Gene ID	Count	Motif	Gene ID	Count	Motif
BnATG1a	2	(GGTT)3, (ATC)4	BnATG13a	2	(AAC)4, (TCC)4
BnATG1b	4	(ATTT)3, (TC)7, (GGA)4, (GGC)4	BnATG13b	1	(TCT)4
BnATG1c	1	(GATG)3	BnATG13d	2	(GAT)4, (TCT)5
BnATG1f	2	(GGTT)3, (ATC)4	BnATG14a	1	(AC)6
BnATG1g	1	(ATTT)3	BnATG14c	1	(AC)6
BnATG1i	2	(GATG)3, (TTTG)3	BnATG14e	1	(GGAAC)3
BnATG3d	1	(GAG)5	BnATG16a	1	(TGATT)3
BnATG4c	1	(GAAGA)3	BnATG16b	1	(TTTGA)6
BnATG4d	1	(TCTA)3	BnATG18a	3	(TTCC)4, (TCT)4, (GGA)5
BnATG5a	2	(CTTT)3, (CCT)5	BnATG18b	1	(GCA)4
BnATG5b	1	(AGA)7	BnATG18e	1	(CTC)4
BnATG5c	1	(TTTC)3	BnATG18g	1	(GGT)4
BnATG6a	3	(GAA)4, (TG)10, (GT)7	BnATG18h	1	(TTTTAT)3
BnATG6b	1	(GAA)4	BnATG18i	1	(CAG)5
BnATG6c	1	(GAA)4	BnATG18l	4	(GAT)4, (AGC)5, (ATG)4, (TTC)4
BnATG6d	1	(GAA)4	BnATG18m	2	(GAT)6, (TTC)4
BnATG7a	1	(TTTC)3	BnATG18r	1	(CTC)4
BnATG8c	2	(TTTG)3, (TTC)4	BnATG18x	3	(ATG)4, (CAT)4, (TTC)5
BnATG8d	2	(TATTT)3, (TTG)5	BnATG20a	2	(AAC)4, (TC)7
BnATG8h	2	(ATTCA)3, (GTT)4	BnATG20b	3	(ATA), (TC)7, (AAC)4
BnATG8j	2	(TCT)5, (AT)12	BnATG101b	1	(TCG)4
BnATG8k	1	(TTTGA)3	BnATG101c	5	(TGGCCT)3, (CTA)6, (TTTA)4, (GATG)3, (CCAT)3
BnATG8l	3	(AAGC)3, (TTGA)3, (TTCT)4	BnATG101d	3	(CCAT)3, (AT)7, (TTC)4
BnATG8o	1	(TA)8	BnATG101e	2	(TCT)4, (TC)6
BnATG8p	1	(CTT)4	BnTORa	5	(TTTA)3, (TC)7, (CT)8, (TC)8, (CTTT)3
BnATG8q	3	(TTTG)3, (TC)7, (TTC)4	BnTORb	3	(TCTT)3, (AT)9, (ATT)4
BnATG8r	1	(TTG)6	BnTORc	2	(TATTT)3, (TTA)5
BnATG8y	2	(AAGC)4, (AT)7	BnVPS15a	2	(TTTC)3, (TTTG)3
BnATG9a	7	(TCAAT)5, (CT)10, (CTC)4, (GAG)5, (GAT)5, (AAGA)3, (GTTA)3	BnVPS15b	2	(TTTC)3, (TTTG)3
BnATG9c	6	(TCAAT)4, (CT)10, (GAT)4, (TATT)3, (AAGA)3, (GTTA)3	BnVPS34a	1	(TA)7
BnATG10a	1	(CGGCAG)3	BnVPS34b	1	(TC)6
BnATG11a	2	(TTCT)3, (TTTA)5	BnVTI12a	1	(AAG)6
BnATG11b	3	(TTAT)3, (AGA)6, (AAC)4	BnVTI12b	3	(CCTT)3, (AAT)5, (TCA)4
BnATG11c	1	(ATC)4	BnVTI12d	1	(AAG)5
BnATG11e	1	(AT)6	BnVTI12e	1	(CCTT)4
BnATG11f	1	(AAC)7	BnVTI12f	2	(ATAC)3, (CTT)4
BnATG12a	1	(TTTGT)3			

2.7. BnATG-Targeted miRNAs Prediction

Due to the present results, 107 miRNAs for 41 *BnATGs* targets were identified (Figure 9 and Supplementary Table S6). miRNA's relationship with their targets was not one by one, and many of them targeted the same gene. For example, *BnATG11d* transcript was co-targeted by seven miRNAs named bna-miR171a, bna-miR171b, bna-miR171c, bna-miR171d, bna-miR171e, and bna-miR6033. In contrast, one miRNA was predicted to target multiple transcripts, such as bna-miR395a, which can suppress the expression of *BnATG18e, BnATG18j, BnATG18o, BnATG18v, BnATG1b,* and *BnATG8s*.

Figure 5. The exon–intron structure of *ATG3–4*, *ATG7–9*, *VPS15*, *VPS34*, *VTI12*, and *TOR* in *B. napus*. Exons and introns were represented by red boxes and green lines, respectively. The exon–intron structure of the *BnATGs* was determined using a gene structure display server (GSDS).

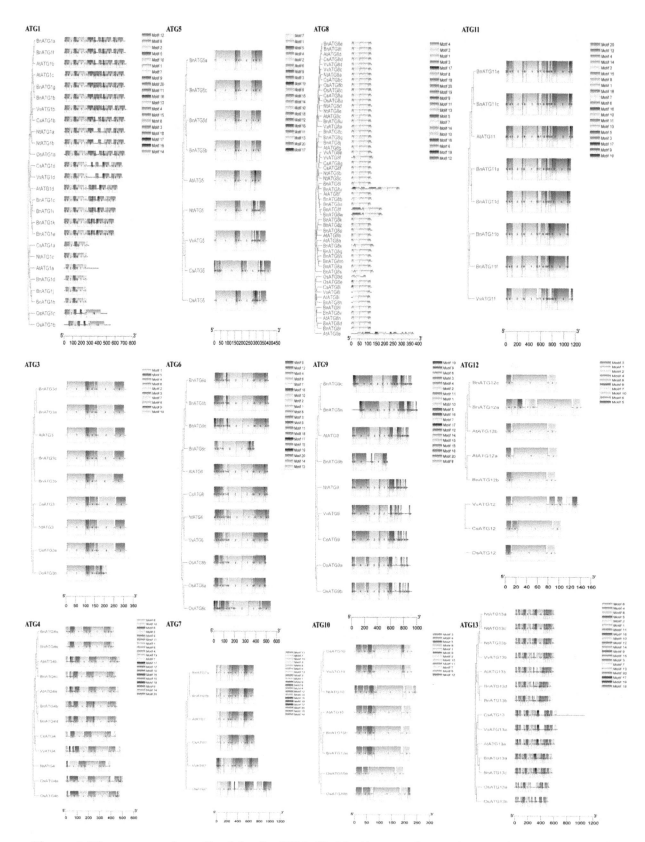

Figure 6. The conserved motifs of the *BnATG1–13*. Motifs were detected using the Multiple Em for Motif Elicitation (MEME) online tool. Different motifs are presented in different colors.

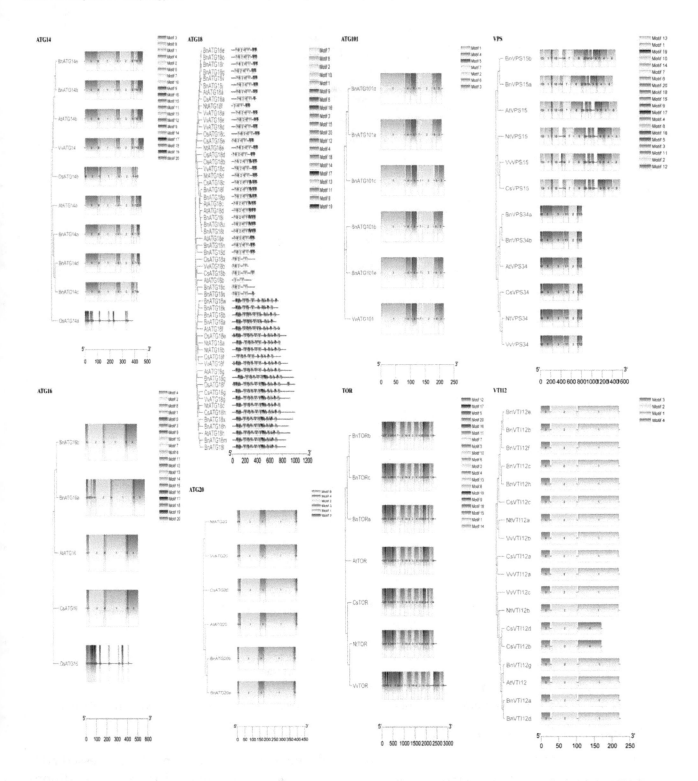

Figure 7. The conserved motifs of the *BnATG14–18*, *BnATG101*, *BnTOR*, *BnVPS*, and *BnVTI12*. Motifs were detected using the MEME online tool. Different motifs are presented in different colors.

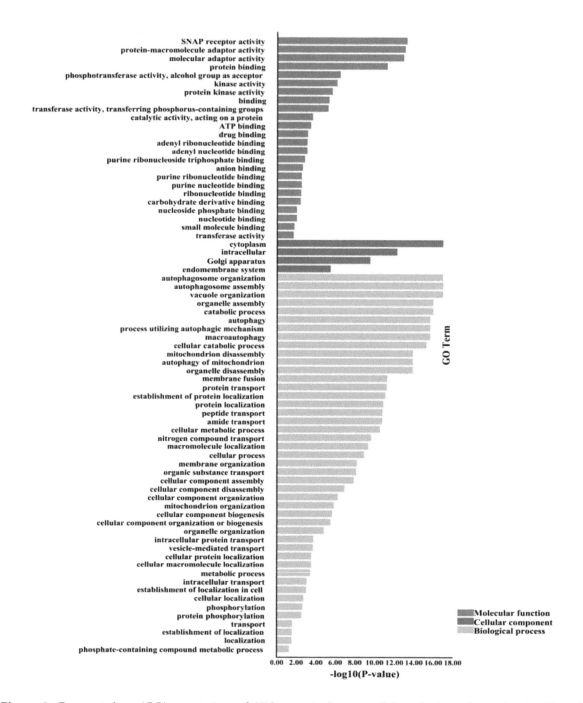

Figure 8. Gene ontology (GO) annotations of *ATG* genes in *B. napus*. GO analysis performed using TBtools.

2.8. The Expression Analysis of BnATGs at Various Developmental Stages

The RNA-seq datasets of Zhang et al. [47] were analyzed to identify differentially expressed ATG genes in roots, stems, leaves, flowers, seeds, and siliques in *B. napus* under normal and stress conditions (Figure 10A). Most of the *BnATGs* were induced in the flowering development, while *BnATG8b, BnATG8s, BnATG18n, BnATGd, BnATGb*, and *BnATG101c* were repressed in this stage. Moreover, all *BnATGs* indicated low-to-high expression in all tissues of *B. napus*, whereas *BnATG1h* (in stem and seed), *BnATG1j* (in root and leaf), *BnATG12b* (in root), *BnATG18d, BnATG18n* (in all tissues except seed), and *BnATG101c* (in leaf and flower) showed no obvious expression.

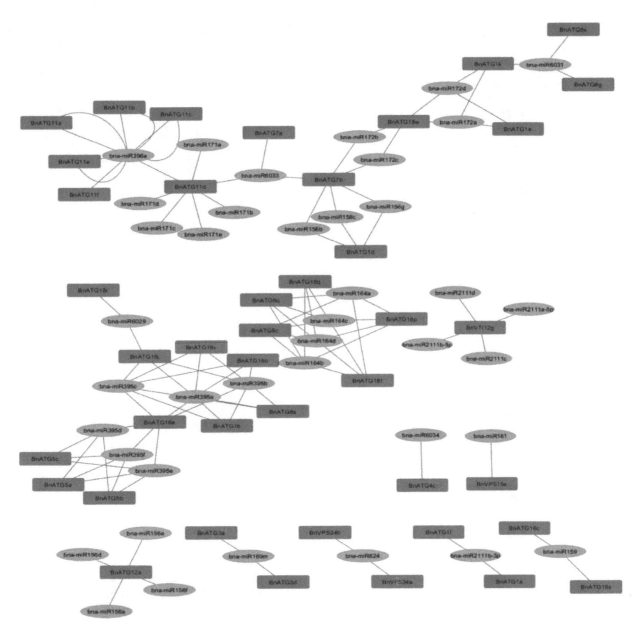

Figure 9. An miRNA-gene network based on interactions between miRNAs and *BnATGs*. Green ellipses represent miRNAs and red rectangles denote genes.

2.9. The Expression Profile of BnATGs under Abiotic Stresses

The RNA-seq datasets of Zhang et al. [47] for *B. napus* related to salt, cold, dehydration, and Abscisic acid (ABA) treatments were analyzed. The results (Figure 10B) indicated that the expression of autophagy genes depends on the stress type. For instance, *BnATG818b* expression was up-regulated under dehydration (after 8 h) and salt stresses (after 4 h), while it showed a moderate-to-low expression under other abiotic stresses. The highest gene expression response to dehydration (after 1 h) was related to *BnATG8p* (increased by 2.74 folds compared to control), while the highest expression in response to dehydration after 8 h was related to *BnATG8q*.

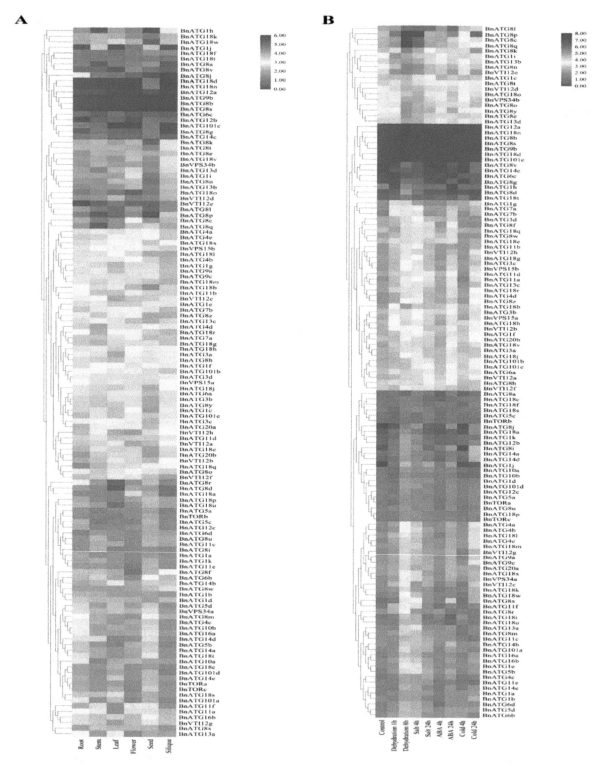

Figure 10. The expression pattern of *BnATG* genes (**A**) in different tissues, (**B**) under diverse abiotic stresses. The color boxes indicate expression values, the lowest (blue), medium (Pale goldenrod), and the highest (red). The heatmap was generated using log10 (TPM + 1) values.

Furthermore, we found that *BnATG8p* expression was up-regulated in response to all abiotic stresses. Indeed, the expression of *BnATG8p* was 2.59, 1.70, 2.00, and 2.26 higher compared to control under dehydration (after 8 h), salinity (after 24 h), ABA (after 24 h), and cold (after 24 h) treatments, respectively. Under dehydration treatment, the expression of all *BnATGs* was increased, whereas the expression of *BnATG8a* and *BnATG12a* was reduced, and the expression of *BnATG8b*, *BnATG8s*,

BnATG9b, and *BnATG101c* unchangeably remained. After 24 h under salinity stress, the expression of the majority of *BnATGs* was up-regulated except *BnATG8s*, *BnATG8v*, *BnATG9b*, *BnATG18n*, and *BnATG101c*, which exhibited a steady expression. Likewise, except *BnATG8b*, *BnATG8s*, and *BnATG9b* genes, which showed no expression, the level of *BnATGs* transcripts increased in response to ABA treatment. Down-regulated expression of *BnATG1h*, *BnATG8d*, *BnATG8v*, *BnATG10b*, *BnATG12a*, *BnATG18c*, *BnATG18i*, *BnATG18k*, *BnATG18s*, *BnATG18u*, and *BnVTI12f* genes was also found. The results of expression analysis were similar in the case of ABA and cold treatments except for *BnATG3d*, *BnATG6b*, *BnATG8f*, *BnATG8h*, *BnATG8l*, *BnATG8m*, *BnATG8o*, *BnATG8w*, *BnATG10a*, *BnATG11c*, *BnATG11e*, *BnATG18q*, *BnATG18t*, *BnATG18u*, *BnATG101a*, and *BnTORb* genes, which exhibited a down-regulation under cold stress and up-regulation under ABA treatment (Supplementary Table S7).

2.10. The Codon Usage Bias Analysis of BnATGs

The results of codon usage bias analysis of *BnATG* are presented in Supplementary Table S8. The GC value for studying genes was in the range of 0.38–0.55, and the GC3s value was calculated to be between 0.32 and 0.66. Due to the significant correlation between GC and GC3, the mutation is the main factor in the formation of codons. Indeed, if the correlation between these two parameters is significant, the mutation is the main factor in the formation of codons. If there had been no correlation between GC and GC3s, the main factor in codon formation would have been a natural selection [48]. codon adaptation index (CAI) is usually used to predict the expression levels of genes, which was in the range of 0.17–0.28 in BnATGs. The closeness of CAI to 1 implies a stronger codon preference and higher gene expression.

A relative synonymous codon usage (RSCU) > 1 shows that the codons are applied more than other synonymous, an RSCU = 1 indicates no preference for codons, and if RSCU < 1, the codons are rarely used by genes [49]. There are 32 codons in *BnVPS34*; 31 codons in *BnATG1*, *BnATG9*, and *BnATG11*; 30 codons in *BnATG3*, *BnATG5*, *BnATG7*, *BnATG8*, and *BnATG14*; 29 codons in *BnVTI12*; 28 codons in *BnATG16*, *BnATG20*, *BnATG101*, and *BnTOR*; 27 codons in *BnATG4*, *BnATG6*, *BnATG10*, and *BnATG13*; and 26 codons in *BnATG12* with RSCU > 1, showing that these codons are the preferred codons for each gene subfamily. The higher RSCU values showed more frequent codons for each subfamily shown in red, while the lower RSCU value is indicated in green color (Figure 11). As shown in Figure 11, *BnATGs* were classified into three clusters based on the RSCU value, including cluster I (*BnATG9*), cluster II (*BnATG1*, *BnATG4*, *BnATG10*, *BnATG14*, *BnATG13*, and *BnVTI12*), and cluster III (*BnTOR*, *BnVPS*, *BnATG101*, *BnATG3*, *BnATG5*, *BnATG6*, *BnATG7*, *BnATG8*, *BnATG11*, *BnATG12*, *BnATG16*, *BnATG18*, and *BnATG20*). Each cluster showed a similar codon preference.

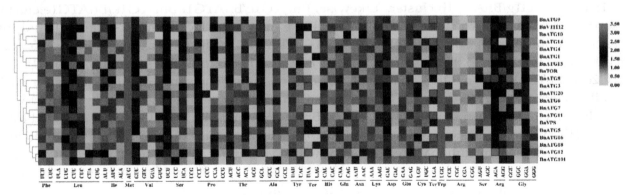

Figure 11. Heat map of relative synonymous codon usage analysis (RSCU) values of BnATGs. The color boxes indicate RSCU values, the lowest (green), and the highest (red) usage of codons. The heatmap was generated using TBtools.

3. Discussion

In present study, a total of 20 core *ATGs* (*ATG1*, *ATG3–4*, *ATG6–14*, *ATG16*, *ATG18*, *ATG20*, *ATG101*, *VPS15*, *VPS34*, *TOR*, and *VTI12*) including 127 members of *ATG* gene family were identified in *B. napus*, which is considerably greater than the number of *ATG* genes in *Z. mays* (45 genes) [29], *Arabidopsis* (40 genes) [42], *S. italica* (37 genes) [38], *V. vinifera* (35 genes) [12], *C. sinensis* (35 genes) [40], *O. sativa* (33 genes) [35], *M. acuminate* (32 genes) [39], *N. tabacum* (30 genes) [36], and *C. annuum* (29 genes) [37]. The *ATG2* subfamily was not identified in the *B. napus*. Due to the number of identified *ATGs* in plants, it can be concluded that there is no significant relationship between genome size and the number of genes. For instance, foxtail millet and tobacco have 37 and 30 *ATG* genes, respectively, while the genome size of these two plants is 490 Mb and 4.5 Gb, respectively. The segmental and tandem duplications can affect the formation of gene families. Therefore, the observed difference in the number of identified *ATGs* may be related to the duplication during plant evolution. The identified *BnATGs* were categorized into ATG1/13 kinase complex, PI3K complex, ATG9/2/18 complex, ubiquitin-like ATG8, and PE conjugation pathway, ubiquitin-like ATG12, and ATG5 conjugation pathway, and SNARE groups (Supplementary Table S1). *BnATG8*, *BnATG18*, and *BnATG1* had multiple copies compared to other core *BnATGs* with 26, 24, and 11 members, respectively. The uneven distribution of *ATGs* has already been observed in rice [35], grapevine [12], and wheat [50]. The *BnATGs* were widely distributed in *B. napus* genome, which can be related to its different ancestors. Studying the gene duplication events is necessary to better understand the expansion of the gene family and the role of genes. In the present survey, both segmental (96.96%) and tandem (3.04%) duplications resulted in the generation of multiple copies of *ATG* genes in *B. napus*. The Ka/Ks ratios of the most duplicated *BnATGs* were <1 except for three duplicated gene pairs (*BnATG16a/16d*, *BnATG8m/BnATG8a*, and *BnATG18g/BnATG18r*) with Ka/Ks > 1 and two duplicated gene pairs (*BnATG1h*, *BnATG1j*, and *BnATG18u*/BnATG18t) with no Ka/Ks value due to the same sequence. It is well known that the alterations in the coding region of duplicated genes during the evolution resulted in different functions associated with amino acid substitution or divergence in exon–intron structure [51]. In general, the importance of the functional role of *ATGs* in *B. napus* was determined due to the strong purifying selection in the *BnATG* gene family.

Considering the phylogenetic tree, a close relation of BnATGs with their counterparts was observed, which related to their sequence conservation and similar function. All members of ATG subfamilies showed similar motif distribution, suggesting that the protein structure was conserved in each subfamily. The phylogenetic distribution of *B. napus* ATG proteins were associated with their motif compositions. BnATG1 subfamily had common motifs 1–2, 5–7, and 10. The difference among the motif compositions of BnATG1a/BnATG1f and BnATG1g/BnATG1b was only the existence of motif 18 in BnATG1g/BnATG1b cluster. Likewise, BnATG1c/BnATG1i/BnATG1k/BnATG1e cluster was separated from the two clusters mentioned above due to the lack of motifs 9 and 20 and the existence of motifs 13 and 17. The BnATG1d/BnATG1j/BnATG1h cluster was completely separated from other BnATG1s due to the lack of motifs 3–4, 8–9, and 11–20. BnATG5s, BnATG7s, BnATG11s, BnATG20s, BnATG101s, BnTORs, and BnVIT12s demonstrated the same motif contents for each subfamily, while the BnATG4b/BnATG4d cluster of BnATG4 subfamily had specific motifs 15, 16, and 18; thus, it clustered in a separated clade. The BnATG6 subfamily, BnATG6c was separated from other BnATGs due to the lack of common motifs 3, 5, 9, 11–12, and 16. BnATG8s were clustered into three clades due to their evolutionary relationships which are similar to their motif contents (Figure 1). Clade I contained motifs 1–4 except for BnATG8o/BnATG8w with motif 5 and BnATG8y with motifs 8–9, 15, 17–18, and 20. The motif contents of clade II were similar to clade I (motifs 1–4) except for BnATG8x and BnATG8s with motif 7. The clade III members contained motifs 1–3, and 6 except for BnATG8h with the lack of motif 6. BnATG9b had only motifs 1, 3, 5–6, 9, 15, and 18; thus, its cluster was separated from other BnATG9s. The difference between BnATG13d/BnATG13b cluster and BnATG13a/BnATG13c clusters was related to the existence of motifs 17–19 in the second cluster. The BnATG14e/BnATG14b cluster was also separated from the BnATG14a/BnATG14d/BnATG14c cluster due to the existence of motifs 5 and 16. BnATg18s were clustered into III groups. Common motifs of Clade I were 1, 2, 5, and 7–10, while clade II contained motifs 1, 2, and 7–8. Likewise, special motifs 3–4, 8, 12–18, and 20 were detected in clade III. BnVPS34 subfamily was completely separated from the BnVPS15 subfamily due to the different motif compositions. Exon–intron structure and splicing phase are important factors in the evolution of gene families [52]. Intron phase 0 and 1 showed the highest and high conservations, respectively, while intron phase 2 indicates the lowest conservation [53,54]. The rate of phases zero and one in all subfamilies including *BnATG1* (84.21%), *BnATG3* (60%), *BnATG4* (66.66), *BnATG5* (89.28%), *BnATG6* (55.88%), *BnATG7* (80%), *BnATG8* (74.31%), *BnATG9* (56.25%), *BnATG10* (100%), *BnATG12* (100%), *BnATG14* 67.3%), *BnATG16* (62.5%), *BnATG18* (76.06%), *BnATG20* (88.8%), *BnVTI12* (100%), *BnTOR* (85.54%), *BnATG101* (83.33%), *BnVPS15* (76%), and *BnVPS34* (75%) was higher than phase two except for *BnATG11* (50%) and *BnATG13* (50%), which ascertained the low diversity in the structure of these genes and high conservation in protein function of the *BnATG* family.

SSRs are short tandem repeats of a simple motif of 1–6 nucleotides, reported to play a significant role in controlling gene expression [55]. In the current paper, tri-nucleotide repeats (45.25%) were higher than other SSRs. The type of dominant SSRs is taxon-dependent, which varies in different plant species, and in general, the frequency of AT repeats is higher in dicot than monocots [56]. In future research, SSR polymorphisms in *BnATGs* may be suitable to select the genotypes with higher levels of resistance to abiotic stresses using marker-assisted selection techniques.

miRNAs are a group of 19–24 bp non-coding small RNAs, playing significant roles in plant growth and response to the environmental stresses through post-transcriptional changes [57]. Therefore, bioinformatics methods have helped predict the target of miRNAs in the shortest possible time. In *BnATG1*, *BnATG3*, *BnATG4*, *BnATG5*, *BnATG8*, *BnATG12*, *BnATG18*, *BnVPS15*, and *BnVTI12* subfamilies, six (*BnATG1a-b*, *BnATG1e-f*, and *BnATG1k*), two (*BnATG3a* and *BnATG3d*), one (*BnATG4c*), five (*BnATG5a-c*), five (*BnATG8c*, *BnATG8g*, *BnATG8q*, *BnATG8s*, and *BnATG8x*), one (*BnATG12a*), 11 (*BnATG18c-f*, *BnATG18j18o-s*, and *BnATG18v-w*), one (*BnVPS15a*), and one (*BnVTI12g*) transcripts were targeted by *B. napus* miRNAs, respectively. Although no target was found in *BnATG6*, *BnATG9*, *BnATG10*, *BnATG13*, *BnATG14*, *BnATG16*, *BnATG101*, and *BnTOR* subfamilies, all members of *BnATG11*, *BnATG7*, and *BnVPS34* were targeted by miRNAs. mir156 is essential to regulate plant transition time from a juvenile to an adult in the vegetative phase [58]. Therefore, it was shown that miR172 is involved in regulating flowering time [59,60]. miR159, miR169, and miR395 play a role in the regulation of seed germination, response to salinity stress, and sulfate starvation, respectively [61–63]. Likewise, miR61 and miR171 are down-regulated under salinity and up-regulated under drought conditions in *B. napus*, respectively [64]. miR164 and miR396 are involved in seed germination and response to abiotic stresses [65–67], and miR824 is progressively accumulated in response to heat exposure [68]. miR2111 is important in response to phosphorus deprivation, which is ascertained in *B. napus* [69]. Finally, miR6029 is necessary for fatty acid biosynthesis during *B. napus* seed development [70].

Considering promoter analysis, the highest number of stress-responsive elements was observed in *BnATG8r* with 12 of 20 stress-related cis-elements. The existence of regulatory elements associated with response to stresses, light, and hormones indicates that *BnATGs* are involved in the plant response to various stresses and biological processes. Likewise, GO annotation found that most *BnATGs* are included in the biological process, most of which are related to stress conditions response. The presence of different regulatory elements, involved in various biological processes, in the promoter of *BnATG* genes and post-transcriptional regulation of these genes by miRNAs show the complex and precise mechanism regulating the expression of these genes that leads to the various functions of *BnATGs*.

An analysis of the expression profile data published by Zhang et al. [47] indicated that the expression profile of genes helps to determine their function. The expression analysis indicated that most *BnATGs* are involved in *B. napus* development and response to abiotic stresses, which is in line with previous researches [9,40,42]. For instance, *BnATG1c*, *BnATG1i*, *BnATG3a-d*, *BnATG6a*, *BnATG7a*, *BnATG8e*, *BnATG8l*, *BnATG8n*, *BnATG8k*, *BnATG8t*, *BnATG8p*, *BnATG8x*, *BnATG8y*, *BnATG8z*, *BnATG13a-d*, *BnATG18e*, *BnATG18o*, *BnATG18v*, *BnATG20a-b*, *Bnvps34b*, *BnVTI12a-b*, and *BnVTI12d-f* revealed a high level of transcripts in seed, suggesting that they may be involved in the regulation of seed development. RNA-seq data analysis revealed that the *ATG* gene expression, in response to abiotic stresses, depends on the stress type and duration. These findings are inconsistent with those of Shangguan et al. [12]. For example, *BnATG18b* and *BnVTI12* genes were up-regulated by salt after 4 h of treatment, while they were repressed after 24 h under salinity condition. Considering the present study, the expression level of *BnATG8p* and *BnATG8q* was the highest among other *BnATGs* under dehydration conditions after 1 and 8 h, respectively. Likewise, the expression of *BnATG8p* and *BnATG8l* was the highest under salinity and ABA treatment. The transcript levels of *BnATG1c* and *BnATG8p* were also higher than other *BnATGs* in response to cold stress after 4 h and 24 h, respectively. These results indicated the potential key role of these genes in *B. napus* response to abiotic stresses, which can be used in future research to develop stress-resistant cultivars. The *BnATGs* exhibited different expression patterns, even in the same subfamily. For instance, in the *BnATG8* subfamily, *BnATG8l*, *BnATG8n*, *BnATG8p*, and *BnATG8q* were significantly up-regulated by dehydration, while no obvious changes in *BnATG8s* and *BnATG8v* expression was observed. In general, the expression of the *BnATG8* subfamily under all abiotic stresses was higher than other subfamilies, which is the following results obtained in *Arabidopsis* [38], foxtail millet [50], and wheat [38,50,71]. ATG8 is a ubiquitin-like protein conjugated to phosphatidylethanolamine (PE) catalyzed by ATG7, ATG3, and ATG12-ATG5 complex [72]. ATG8-PE complex is essential for autophagosome formation through the membrane connection and remodeling [73]. The expression of *ATG8* in wheat and *Arabidopsis* increases plant tolerance to osmotic stress [13,74]. The expression of *GmATG8c* in *Arabidopsis* and *SiATG8a* in rice also improved plant tolerance to nitrogen deficiency [38,75]. Likewise, *ATG8* showed a response to leaf senescence due to nitrate conditions in barley [76].

In this study, the factors involved in codon usage bias (CUB), including expression level, GC value, and mutation were investigated. CUB index varies among genes in each genome. Effective codon number (ENC) value was used to evaluate the effective codon number between 20 and 61.20 [77]. In this study, the ENC value was between 43.8 and 61, indicating that there are various synonymous codons among studying genes. Highly expressed genes have a higher codon preference with higher CAI and lower NC values. However, the genes with low expression have more types of rare codons; thus, they have a lower codon preference and a lower CAI but a higher NC. *BnATG8p*, *BnATG8q*, and *BnATG8l* genes, which had higher expression under different abiotic stresses, revealed almost higher CAI and relatively lower NC. Most of the *BnATGs* showed GC content < 0.5, indicating that *B. napus* ATGs have no perceptible preference for GC nucleotides. Only 12.59% of *BnATGs* revealed GC3s value > 0.5, indicating that codons with A/T end are preferred.

4. Materials and Methods

4.1. In Silico Identification of BnATG Genes

To determine the *BnATG* gene family in *B. napus*, related protein sequences in *Arabidopsis* [42], *O. sativa* [35], *C. sinensis* [40], *N. tabacum* [36], and *V. vinifera* [12] were obtained from Phytozome 12.1.6 database [78]. Then, the protein sequence of *ATG* genes related to the mentioned plants was used to identify *BnATG* genes in the BRAD database [43] using the BlastP algorithm (E.Value < 1e^{-5}) [41]. The existence of ATGs domains in the obtained sequences was assessed using HMMscan databases [79]. Sequences without ATG domains were deleted, and the remaining genes were classified into different ATG groups based on the specific domain of each group. The genes were named by first adding Bn (*Brassica napus*), and then adding the name of the group and the English letters based on the chromosomal location of the genes. Molecular weight (M$_W$), length, and theoretical isoelectric points (pI) of *B. napus* ATGs were calculated using the ProtParam tool of the ExPASY database [80]. To identify the cellular localization of proteins, subCELlular LOcalization (CELLO) has been applied [81].

4.2. Phylogenetic Analysis of B. napus ATG Gene Family

Full-length protein sequence alignment of *B. napus*, *Arabidopsis*, rice, sweet orange, tobacco, and the grapevine was handled using ClustalX 2.0.8 program to evaluate the evolutionary relationships of the *ATG* gene family. A phylogenetic tree of ATG proteins with 1000 bootstraps [82] constructed using MEGA 7 [83] by the neighbor-joining (NJ) method.

4.3. Chromosome Localization, Gene Duplication, and Selection Pressure

ATGs location on *B. napus* chromosomes was obtained from the *Brassica* database (BRAD) database [43]. Genes with a maximum interval of 10 genes on the same chromosome were regarded as tandem duplication [84]. Two criteria are considered to recognize segmental duplication, including more than 80% identification of the aligned region and more than 80% alignment coverage compared to the longer genes [85]. DnaSP ver. 5 software [86] was applied to compute the substitution rates of non-synonymous (Ka) per synonymous (Ks) of the duplicated genes to determine the type of selection pressure. The location of genes on chromosomes and the duplication relationship among them were presented using TBtools [46].

4.4. Exon–Intron Structure and Conserved Motifs

The gene structure of *BnATGs* was assessed using a gene structure display server (GSDS 2.0) [87]. This server evaluates the genomic DNA sequence of each gene based on its coding sequence and presents the exon, intron, 3'-UTR, 5'-UTR, and intron phase of the gene. Multiple Em for Motif Elicitation (MEME 5.0.5) was used to identify the conserved motifs of the *ATG* gene family [88]. A limit of 20 motifs, and minimum and maximum motifs length were 10 and 200, respectively.

4.5. Gene Ontology Annotations and Cis-Regulatory Element Identification

To investigate the functional role of *ATGs* in *B. napus* biology, gene ontology (GO) analysis was performed using TBtools [46]. Three levels of GO classification, including molecular functions, biological processes, and cellular components, were applied to present genes. An amount of 1.5 kb upstream of the initiation codon of *ATG* genes was retrieved from the Ensemble Plants database [89], and cis-acting regulatory elements were identified using PlantCare [90]. Only cis-elements with scores ≥ 6 were examined.

4.6. The Prediction of Simple Sequence Repeats (SSR) Markers and BnATG-Targeted miRNAs

SSR markers were detected in *BnATG* genes using the BatchPrimer3v1.0 server [91]. In the psRNATarget database, CD sequences of them were examined by considering default parameters to find *BnATG*-targeted miRNAs. The relationship between *BnATGs* and identified miRNA molecules was visualized using Cytoscape software [92].

4.7. Analysis of Previously Published B. napus RNA-Seq Data

Transcript data for silique, stem, leaf, flower, and root tissues as well as salinity (200 mM), ABA (25 μM), and cold (4C) stresses at 4 and 24 h after treatment and dehydration stress at 1 and 8 h after treatment are related to the study of Zhang et al. [47]. These data are available with the project ID CRA001775 at [93]. Initial quality analysis was performed on FastQ files using FastQC software [94], then preprocessing of raw sequence data was conducted and low quality reads, adapter sequences, and duplicate mapping reads were filtered using Trimmomatic on Linux [95]. The preprocessed FastQ files were aligned to the Brassica napus reference genome using STAR [96]. The counts obtained from STAR normalized to transcript per million (TPM). Log2 (TPM + 1) used to generate the heatmap utilizing TBtools [46]. Clustering the data was performed using the Pearson correlation coefficient and the complete linkage method.

4.8. Codon Usage Bias Analysis

The sequences were evaluated for the frequency of optimal codons (FOP), GC content, GC content at the third site position of codon (GC3s), codon adaptation index (CAI), effective codon number (ENC), and relative synonymous codon usage (RSCU) for the BnATGs using CodonW 1.4.2 [97]. Statistical analysis was performed using Excel software. Clustering the data was performed using the Pearson correlation coefficient and the complete linkage method using TBtools [46].

5. Conclusions

In recent years, bioinformatics tools have been used to identify important genes in increasing plant tolerance to biotic and abiotic stresses. On the other hand, autophagy-related genes have important roles in plant growth, development, and responding to environmental stresses. Therefore, in the present study, 127 BnATGs were detected using bioinformatics tools in rapeseed. The reason for expanding this gene family was the tandem and segmental duplications. Ka/Ks for most of the paired genes were <1, indicating a negative selection during the evolution of *B. napus* plants to maintain their function. Promoter analysis showed hormone- and stress-responsive elements in the *BnATGs* promoters, which is in line with gene ontology results suggesting their role in various plant biological processes. Likewise, the expression patterns of *BnATGs* ascertained their roles in different tissues under various environmental stresses in *B. napus*, which can be applied to develop stress-resistant cultivars in future studies. Besides, the mutation was the main factor in the formation of *BnATGs* codons. In addition, they are more likely to have A/T at the 3' end of their codons. This study was performed to detect the molecular evolution and the possible function of *BnATGs*, and it has provided useful information about *BnATGs* for future studies.

Supplementary Materials
Table S1: Features of rapeseed ATG proteins, ATG1/13 kinase complex, PI3K complex, ATG9/2/18 complex, ubiquitin-like ATG8, and PE conjugation pathway, ubiquitin-like ATG12, and ATG5 conjugation pathway, and SNARE groups shown in yellow, green, gray, red, blue, and pink colors, respectively; Table S2: Ka/Ks analysis of the BnATG duplicated paired genes, positive selection, segmental, and tandem duplications shown in yellow, blue, and green colors, respectively; Table S3: ATG protein domain. The existence of ATGs domains in the Arabidopsis, *O. sativa*, *C. sinensis*, *N. tabaccum*, and *V. vinifera* sequences was assessed using HMMscan databases; Table S4: Gene annotations of BnATGs, biological process, cellular component, and molecular function are shown in yellow, brown, and blue colors, respectively; Table S5: List of cis-acting elements in BnATGs promoters, cell-cycle- and tissue-specific, hormone-responsive, light-responsive, regulatory, stress-responsive, and unknown function elements are shown in red, yellow, blue, brown, green, and gray colors, respectively; Table S6: Putative *BnATGs*-targeted miRNA; Table S7: RNA-seq data analysis of *BnATGs*, the color boxes indicate expression values, the lowest (blue), medium (Pale goldenrod), and the highest (red); Table S8: Relative synonymous codon usage analysis (RSCU) of BnATGs.

Author Contributions: Conceptualization, C.J., A.A. and E.M.E.; methodology, E.M.E.; software, A.A. and E.M.E.; validation, C.J. and A.A.; formal analysis, M.K.; investigation, E.M.E., and A.A. resources, A.A.; data curation, Z.H. and A.A.; writing—original draft preparation, Z.H. and E.M.E.; writing—review and editing, C.J. and M.K.; visualization, A.A. and E.M.E.; supervision, C.J. and M.K.; project administration, M.K. All authors have read and agreed to the published version of the manuscript.

References

1. Li, F.; Vierstra, R.D. Autophagy: A multifaceted intracellular system for bulk and selective recycling. *Trends Plant Sci.* **2012**, *17*, 526–537. [CrossRef] [PubMed]
2. Klionsky, D.J.; Meijer, A.J.; Codogno, P.; Neufeld, T.P.; Scott, R.C. Autophagy and p70S6 Kinase. *Autophagy* **2005**, *1*, 59–61. [CrossRef] [PubMed]
3. Wen, X.; Klionsky, D.J. An overview of macroautophagy in yeast. *J. Mol. Biol.* **2016**, *428*, 1681–1699. [CrossRef] [PubMed]
4. Li, W.-W.; Li, J.; Bao, J. Microautophagy: Lesser-known self-eating. *Cell. Mol. Life Sci.* **2011**, *69*, 1125–1136. [CrossRef] [PubMed]

5. Avin-Wittenberg, T.; Baluška, F.; Bozhkov, P.V.; Elander, P.H.; Fernie, A.R.; Galili, G.; Hassan, A.; Hofius, D.; Isono, E.; Le Bars, R.; et al. Autophagy-related approaches for improving nutrient use efficiency and crop yield protection. *J. Exp. Bot.* **2018**, *69*, 1335–1353. [CrossRef] [PubMed]

6. Chen, L.; Liao, B.; Qin-Fang, C.; Xie, L.-J.; Huang, L.; Tan, W.-J.; Zhai, N.; Yuan, L.-B.; Zhou, Y.; Yu, L.-J.; et al. Autophagy contributes to regulation of the hypoxia response during submergence in Arabidopsis thaliana. *Autophagy* **2015**, *11*, 2233–2246. [CrossRef] [PubMed]

7. Guiboileau, A.; Yoshimoto, K.; Soulay, F.; Bataillé, M.-P.; Avice, J.-C.; Masclaux-Daubresse, C. Autophagy machinery controls nitrogen remobilization at the whole-plant level under both limiting and ample nitrate conditions in Arabidopsis. *New Phytol.* **2012**, *194*, 732–740. [CrossRef] [PubMed]

8. Li, W.-W.; Chen, M.; Zhong, L.; Liu, J.-M.; Xu, Z.-S.; Li, L.-C.; Zhou, Y.-B.; Guo, C.-H.; Ma, Y.-Z. Overexpression of the autophagy-related gene SiATG8a from foxtail millet (Setaria italica L.) confers tolerance to both nitrogen starvation and drought stress in Arabidopsis. *Biochem. Biophys. Res. Commun.* **2015**, *468*, 800–806. [CrossRef] [PubMed]

9. Luo, L.; Zhang, P.; Zhu, R.; Fu, J.; Su, J.; Zheng, J.; Wang, Z.; Wang, D.; Gong, Q. Autophagy Is Rapidly Induced by Salt Stress and Is Required for Salt Tolerance in Arabidopsis. *Front. Plant Sci.* **2017**, *8*, 1459. [CrossRef]

10. Slobodkin, M.R.; Elazar, Z. The Atg8 family: Multifunctional ubiquitin-like key regulators of autophagy. *Essays Biochem.* **2013**, *55*, 51–64. [CrossRef] [PubMed]

11. Yoshimoto, K.; Hanaoka, H.; Sato, S.; Kato, T.; Tabata, S.; Noda, T.; Ohsumi, Y. Processing of ATG8s, Ubiquitin-Like Proteins, and Their Deconjugation by ATG4s Are Essential for Plant Autophagy. *Plant Cell* **2004**, *16*, 2967–2983. [CrossRef] [PubMed]

12. Shangguan, L.; Fang, X.; Chen, L.; Cui, L.; Fang, J. Genome-wide analysis of autophagy-related genes (ARGs) in grapevine and plant tolerance to copper stress. *Planta* **2018**, *247*, 1449–1463. [CrossRef] [PubMed]

13. Kuzuoğlu-Öztürk, D.; Yalcinkaya, O.C.; Akpınar, B.A.; Mitou, G.; Korkmaz, G.; Gozuacik, D.; Budak, H.; Akpinar, B.A. Autophagy-related gene, TdAtg8, in wild emmer wheat plays a role in drought and osmotic stress response. *Planta* **2012**, *236*, 1081–1092. [CrossRef] [PubMed]

14. Liu, Y.; Xiong, Y.; Bassham, D.C. Autophagy is required for tolerance of drought and salt stress in plants. *Autophagy* **2009**, *5*, 954–963. [CrossRef]

15. Shin, J.-H.; Yoshimoto, K.; Ohsumi, Y.; Jeon, J.-S.; An, G. OsATG10b, an autophagosome component, is needed for cell survival against oxidative stresses in rice. *Mol. Cells* **2009**, *27*, 67–74. [CrossRef]

16. Minina, E.A.; Moschou, P.N.; Vetukuri, R.R.; Sanchez-Vera, V.; Cardoso, C.; Liu, Q.; Elander, P.H.; Dalman, K.; Beganovic, M.; Yilmaz, J.L.; et al. Transcriptional stimulation of rate-limiting components of the autophagic pathway improves plant fitness. *J. Exp. Bot.* **2018**, *69*, 1415–1432. [CrossRef]

17. Wang, P.; Sun, X.; Jia, X.; Ma, F. Apple autophagy-related protein MdATG3s afford tolerance to multiple abiotic stresses. *Plant Sci.* **2017**, *256*, 53–64. [CrossRef]

18. Hanaoka, H.; Noda, T.; Shirano, Y.; Kato, T.; Hayashi, H.; Shibata, D.; Tabata, S.; Ohsumi, Y. Leaf Senescence and Starvation-Induced Chlorosis Are Accelerated by the Disruption of an Arabidopsis Autophagy Gene. *Plant Physiol.* **2002**, *129*, 1181–1193. [CrossRef]

19. Islam, S.; Proshad, R.; Kormoker, T.; Tusher, T.R. Autophagy-mediated Nutrient Recycling and Regulation in Plants: A Molecular View. *J. Plant Biol.* **2019**, *62*, 307–319. [CrossRef]

20. Klionsky, D.J.; Abdelmohsen, K.; Abe, A.; Abedin, J.; Abeliovich, H.; Arozena, A.A.; Adachi, H.; Adams, C.M.; Adams, P.D.; Adeli, K.; et al. Guidelines for the use and interpretation of assays for monitoring autophagy 3rd ed. *Autophagy* **2016**, *12*, 1–222. [CrossRef]

21. Tang, J.; Bassham, D.C. Autophagy in crop plants: What's new beyond Arabidopsis? *Open Biol.* **2018**, *8*, 180162. [CrossRef] [PubMed]

22. Su, T.; Li, X.; Yang, M.; Shao, Q.; Zhao, Y.; Ma, C.; Wang, P. Autophagy: An Intracellular Degradation Pathway Regulating Plant Survival and Stress Response. *Front. Plant Sci.* **2020**, *11*, 164. [CrossRef] [PubMed]

23. Li, S.; Yan, H.; Mei, W.; Tse, Y.C.; Wang, H. Boosting autophagy in sexual reproduction: A plant perspective. *New Phytol.* **2020**, *226*, 679–689. [CrossRef] [PubMed]

24. Chung, T.; Suttangkakul, A.; Vierstra, R.D. The ATG Autophagic Conjugation System in Maize: ATG Transcripts and Abundance of the ATG8-Lipid Adduct Are Regulated by Development and Nutrient Availability. *Plant Physiol.* **2008**, *149*, 220–234. [CrossRef] [PubMed]

25. Breeze, E.; Harrison, E.; McHattie, S.; Hughes, L.; Hickman, R.; Hill, C.; Kiddle, S.; Kim, Y.-S.; Penfold, C.A.; Jenkins, D.; et al. High-Resolution Temporal Profiling of Transcripts during Arabidopsis Leaf Senescence Reveals a Distinct Chronology of Processes and Regulation. *Plant Cell* **2011**, *23*, 873–894. [CrossRef]

26. Di Berardino, J.; Marmagne, A.; Berger, A.; Yoshimoto, K.; Cueff, G.; Chardon, F.; Masclaux-Daubresse, C.; Reisdorf-Cren, M. Autophagy controls resource allocation and protein storage accumulation in Arabidopsis seeds. *J. Exp. Bot.* **2018**, *69*, 1403–1414. [CrossRef]

27. Kurusu, T.; Koyano, T.; Hanamata, S.; Kubo, T.; Noguchi, Y.; Yagi, C.; Nagata, N.; Yamamoto, T.; Ohnishi, T.; Okazaki, Y.; et al. OsATG7 is required for autophagy-dependent lipid metabolism in rice postmeiotic anther development. *Autophagy* **2014**, *10*, 878–888. [CrossRef]

28. Wada, S.; Hayashida, Y.; Izumi, M.; Kurusu, T.; Hanamata, S.; Kanno, K.; Kojima, S.; Yamaya, T.; Kuchitsu, K.; Makino, A.; et al. Autophagy Supports Biomass Production and Nitrogen Use Efficiency at the Vegetative Stage in Rice. *Plant Physiol.* **2015**, *168*, 60–73. [CrossRef]

29. Li, F.; Chung, T.; Pennington, J.G.; Federico, M.L.; Kaeppler, H.F.; Kaeppler, S.M.; Otegui, M.S.; Vierstra, R.D. Autophagic Recycling Plays a Central Role in Maize Nitrogen Remobilization. *Plant Cell* **2015**, *27*, 1389–1408. [CrossRef]

30. Zomorodian, A.; Kavoosi, Z.; Momenzadeh, L. Determination of EMC isotherms and appropriate mathematical models for canola. *Food Bioprod. Process.* **2011**, *89*, 407–413. [CrossRef]

31. Selvam, A.; Wong, J.W. Cadmium uptake potential of Brassica napus cocropped with Brassica parachinensis and Zea mays. *J. Hazard. Mater.* **2009**, *167*, 170–178. [CrossRef]

32. Wang, B.; Liu, L.; Gao, Y.; Chen, J. Improved phytoremediation of oilseed rape (Brassica napus) by Trichoderma mutant constructed by restriction enzyme-mediated integration (REMI) in cadmium polluted soil. *Chemosphere* **2009**, *74*, 1400–1403. [CrossRef] [PubMed]

33. Fiebelkorn, D.; Rahman, M. Development of a protocol for frost-tolerance evaluation in rapeseed/canola (*Brassica napus* L.). *Crop. J.* **2016**, *4*, 147–152. [CrossRef]

34. Bassham, D.C.; Laporte, M.; Marty, F.; Moriyasu, Y.; Ohsumi, Y.; Olsen, L.J.; Yoshimoto, K. Autophagy in Development and Stress Responses of Plants. *Autophagy* **2006**, *2*, 2–11. [CrossRef] [PubMed]

35. Xia, K.; Liu, T.; Ouyang, J.; Wang, R.; Fan, T.; Zhang, M. Genome-Wide Identification, Classification, and Expression Analysis of Autophagy-Associated Gene Homologues in Rice (*Oryza sativa* L.). *DNA Res.* **2011**, *18*, 363–377. [CrossRef] [PubMed]

36. Zhou, X.-M.; Zhao, P.; Wang, W.; Zou, J.; Cheng, T.-H.; Peng, X.-B.; Sun, M.-X. A comprehensive, genome-wide analysis of autophagy-related genes identified in tobacco suggests a central role of autophagy in plant response to various environmental cues. *DNA Res.* **2015**, *22*, 245–257. [CrossRef] [PubMed]

37. Ezhai, Y.; Eguo, M.; Ewang, H.; Elu, J.; Eliu, J.; Ezhang, C.; Egong, Z.-H.; Lu, M.-H. Autophagy, a Conserved Mechanism for Protein Degradation, Responds to Heat, and Other Abiotic Stresses in Capsicum annuum L. *Front. Plant Sci.* **2016**, *7*, 131. [CrossRef]

38. Li, W.; Chen, M.; Wang, E.; Hu, L.; Hawkesford, M.J.; Zhong, L.; Chen, Z.; Xu, Z.; Li, L.; Zhou, Y.; et al. Genome-wide analysis of autophagy-associated genes in foxtail millet (*Setaria italica* L.) and characterization of the function of SiATG8a in conferring tolerance to nitrogen starvation in rice. *BMC Genom.* **2016**, *17*, 797. [CrossRef]

39. Wei, Y.; Liu, W.; Hu, W.; Liu, G.; Wu, C.; Liu, W.; Zeng, H.; He, C.; Shi, H. Genome-wide analysis of autophagy-related genes in banana highlights MaATG8s in cell death and autophagy in immune response to Fusarium wilt. *Plant Cell Rep.* **2017**, *36*, 1237–1250. [CrossRef]

40. Fu, X.-Z.; Zhou, X.; Xu, Y.-Y.; Hui, Q.-L.; Chang-Pin, C.; Li-Li, L.; Peng, L.-Z. Comprehensive Analysis of Autophagy-Related Genes in Sweet Orange (Citrus sinensis) Highlights Their Roles in Response to Abiotic Stresses. *Int. J. Mol. Sci.* **2020**, *21*, 2699. [CrossRef]

41. Han, B.; Xu, H.; Feng, Y.; Xu, W.; Cui, Q.; Liu, A. Genomic Characterization and Expressional Profiles of Autophagy-Related Genes (ATGs) in Oilseed Crop Castor Bean (*Ricinus communis* L.). *Int. J. Mol. Sci.* **2020**, *21*, 562. [CrossRef] [PubMed]

42. Han, S.; Yu, B.; Wang, Y.; Liu, Y. Role of plant autophagy in stress response. *Protein Cell* **2011**, *2*, 784–791. [CrossRef] [PubMed]

43. Cheng, F.; Liu, S.; Wu, J.; Fang, L.; Sun, S.; Liu, B.; Li, P.; Hua, W.; Wang, X. BRAD, the genetics and genomics database for Brassica plants. *BMC Plant Biol.* **2011**, *11*, 136. [CrossRef] [PubMed]

44. Sharp, P.A. Speculations on RNA splicing. *Cell* **1981**, *23*, 643–646. [CrossRef]

45. Altenhoff, A.M.; Studer, R.A.; Robinson-Rechavi, M.; Dessimoz, C. Resolving the Ortholog Conjecture: Orthologs Tend to Be Weakly, but Significantly, More Similar in Function than Paralogs. *PLoS Comput. Biol.* **2012**, *8*, e1002514. [CrossRef]

46. Chen, C.; Chen, H.; He, Y.; Xia, R. TBtools, a toolkit for biologists integrating various biological data handling tools with a user-friendly interface. *BioRxiv* **2018**, 289660. [CrossRef]

47. Zhang, Y.; Ali, U.; Zhang, G.; Yu, L.; Fang, S.; Iqbal, S.; Li, H.; Lu, S.; Guo, L. Transcriptome analysis reveals genes commonly responding to multiple abiotic stresses in rapeseed. *Mol. Breed.* **2019**, *39*, 158. [CrossRef]

48. Sueoka, N.; Kawanishi, Y. DNA G+C content of the third codon position and codon usage biases of human genes. *Gene* **2000**, *261*, 53–62. [CrossRef]

49. Sharp, P.M.; Tuohy, T.M.; Mosurski, K.R. Codon usage in yeast: Cluster analysis clearly differentiates highly and lowly expressed genes. *Nucl. Acids Res.* **1986**, *14*, 5125–5143. [CrossRef]

50. Yue, W.; Nie, X.; Cui, L.; Zhi, Y.; Zhang, T.; Du, X.; Song, W. Genome-wide sequence and expressional analysis of autophagy Gene family in bread wheat (*Triticum aestivum* L.). *J. Plant Physiol.* **2018**, *229*, 7–21. [CrossRef]

51. Xu, G.; Guo, C.; Shan, H.; Kong, H. Divergence of duplicate genes in exon-intron structure. *Proc. Natl. Acad. Sci. USA* **2012**, *109*, 1187–1192. [CrossRef] [PubMed]

52. Wang, G.-M.; Yin, H.; Qiao, X.; Tan, X.; Gu, C.; Wang, B.-H.; Cheng, R.; Wang, Y.-Z.; Zhang, S.-L. F-box genes: Genome-wide expansion, evolution and their contribution to pollen growth in pear (Pyrus bretschneideri). *Plant Sci.* **2016**, *253*, 164–175. [CrossRef] [PubMed]

53. Bai, C.; Sen, P.; Hofmann, K.; Ma, L.; Goebl, M.; Harper, J.; Elledge, S.J. SKP1 Connects Cell Cycle Regulators to the Ubiquitin Proteolysis Machinery through a Novel Motif, the F-Box. *Cell* **1996**, *86*, 263–274. [CrossRef]

54. Dinant, S.; Clark, A.M.; Zhu, Y.; Vilaine, F.; Palauqui, J.-C.; Kusiak, C.; Thompson, G.A. Diversity of the Superfamily of Phloem Lectins (Phloem Protein 2) in Angiosperms. *Plant Physiol.* **2003**, *131*, 114–128. [CrossRef] [PubMed]

55. Haasl, R.J.; Payseur, B.A. Microsatellites as targets of natural selection. *Mol. Biol. Evol.* **2012**, *30*, 285–298. [CrossRef]

56. Qin, Z.; Wang, Y.; Wang, Q.; Li, A.; Hou, F.; Zhang, L. Evolution Analysis of Simple Sequence Repeats in Plant Genome. *PLoS ONE* **2015**, *10*, e0144108. [CrossRef] [PubMed]

57. Wang, Z.; Qiao, Y.; Zhang, J.; Shi, W.; Zhang, J. Genome wide identification of microRNAs involved in fatty acid and lipid metabolism of Brassica napus by small RNA and degradome sequencing. *Gene* **2017**, *619*, 61–70. [CrossRef] [PubMed]

58. Wu, G.; Park, M.Y.; Conway, S.R.; Wang, J.-W.; Weigel, D.; Poethig, R.S. The Sequential Action of miR156 and miR172 Regulates Developmental Timing in Arabidopsis. *Cell* **2009**, *138*, 750–759. [CrossRef]

59. Chuck, G.; Cigan, A.M.; Saeteurn, K.; Hake, S. The heterochronic maize mutant Corngrass1 results from overexpression of a tandem microRNA. *Nat. Genet.* **2007**, *39*, 544–549. [CrossRef]

60. Zhao, L.; Kim, Y.; Dinh, T.T.; Chen, X. miR172 regulates stem cell fate and defines the inner boundary of APETALA3 and PISTILLATA expression domain in Arabidopsis floral meristems. *Plant J.* **2007**, *51*, 840–849. [CrossRef]

61. Jones-Rhoades, M.W.; Bartel, D.P. Computational Identification of Plant MicroRNAs and Their Targets, Including a Stress-Induced miRNA. *Mol. Cell* **2004**, *14*, 787–799. [CrossRef] [PubMed]

62. Nonogaki, H. MicroRNA Gene Regulation Cascades During Early Stages of Plant Development. *Plant Cell Physiol.* **2010**, *51*, 1840–1846. [CrossRef]

63. Zhao, B.; Ge, L.; Liang, R.-Q.; Li, W.; Ruan, K.; Lin, H.; Jin, Y. Members of miR-169 family are induced by high salinity and transiently inhibit the NF-YA transcription factor. *BMC Mol. Biol.* **2009**, *10*, 29. [CrossRef] [PubMed]

64. Jian, H.; Wang, J.; Wang, T.; Wei, L.; Li, J.; Liu, L. Identification of Rapeseed MicroRNAs Involved in Early Stage Seed Germination under Salt and Drought Stresses. *Front. Plant Sci.* **2016**, *7*, 658. [CrossRef] [PubMed]

65. Liu, H.-H.; Tian, X.; Li, Y.-J.; Wu, C.-A.; Zheng, C.-C. Microarray-based analysis of stress-regulated microRNAs in Arabidopsis thaliana. *RNA* **2008**, *14*, 836–843. [CrossRef] [PubMed]

66. Nakashima, K.; Takasaki, H.; Mizoi, J.; Shinozaki, K.; Yamaguchi-Shinozaki, K. NAC transcription factors in plant abiotic stress responses. *Biochim. Biophys. Acta (BBA) Bioenerg.* **2012**, *1819*, 97–103. [CrossRef] [PubMed]

67. Zhang, J.; Zhang, S.; Han, S.; Li, X.; Tong, Z.; Qi, L. Deciphering Small Noncoding RNAs during the Transition from Dormant Embryo to Germinated Embryo in Larches (Larix leptolepis). *PLoS ONE* **2013**, *8*, e81452. [CrossRef]

68. Szaker, H.M.; Darkó, É.; Medzihradszky, A.; Janda, T.; Liu, H.-C.; Charng, Y.-Y.; Csorba, T. miR824/AGAMOUS-LIKE16 Module Integrates Recurring Environmental Heat Stress Changes to Fine-Tune Poststress Development. *Front. Plant Sci.* **2019**, *10*, 10. [CrossRef]

69. Pant, B.D.; Musialak-Lange, M.; Nuc, P.; May, P.; Buhtz, A.; Kehr, J.; Walther, D.; Scheible, W.-R. Identification of Nutrient-Responsive Arabidopsis and Rapeseed MicroRNAs by Comprehensive Real-Time Polymerase Chain Reaction Profiling and Small RNA Sequencing. *Plant Physiol.* **2009**, *150*, 1541–1555. [CrossRef]

70. Wang, J.; Jian, H.; Wang, T.; Wei, L.; Li, J.; Li, C.; Liu, L. Identification of microRNAs Actively Involved in Fatty Acid Biosynthesis in Developing Brassica napus Seeds Using High-Throughput Sequencing. *Front. Plant Sci.* **2016**, *7*, 1570. [CrossRef]

71. Slávikova, S.; Shy, G.; Yao, Y.; Glozman, R.; Levanony, H.; Pietrokovski, S.; Elazar, Z.; Egalili, G. The autophagy-associated Atg8 gene family operates both under favourable growth conditions and under starvation stresses in Arabidopsis plants. *J. Exp. Bot.* **2005**, *56*, 2839–2849. [CrossRef] [PubMed]

72. Bu, F.; Yang, M.; Guo, X.; Huang, W.; Chen, L. Multiple Functions of ATG8 Family Proteins in Plant Autophagy. *Front. Cell Dev. Biol.* **2020**, *8*, 466. [CrossRef] [PubMed]

73. Nakatogawa, H.; Ichimura, Y.; Ohsumi, Y. Atg8, a Ubiquitin-like Protein Required for Autophagosome Formation, Mediates Membrane Tethering and Hemifusion. *Cell* **2007**, *130*, 165–178. [CrossRef] [PubMed]

74. Slavikova, S.; Ufaz, S.; Avin-Wittenberg, T.; Levanony, H.; Egalili, G. An autophagy-associated Atg8 protein is involved in the responses of Arabidopsis seedlings to hormonal controls and abiotic stresses. *J. Exp. Bot.* **2008**, *59*, 4029–4043. [CrossRef]

75. Xia, T.; Xiao, D.; Liu, N.; Chai, W.; Gong, Q.; Wang, N.N. Heterologous Expression of ATG8c from Soybean Confers Tolerance to Nitrogen Deficiency and Increases Yield in Arabidopsis. *PLoS ONE* **2012**, *7*, e37217. [CrossRef]

76. Avila-Ospina, L.; Marmagne, A.; Soulay, F.; Masclaux-Daubresse, C. Identification of Barley (*Hordeum vulgare* L.) Autophagy Genes and Their Expression Levels during Leaf Senescence, Chronic Nitrogen Limitation and in Response to Dark Exposure. *Agronomy* **2016**, *6*, 15. [CrossRef]

77. He, H.; Liang, G.; Lu, S.; Wang, P.; Liu, T.; Ma, Z.; Zuo, C.; Sun, X.; Chen, B.; Mao, J. Genome-Wide Identification and Expression Analysis of GA2ox, GA3ox, and GA20ox Are Related to Gibberellin Oxidase Genes in Grape (*Vitis Vinifera* L.). *Genes* **2019**, *10*, 680. [CrossRef]

78. Goodstein, D.M.; Shu, S.; Howson, R.; Neupane, R.; Hayes, R.D.; Fazo, J.; Mitros, T.; Dirks, W.; Hellsten, U.; Putnam, N.; et al. Phytozome: A comparative platform for green plant genomics. *Nucleic Acids Res.* **2011**, *40*, D1178–D1186. [CrossRef]

79. Finn, R.D.; Bateman, A.; Clements, J.; Coggill, P.; Eberhardt, R.Y.; Eddy, S.R.; Heger, A.; Hetherington, K.; Holm, L.; Mistry, J.; et al. Pfam: The protein families database. *Nucleic Acids Res.* **2013**, *42*, D222–D230. [CrossRef]

80. Artimo, P.; Jonnalagedda, M.; Arnold, K.; Baratin, D.; Csardi, G.; De Castro, E.; Duvaud, S.; Flegel, V.; Fortier, A.; Gasteiger, E.; et al. ExPASy: SIB bioinformatics resource portal. *Nucleic Acids Res.* **2012**, *40*, W597–W603. [CrossRef]

81. Yu, C.-S.; Chen, Y.-C.; Lu, C.-H.; Hwang, J.-K. Prediction of protein subcellular localization. *Proteins Struct. Funct. Bioinform.* **2006**, *64*, 643–651. [CrossRef]

82. Felsenstein, J. Confidence Limits on Phylogenies: An Approach Using the Bootstrap. *Evolution* **1985**, *39*, 783. [CrossRef]

83. Kumar, S.; Stecher, G.; Tamura, K. MEGA7: Molecular Evolutionary Genetics Analysis Version 7.0 for Bigger Datasets. *Mol. Biol. Evol.* **2016**, *33*, 1870–1874. [CrossRef] [PubMed]

84. Wei, K.; Pan, S.; Li, Y. Functional Characterization of Maize C2H2 Zinc-Finger Gene Family. *Plant Mol. Biol. Rep.* **2015**, *34*, 761–776. [CrossRef]

85. Wu, C.; Ding, X.; Ding, Z.; Tie, W.; Yan, Y.; Wang, Y.; Yang, H.; Hu, W. The Class III Peroxidase (POD) Gene Family in Cassava: Identification, Phylogeny, Duplication, and Expression. *Int. J. Mol. Sci.* **2019**, *20*, 2730. [CrossRef]

86. Librado, P.; Rozas, J. DnaSP v5: A software for comprehensive analysis of DNA polymorphism data. *Bioinformatics* **2009**, *25*, 1451–1452. [CrossRef]

87. Hu, B.; Jin, J.; Guo, A.-Y.; Zhang, H.; Luo, J.; Gao, G. GSDS 2.0: An upgraded gene feature visualization server. *Bioinformatics* **2014**, *31*, 1296–1297. [CrossRef]

88. Bailey, T.L.; Williams, N.; Misleh, C.; Li, W.W. MEME: Discovering and analyzing DNA and protein sequence motifs. *Nucleic Acids Res.* **2006**, *34*, W369–W373. [CrossRef]

89. Howe, K.L.; Contreras-Moreira, B.; De Silva, N.; Maslen, G.; Akanni, W.; Allen, J.; Alvarez-Jarreta, J.; Barba, M.; Bolser, D.M.; Cambell, L.; et al. Ensembl Genomes 2020—enabling non-vertebrate genomic research. *Nucleic Acids Res.* **2019**, *48*, D689–D695. [CrossRef]

90. Lescot, M. PlantCARE, a database of plant cis-acting regulatory elements and a portal to tools for in silico analysis of promoter sequences. *Nucleic Acids Res.* **2002**, *30*, 325–327. [CrossRef]

91. You, F.M.; Huo, N.; Gu, Y.Q.; Luo, M.-C.; Ma, Y.; Hane, D.; Lazo, G.R.; Dvorák, J.; Anderson, O.D. BatchPrimer3: A high throughput web application for PCR and sequencing primer design. *BMC Bioinform.* **2008**, *9*, 1–13. [CrossRef]

92. Shannon, P.; Markiel, A.; Ozier, O.; Baliga, N.S.; Wang, J.T.; Ramage, D.; Amin, N.; Schwikowski, B.; Ideker, T. Cytoscape: A Software Environment for Integrated Models of Biomolecular Interaction Networks. *Genome Res.* **2003**, *13*, 2498–2504. [CrossRef] [PubMed]

93. Available online: https://digd.big.ac.cn/ (accessed on 18 October 2020).

94. Andrews, S. *FastQC: A Quality Control Tool for High Throughput Sequence Data*; Babraham Bioinformatics, Babraham Institute: Cambridge, UK, 2010. Available online: http://www.bioinformatics.babraham.ac.uk/projects/fastqc (accessed on 18 October 2020).

95. Bolger, A.M.; Lohse, M.; Usadel, B. Trimmomatic: A flexible trimmer for Illumina sequence data. *Bioinformatics* **2014**, *30*, 2114–2120. [CrossRef] [PubMed]

96. Dobin, A.; Gingeras, T.R. Mapping RNA-seq Reads with STAR. *Curr. Protoc. Bioinform.* **2015**, *51*, 11–14. [CrossRef]

97. Peden, J.F. Analysis of Codon Usage. Ph.D. Thesis, Department of Genetics, University of Nottingham, Nottingham, UK, 1999.

3

Main Root Adaptations in Pepper Germplasm (*Capsicum* spp.) to Phosphorus Low-Input Conditions

Leandro Pereira-Dias [1], Daniel Gil-Villar [1], Vincente Castell-Zeising [2], Ana Quiñones [3], Ángeles Calatayud [3], Adrián Rodríguez-Burruezo [1] and Ana Fita [1,*]

[1] Instituto de Conservación y Mejora de la Agrodiversidad Valenciana (COMAV), Universitat Politècnica de València, 46022 Valencia, Spain; leapedia@etsiamn.upv.es (L.P.-D.); dagivil@upv.es (D.G.-V.); adrodbur@upv.es (A.R.-B.)
[2] Departamento de Producción Vegetal, Universitat Politècnica de València, 46022 Valencia, Spain; vcastell@prv.upv.es
[3] Instituto Valenciano de Investigaciones Agrarias (IVIA), 46113 Moncada, Valencia, Spain; quinones_ana@gva.es (A.Q.); calatayud_ang@gva.es (Á.C.)
* Correspondence: anfifer@btc.upv.es

Abstract: Agriculture will face many challenges regarding food security and sustainability. Improving phosphorus use efficiency is of paramount importance to face the needs of a growing population while decreasing the toll on the environment. Pepper (*Capsicum* spp.) is widely cultivated around the world; hence, any breakthrough in this field would have a major impact in agricultural systems. Herein, the response to phosphorus low-input conditions is reported for 25 pepper accessions regarding phosphorus use efficiency, biomass and root traits. Results suggest a differential response from different plant organs to phosphorus starvation. Roots presented the lowest phosphorus levels, possibly due to mobilizations towards above-ground organs. Accessions showed a wide range of variability regarding efficiency parameters, offering the possibility of selecting materials for different inputs. Accessions bol_144 and fra_DLL showed an interesting phosphorus efficiency ratio under low-input conditions, whereas mex_scm and sp_piq showed high phosphorus uptake efficiency and mex_pas and sp_bola the highest values for phosphorus use efficiency. Phosphorus low-input conditions favored root instead of aerial growth, enabling increases of root total length, proportion of root length dedicated to fine roots and root specific length while decreasing roots' average diameter. Positive correlation was found between fine roots and phosphorus efficiency parameters, reinforcing the importance of this adaptation to biomass yield under low-input conditions. This work provides relevant first insights into pepper's response to phosphorus low-input conditions.

Keywords: *Capsicum annuum*; root structure; root hairs; phosphorus use efficiency; P-starvation; abiotic stress; macrominerals; nutrient; breeding

1. Introduction

Agriculture will face many challenges in the next generations, especially those related to food security and agricultural sustainability [1,2]. On one hand, intensive agriculture has a significant impact on the environment, contributing to soil erosion, soil salinization, eutrophication and contamination of water bodies, and biodiversity reduction [3,4]. On the other hand, agricultural systems need to be improved in order to cope with requirements of an increasing population as well as the impact of climate change consequences [1,5].

In both cases, one of the most critical resources involved is phosphorus (P), an inorganic mineral with a major role within the physiochemical processes of plants [6,7]. Since almost 40% of the world's arable land lacks of P or the soil properties to make it available for crops, P absence is a major constraint to food production all around the world [8–10]. Until now, application of P-enriched fertilizers has been the main strategy to face its deficiency in soils despite the severe contaminants emissions associated to its production [3,9,11]. In addition, only 15 to 40% of the added P is taken up by crops [3,9,12], while the remaining ends up being washed down through the soil, contributing to eutrophication of water bodies [13,14]. Furthermore, as costs of extraction increase and rock-phosphate reserves decline, P is becoming an extremely expensive resource that is already unaffordable in many regions of the globe [10]. As demand for P-enriched fertilizers is going to increase in the next decades, the control for P supply will be a source of conflicts [7,9]. Therefore, there is a need for P low-input adapted varieties.

The response to P-starvation conditions has been studied for a few model organisms and some economically important crops, such as soybean, maize, sunflower, brassica or melon over the last decades [15–19]. As a result, researchers have linked several root traits to a greater performance under low P conditions [20]. Thus, morphological changes, such as the increment of number of root hairs and higher root branching [15,18,21], as well as physiological changes, such as cellular structure alteration, enhanced phosphatases enzyme activity and organic acids production and root P transporters enhanced expression [12,16,22,23], are adaptations expressed under P-starvation conditions. The exploitation of these plant adaptations could have a remarkable impact on the reduction of chemical fertilizers inputs in agricultural systems [12,24].

Peppers (*Capsicum* spp.) are one of the most relevant vegetables, grown in almost all temperate and tropical regions of the world [25]. Food and Agriculture Organization of the United Nations (FAO) last available data estimates around 40×10^6 t of pepper produced each year [26]. Therefore, improving pepper for its uptake and use of P would significantly reduce the need for P-fertilizer applications [3,12]. Notwithstanding, the development of improved *Capsicum* varieties for P low-input conditions is a challenging goal and is conditioned by both the availability of genetic variability within *Capsicum* and the understanding of the mechanisms underlying the response. Regarding the first point, *Capsicum* spp., particularly *Capsicum annuum* L., is remarkably diverse, as well as adapted to a wide range of environments and, therefore, tolerant to several abiotic stresses [27–30]. However, pepper fundamentals regarding this subject have never been studied. Hence, we believe that an exhaustive characterization of pepper germplasm for its responses under P low-input conditions is of paramount importance in order to recognize the variability within the genus, to enhance our understanding regarding the responses activated under such conditions and, finally, to link those responses to the genomic regions controlling them. Herein, the characterization of the main root adaptations of pepper accessions to low P conditions was established as a main goal, as a first step towards the identification of elite individuals for future pepper breeding programs.

2. Materials and Methods

2.1. Germplasm

A collection of 25 pepper accessions, encompassing 22 *Capsicum annuum*, two *Capsicum chinense* and one *Capsicum frutescens* accessions, comprising a wide range of variability for fruit shape, fruit pungency, fruit color, biotic resistances and adaptation to the environments, was studied herein [31] (Table 1). The considered collection belongs to the Instituto Universitario de Conservación y Mejora de la Agrodiversidad Valenciana (COMAV) Germplasm Bank (Universitat Politècnica de València, Spain) and to the COMAV *Capsicum* breeding group, and was selected based on previous experiments, where an interesting performance and diversity for several relevant root and P uptake traits was observed [32].

Table 1. List of the 25 accessions and corresponding abbreviation, species, varietal status, origin, fruit shape, fruit taste, fruit color and trial year.

Abbreviation	Species	Accession (UPV Genebank Code)	Origin	Fruit Shape	Fruit Taste	Fruit Color	Trial
Traditional varieties							
fra_DLL	Capsicum annuum	Doux Long des Landes	France (INRA-GEVES, F. Jourdan)	Cayenne, long-sized	Sweet	Red	Trial 2
mex_096D	Capsicum annuum	Chile Ancho Poblano	Mexico, Aguascalientes	Triangular, Pochard's C4 type	Hot	Red	Trial 2
mex_103B	Capsicum annuum	Chile Ancho Poblano	Mexico, Aguascalientes	Triangular, Pochard's C4 type	Hot	Red	Trial 2
mex_pas	Capsicum annuum	Pasilla Bajío	Mexico, Reymer Seeds	Cayenne, long-sized	Hot	Brown	Trial 1 and Trial 2
mex_ng	Capsicum annuum	Numex Garnet	Mexico, Aguascalientes	Elongated, Pochard's C2 type	Sweet	Red	Trial 2
mu_esp	Capsicum annuum	Jalapeno Espinalteco	Mexico/USA (P. W. Bosland)	Jalapeno	Hot	Red	Trial 1 and Trial 2
sp_060	Capsicum annuum	Pimiento morrón de bola (BGV00060)	Spain, Zamora	Round, Pochard's F type	Sweet	Red	Trial 2
sp_11814	Capsicum annuum	Dulce Italiano (BGV11814)	Spain, León	Elongated, Pochard's C2 type	Sweet	Red	Trial 2
sp_bola	Capsicum annuum	Pimiento de bola, ñora	Spain, Murcia (P.D.O. Pimentón Murcia)	Round, Pochard's N type	Sweet	Red	Trial 1 and Trial 2
sp_lam	Capsicum annuum	Lamuyo	Spain, Valencia	Blocky, Pochard's B1 or B2 type	Sweet	Red	Trial 2
sp_piq	Capsicum annuum	Pimiento Piquillo de Lodosa	Spain, Navarra (P.D.O. Piquillo Lodosa)	Triangular, Pochard's C4 type	Sweet	Red	Trial 1 and Trial 2
usa_chi	Capsicum annuum	Chimayó	USA, New Mexico (P. W. Bosland)	Blocky small-sized, Pochard's B4 type	Hot	Red	Trial 1
usa_conq	Capsicum annuum	Numex Conquistador	USA, New Mexico	Elongated, Pochard's C2 type	Sweet	Red	Trial 2
usa_jap	Capsicum annuum	Chile Japonés	USA, New Mexico	Cayenne, very short-sized	Hot	Red	Trial 2
usa_numex	Capsicum annuum	Numex X	USA, New Mexico	Elongated, Pochard's C2 type	Hot	Red	Trial 2
usa_sandia	Capsicum annuum	Numex Sandia (BGV13293)	USA, New Mexico	Elongated, Pochard's C2 type	Hot	Red	Trial 2
Experimental lines							
mex_scm	Capsicum annuum	Serrano Criollo de Morellos (SCM334)	Mexico	Serrano	Hot	Red	Trial 1 and Trial 2
sp_cwr	Capsicum annuum	California Wonder	Spain, Valencia (COMAV)	Blocky, Pochard's A type	Sweet	Red	Trial 1
Commercial hybrids (F1)							
sp_anc	Capsicum annuum	Ancares	Spain (Ramiro Arnedo)	Blocky, Pochard's B1 or B2 type	Sweet	Red	Trial 2
sp_cat	Capsicum annuum	Catedral	Spain (Zeraim Ibérica)	Blocky, Pochard's A type	Sweet	Red	Trial 1
sp_lobo	Capsicum annuum	El Lobo	Spain (Zeraim Ibérica)	Blocky, Pochard's A type	Sweet	Red	Trial 2
sp_mel	Capsicum annuum	Melchor	Spain (Ramiro Arnedo)	Blocky, Pochard's A type	Sweet	Red	Trial 1
Other Capsicums							
bol_037	Capsicum chinense	Bol-37R (BGV007644)	Chuquisaca, Bolivia	Triangular, small-sized, thin flesh	Hot	Red	Trial 1
bol_144	Capsicum baccatum	Bol-144 (BGV007751)	Bolivia, Santa Cruz	Cayenne, very short-sized	Hot	Red	Trial 1
eq_973	Capsicum chinense	ECU-973	Ecuador, Napo	Triangular, small-sized, thin flesh	Hot	Red	Trial 1

2.2. Germination and Cultivation Conditions

Seeds were surface sterilized with a 30% NaClO solution (v:v) for five minutes, followed by rinsing with steril deionized water, and transfered to individual Petri dishes containing a wet layer of cotton under a filter paper disk. Two drops of 2% Tetramethylthiuram disulphide solution were added to each Petri dish to prevent fungal proliferation. Petri dishes were kept under germination chamber conditions until two-cotyledon stage. Seedlings were then transferred to seedling trays filled with Neuhaus N3 substrate (Klasmann-Dellmann GmbH, Geeste, Germany), kept under heated nursery conditions until the five leaves stage and, finally, transplanted to the greenhouse.

The experiment was carried out in two years. In the first year (from now on Trial 1), 12 accessions were trialed and the five most interesting genotypes were re-trialed in the second year (from now on Trial 2), against 13 new accessions (Table 1). In both trial years, plants were grown for 60 days under a mesh greenhouse, during the spring-summer cycle, on COMAV experimental fields (Universitat Politècnica de València Vera Campus GPS coordinates: 39°28'56.33" N; 0°20'10.88" W). Transplant was carried out in June and the experiment was finished in August. Nine (Trial 1) and six (Trial 2) plants, per accession and treatment, were grown in 15 L plastic pots filled with substrate made by mixing a part of soil with a part of sand (1:1) and arranged into a completely randomized design with six rows. Pots were spaced 1.2 m between rows and 0.40 m inside rows, while a drip irrigation system provided water and nutrient solutions to cover the plants' water and nutritional requirements. Individual plants were trained with vertical strings, according to standard local practices for pepper. Plants were not pruned during the experiment in order to avoid interference with biomass yield. Likewise, phytosanitary treatments against whiteflies, spider mites, aphids and caterpillars were applied accordingly to population levels.

Plants were subjected to two treatments. On one hand, control treatment was applied using a standard solution providing all elements (Table S1). On the other hand, stress treatment (from now on NoP) was applied using similar solution to the control treatment except for P carrying ions, which were removed from the formulation of the solution (Table S1).

2.3. Sample Preparation

After the 60 days period plants were harvested for processing. Shoot and fruits were processed separately in order to assess effects of P deprivation on both tissues. Each tissue was put into individual paper bags and dried at 70 °C, until constant weight was achieved, in a Raypa ID-150 oven (R. Espinar S.L., Barcelona, Spain). At this point, shoot (SW, g) and fruit (FW, g) dry weights were determined, and those tissues were ground into a thin powder, using a domestic Taurus coffee grinder (Taurus Group, Oliana, Spain), for later mineral content analysis. Furthermore, all plants' roots were separated from substrate by gently washing them with running tap water and processed separately from other tissues [33]. This was done by hand, one root at a time (Figure 1).

For Trial 1 ($n = 9$), root hairs (Ø < 0.5 mm) were separated from lateral roots (Ø > 0.5 mm) and dried at 70 °C in order to obtain root hairs dry weight (RHW, g) (Figure 1). It is important to note that what is referred here as root hairs does not correctly translate to the root anatomical definition of root hairs; instead, it includes root hairs and some fine tertiary and lower order roots. However, herein it is useful to differentiate between the evaluated root parts. In the same way, lateral roots are mainly secondary roots; however, as can be seen in the picture Figure 1C, they can also include a portion of tertiary roots, as it was impossible to separate all in such a large amount of samples. Lateral roots were scanned, using an Epson Expression 1640XL G650C scanner (Seiko Epson Corp., Suwa, Japan), and resulting images were analyzed by WinRIZHO™ Pro 2.3 software (Regent Instruments Inc., Québec, QC, Canada). Lateral root total length (LRL, m), lateral root average diameter (LRAD, mm) and total length of lateral roots with diameter under (LRL$_{<1mm}$, m) and above (LRL$_{>1mm}$, m) 1 mm were determined based on said images for each plant included in the experiment. Finally, scanned lateral roots were dried in order to obtain lateral roots dry weight (LRW, g) and ground for mineral content determination (Figure 1). From those measurements, several parameters were calculated in order to better characterize plants' performance. Hence, for trial 1, total root dry weight (RW, g) was determined as the sum of RHW and LRW and, therefore, total biomass dry weight (BW, g) was calculated as the sum of RW, SW and FW. In addition, root to shoot weight ratio (R/S) was calculated by dividing RW by SW; the percentage of root dry weight devoted to root hairs (RHW%) was calculated by the division of RHW by RW. Furthermore, the proportion of root length devoted to fine lateral roots (PLFR, %) was defined as the ratio between $LRL_{<1mm}$ and LRL. Finally, lateral root specific length (LRSL, m/g) was calculated by dividing LRL by LRW.

For Trial 2 ($n = 6$), roots were entirely scanned (Figure 1). In order to fully capture a root's morphometrics, individual roots were properly spread over several transparent acetate sheets (Figure 1) and analyzed by WinRIZHO™ Pro 2.3 software (Regent Instruments Inc., Canada). Root total length (RTL, m), total root average diameter (TAD, mm) and total length of roots with diameters under ($RL_{<1mm}$, m) and above ($RL_{>1mm}$, m) 1 mm, were determined for each plant. Finally, the scanned roots were dried until constant weight was achieved and ground to a powder as in Trial 1. Root hairs dry weight (RHW, g), lateral roots dry weight (LRW, g), total root dry weight (RW, g), total biomass dry weight (BW, g), root to shoot ratio (R/S), percentage of root dry weight devoted to root hairs (RHW%) and root specific length (RSL, m/g) were determined as in Trial 1. Finally, the proportion of root length devoted to fine lateral roots was determined, that is, including root hairs and roots below 1 mm (PFR, %), as the ratio between $RL_{<1mm}$ and RTL.

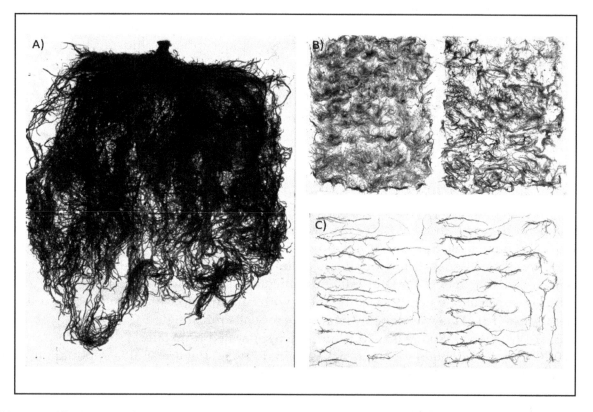

Figure 1. Illustration of the roots along the scanning process (from a representative sample). Individual root systems were separated from the soil with running tap water and taken to the laboratory to be scanned and dried. In Trial 1, whole roots (**A**) were separated into (**B**) root hairs (Ø < 0.5 mm) and (**C**) lateral roots (Ø > 0.5 mm). Root hairs (**B**) were only weighed while lateral roots (**C**) were scanned and weighed. In Trial 2, whole roots (**A**) were also separated into root hairs (**B**) and lateral roots (**C**) and both were scanned and weighed.

2.4. Tissue Mineral Concentration Assessment

Before mineral content determination, samples were mineralized [34]. Thus, 2 g of powdered plant tissue were calcined for 2 h in a muffle at 450 °C. Ashes resulting from mineralization were let to cool down, weighted and then hydrated with 2 mL of distilled water followed by addition of 2 mL of concentrated HCl (Scharlau, Valencia, Spain). At this point, the solution was heated on a hot plate, until first fumes appeared, and then filtered with Whatman filter paper (Sigma-Aldrich, St. Louis, MI, USA). Finally, distilled water was added in order to make up to 100 mL volume [34].

In Trial 1 ($n = 4$), phosphorus (P), potassium (K), calcium (Ca), magnesium (Mg), sodium (Na) and sulfur (S) concentration (g 100 g^{-1} DW) in different plant tissues (root, shoot and fruits, [Mineral]$_{Tissue}$) was determined by Inductively Coupled Plasma-Atomic Emission Spectrometry (ICP-AES; iCAp-AES 6000, Thermo Scientific, Cambridge UK). Samples were digested for 24 h by adding 10 mL 65% HNO_3 solution (Panreac Quimica S.A.U., Barcelona, Spain) to 0.5 g dried material, in a 25 mL open vessel. Digested samples were then boiled at 120 °C for 10 min followed by another 25 min at 170 °C. Finally, samples were cooled, 2 mL of 70% $HClO_4$ was added (Panreac Quimica S.A.U., Barcelona, Spain) and were then heated at 200 °C for 40 min. At this point, samples were transferred to a flask and volume was brought up to 25 mL with distilled water.

For Trial 2 ($n = 6$), leaves' P-concentration ([P]$_{Shoot}$) was determined by colorimetric reaction (MAPA, 1994). This method is based on absorbance measurement at 430 nm of each sample in acid solution and on the presence of vanadium (V^{5+}) and molybdenum (Mo^{6+}) ions. Under these conditions, phosphoric acid forms a phosphomolybvanadate complex that gives yellow coloration. Hence, 5 mL of mineralized solution were pipetted into a new 25 mL volumetric flask, followed by the addition of 5 mL of nitro-vanado-molybdic reagent. Volume was then brought up to 25 mL with distilled water. Prior to mineral concentration determination, a standard curve was constructed with standards 0, 2, 4, 6, 8, 10 and 12 µg of P mL^{-1} prepared from an initial solution of 20 µg of P mL^{-1}. Sample P concentration was determined using a 6305 model UV/V spectrophotometer (Jenway, Gransmore Green, England, UK) at 430 nm against a standard curve.

2.5. Phosphorus Uptake and Use Efficiency Parameters

In order to better characterize treatment effect on accessions performance, several widely-used P uptake and P use efficiency parameters (PUE) were calculated based on previous works [18,21] (Table 2).

Table 2. P uptake and P use efficiency (PUE) parameters used in this experiment and corresponding abbreviation, formula and expressed units. Dry weight (DW), total biomass dry weight biomass weight (BW).

Parameter	Abbreviation	Formula [1]	Units
Tissue total P content	RootP, ShootP, FruitP	$[P]_{Tissue} \times DW_{Tissue}$	G
Plant total P content	PTP [2]	$[P]_{Root} \times DW_{Root} + [P]_{Shoot} \times DW_{Shoot} + [P]_{Fruit} \times DW_{Fruit}$	mg P
P uptake efficiency	PUpE [3]	$([P]_{Control} \times BW_{Control}) - ([P]_{NoP} \times BW_{NoP})$	mg P
P utilization efficiency	PUtE [3]	$(BW_{Control} - BW_{NoP}) / (([P]_{Control} \times BW_{Control}) - ([P]_{NoP} \times BW_{NoP}))$	g DW g^{-1} P
Physiological P use efficiency	PPUE	$BW_{Control}/[P]_{Control}$ and $BW_{NoP}/[P]_{NoP}$	g^2 DW g^{-1} P
P efficiency ratio	PER	$BW_{Control}/ ([P]_{Control} \times BW_{Control})$ and $BW_{NoP}/ ([P]_{NoP} \times BW_{NoP})$	g DW g^{-1} P

[1] P concentration ([P]), Dry weight (DW), total biomass dry weight (BW) [2] Note that for Trial 2 only [P]$_{shoot}$ was measured, therefore PTP was obtained as [P]$_{shoot}$ × BW; [3] Note that [P] in Trial 1 is the weighted average [P] among different tissues, whereas in Trial 2 [P] = [P]$_{shoot}$.

2.6. Statistical Analysis

Two-way factorial analysis of variance (ANOVA) was performed using individual plant values in order to assess accession and treatment effects and interaction significance [35]. In addition,

Student-Newman-Keuls post-hoc multiple range test ($p < 0.05$) was used to detect significant differences among accession means for all evaluated traits. Finally, trait differences between treatments (μ_{NoP}-$\mu_{Control}$) were used to perform multivariate Principal Component Analysis (PCA) using Euclidean pairwise distances. In addition, traits variation (%) between control and NoP conditions was calculated as $\left(\frac{\mu_{NoP} - \mu_{Control}}{\mu_{Control}} \right) \times 100\%$. All statistical analysis were performed using Statgraphics Centurion XVII (StatPoint Technologies, Warrenton, VA, USA) and plotted using R package ggplot2 [36,37].

3. Results

3.1. General Treatment Effect on P and Other Minerals Concentrations for Trial 1

P concentration ([P]) in plant tissues is an important indicator of both treatment effectiveness and accession's capability to make the most with the available resources. In Trial 1 ($n = 4$), plants cultivated under NoP conditions showed significantly lower [P] compared to control plants. This behavior was statistically significant for all three sampled tissues (Table S2). For [P]$_{Roots}$, there was a reduction from 0.56 g P 100 g^{-1} DW, when cultivated under control conditions, to 0.10 g P 100 g^{-1} DW (−81.76%) when cultivated under NoP conditions (Table S2). For [P]$_{Shoot}$, values decreased from 0.18 g 100 g^{-1} DW to 0.12 g P 100 g^{-1} DW (−29.31%) for control and NoP conditions, respectively; this the organ is less affected by the treatment (Table S2). Finally, fruit P levels dropped from 0.26 g P 100 g^{-1} DW, when irrigated with control solution, to 0.17 g P 100 g^{-1} DW (−35.18%) when NoP solution treatment was applied (Table S2).

Concentration of other macrominerals was determined in order to assess possible deficiencies induced by the applied treatments. Regarding that, significant differences between treatments were observed, particularly for K and Mg, probably due to the differences in the nutrient solutions and as a result of plant ionic adjustments. Despite that, mineral concentrations were within the normal range for pepper (Table S2) [6,38].

3.2. Treatment Effect on P Accumulation and Efficiency Parameters for Trial 1 by Accessions

Accessions were significantly affected by the NoP treatment, but not all to the same extent, as shown by the two-way ANOVA (Table S3) and the accession mean values for the evaluated traits (Table S4). For example, [P]$_{Root}$ dropped as much as 91.62% and 90.53% for bol_037 and sp_cwr accessions, respectively, while for bol_144 the reduction was considerably lower (74.12%) (Table 3). Regarding [P]$_{Shoot}$, the most affected accessions were sp_piq (−45.26%) and sp_cwr (−45.03%), while some accessions experienced no statistically significant reduction of their shoot P concentration, e.g., bol_037, bol_144, eq_973, mex_pas, sp_bola and sp_cat (Table 3 and Table S4). Finally, for [P]$_{Fruit}$, only sp_bola showed no statistical difference between both treatments, whereas the remaining accessions represented significant reductions of around 35% (Table 3).

In terms of P tissue accumulation, significant differences were found among accessions. Thus, despite P-deficient plants had on average 86.16% less accumulated P in the root (RootP) than control plants, 56.12% less P in the shoots (ShootP), and 34.32% less in fruits (FruitP), some genotypes, such as bol_144 or eq_973, increased the amount of fruit accumulated P (although this was not statistically significant). Overall, plant total P (PTP) was reduced by 63.30%; however, several genotypes experienced no statistically significant reduction of this trait, e.g., bol_037, bol_144_eq_973 and sp_bola, while others, like sp_mel, were highly affected (Table 3 and Table S4).

Furthermore, in order to evaluate how efficient genotypes were under these conditions parameters of physiological P use efficiency (PPUE), P efficiency ratio (PER), P uptake efficiency (PUpE) and P utilization efficiency (PUtE), were calculated [18,21]. Overall, PPUE was on average 36.26% higher under NoP conditions (Table 3). However, accessions' behavior for this parameter was extremely variable and significant differences between treatments were only found for two accessions, mu_esp (41.31%) and sp_cwr (97.53%) (Table 3 and Table S4). On the other hand, PER's behavior was more consistent and NoP treatment produced a generalized increase, averaging at 87.64% (Table 3). In this case, only two accessions showed no significant differences, bol_144 and sp_bola (Table 3 and Table S4). Interestingly, the best performers in terms of increasing PER from control to NoP conditions were California type accessions (sp_cat, sp_mel and sp_cwr) and *Capsicum chinense* accession eq_973.

In addition, an interesting amount of variability was observed for P uptake efficiency (PUpE) and P utilization efficiency (PUtE) parameters (Figure 2). PUpE refers to the increase of internal P when it is available in the environment, whereas PUtE measures the ability of a genotype to increase its biomass per unit of internal P. Both measures always compare two conditions differing in P levels. PUpE averaged 96.56 mg P for the whole collection, where accessions mex_scm (128.77 mg P), mu_esp (144.43 mg P), sp_cat (130.06 mg P), sp_piq (124.53 mg P) and usa_chi (108.70 mg P) showed the highest values of the experiment (Figure 2). PUtE values ranged from 29.77 g DW g^{-1} P (bol_144) to 404.95 g DW g^{-1} P (mex_pas) and averaged 186.55 g DW g^{-1} P. Accessions mex_pas (404.95 g DW g^{-1} P), usa_chi (316.77 g DW g^{-1} P) and sp_bola (273.16 g DW g^{-1} P) presented the most interesting results (Figure 2).

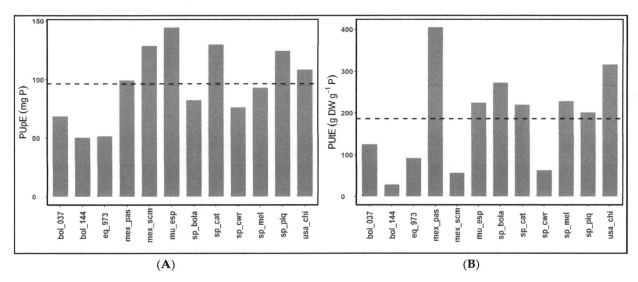

Figure 2. (**A**) Average PUpE (P Uptake Efficiency) and PUtE (P Utilisation Efficiency) values for the 12 accessions (*n* = 4) studied in Trial 1. (**B**) The black dashed line represents the average value for the whole collection for both PUpE (96.56 mg P) and PUtE (186.55 g DW g^{-1} P) parameters.

3.3. Treatment Effect on Root and Shoot Biomass and Morphometrics for Trial 1 by Accessions

P is a major factor controlling root structure and architecture [20]. Hence, in order to understand the possible effects on plant morphology, root structure and architecture resulting from lack of P, we compared several biomass and root traits.

Thus, roots dry weight (RW) showed a significant generalized decrease (−24.52%) when genotypes were cultivated under NoP conditions compared to control plants (Table 3, Tables S3 and S4). This weight difference was more evident in lateral roots (LRW), which, on average, weighed 26.07% less, while root hairs (RHW) weight was 18.34% lower than under control conditions. Notwithstanding, the greatest weight difference was observed for shoot dry weight (SW), −36.04% under NoP conditions (Table 3, Table S3 and Table S4). Despite that, taking a closer look at the treatment effect on biomass by accession, it is observed that only three accessions reduced it significantly in all organs: mex_pas, mu_esp and usa_chi. The rest of the accessions also reduced their biomass but not so systematically (Table 3 and Table S4). Interestingly, accession mex_scm, presented similar RW and SW under both treatments, while presenting the heaviest root system and shoot within the collection under NoP conditions (Table S4). Finally, root to shoot ratio (R/S) was positively affected under NoP treatment. This parameter increased by 20.94%, on average, although only usa_chi showed statistically significant differences between treatments (+22.73%) (Table 3), apparently achieved by reducing a shoot's weight instead of increasing a root's weight (Table S4).

Table 3. Accession behavior given by differences (%) between control and NoP conditions for Trial 1. Twenty different P accumulation and efficiency ($n = 4$), biomass and root traits ($n = 9$) and parameters were considered.

Accession	P Accumulation and Efficiency Traits									Biomass Traits						Root Traits				
	$[P]_{Root}$	$[P]_{Shoot}$	$[P]_{Fruit}$	RootP	ShootP	FruitP	PTP	PPUE	PER	RW	LRW	RHW	SW	BW	R/S	LRL	LRAD	RHW%	PLFR	LRSL
bol_037	-91.62 *	-32.10	-32.56 *	-93.37 *	-37.32	-25.00	-54.72	95.23	106.37 *	-12.91	-8.50	-18.68	-22.16	-21.67	25.41	33.30	-8.59	-6.42	5.21	38.14
bol_144	-74.12 *	-34.29	-27.41 *	-86.11 *	-55.45	98.52	-51.66	90.59	78.15	-26.94	-36.58	-1.37	-8.13	-4.08	-16.22	15.14	-21.81 *	51.06	9.56	112.94
eq_973	-87.67 *	-34.32	-35.19 *	-87.60 *	-41.15	55.56	-53.92	112.72	115.10 *	-21.00	-30.62	-3.38	-16.21	-14.17	6.67	49.55	-16.80	39.34	6.02	78.20
mex_pas	-78.01 *	1.51	-19.10 *	-86.13 *	-52.54 *	-59.45	-63.36 *	-27.82	36.65 *	-36.24 *	-42.38 *	-26.69	-59.65 *	-54.20 *	27.89	-3.07	-9.83	12.06	6.63	70.84
mex_scm	-84.98 *	-44.35 *	-37.94 *	-85.62 *	-62.45 *	-24.63	-57.57 *	50.86	96.13 *	0.65	9.15	-8.95	-18.18	-12.62	21.37	8.86	7.98	-11.28	1.63	16.26
mu_esp	-77.38 *	-32.15 *	-33.93 *	-88.61 *	-77.91 *	-72.38 *	-76.17 *	-41.31 *	51.20 *	-46.63 *	-29.39	-52.87 *	-56.71 *	-55.84 *	9.88	26.88	-8.57	-12.76	7.36	115.25
sp_bola	-77.51 *	-4.11	-17.47	-75.94 *	-41.77	-53.31	-55.46	-0.75	38.47	-19.85	-30.22 *	-5.64	-43.10	-40.06	27.73	2.12	-10.84	2.89	8.55	57.72
sp_cat	-83.53 *	-24.27	-45.63 *	-83.02 *	-56.11	-78.90 *	-73.96 *	27.80	108.33 *	-2.79	-23.62	31.54	-33.26	-41.92	50.00	3.73	-13.32	25.62	6.58	20.20
sp_cwr	-90.53 *	-45.03 *	-47.21 *	-92.86 *	-55.30 *	-46.11	-61.87 *	97.53 *	132.69 *	-19.90 *	-14.68	-26.20	-16.17	-11.06	-6.58	65.62 *	-7.13	-6.42	4.17	91.05
sp_mel	-83.67 *	-35.14 *	-49.43 *	-87.27 *	-66.67	-74.04	-77.17 *	24.96	137.24 *	-30.17	-35.41	-21.27	-58.18 *	-48.90	63.81	-10.47	-3.90	11.05	3.12	54.84
sp_piq	-75.52 *	-45.26 *	-44.11 *	-78.51 *	-59.93 *	-70.57	-67.78 *	24.31	98.40 *	-25.14	-17.60	-32.96	-39.38 *	-38.84 *	18.59	-19.76	4.76	-14.57	0.98	6.84
usa_chi	-76.56 *	-22.25	-32.20 *	-88.89 *	-66.88 *	-61.56 *	-65.95 *	-18.94	52.95	-53.35 *	-53.03 *	-53.63 *	-61.36 *	-59.42 *	22.73 *	27.85	12.59	-1.07	-1.29	142.63
Global mean	-81.76 *	-29.31 *	-35.18 *	-86.16 *	-56.12 *	-34.32 *	-63.30 *	36.26 *	87.64 *	-24.52 *	-26.07 *	-18.34 *	-36.04 *	-33.56 *	20.94 *	16.65 *	-6.29 *	7.46	4.88	67.08

Root P concentration ($[P]_{Root}$), shoot P concentration ($[P]_{Shoot}$), fruit P concentration ($[P]_{Fruit}$), root total P content (Root P), shoot total P content (ShootP), fruit total P content (FruitP), plant total P content (PTP), physiological P use efficiency (PPUE), P efficiency ratio (PER), total root dry weight (RW), lateral root dry weight (LRW), root hairs dry weight (RHW), shoot dry weight (SW), total biomass dry weight (BW), root to shoot ratio (R/S), lateral root total length (LRL), lateral root average diameter (LRAD), percentage of root dry weight devoted to root hairs (RHW%), proportion of root length devoted to fine lateral roots (PLFR) and lateral root specific length (LRSL). * Indicates significant differences between treatments for that accession and trait.

Regarding root morphology traits, treatment and accession effects showed significant influence over most traits, except for the percentage of root dry weight devoted to root hairs (RHW%), for which significant differences between treatments were not detected (Table S3), despite there being differences among accessions. In addition, for lateral root specific length (LRSL), there was a significant accession per treatment interaction (Table S3). The significant effects of the NoP conditions on pepper's roots where to increase: the lateral root length (LRL), by 16.65%, the proportion of root length devoted to fine lateral roots (PLFR), by 4.88%, and the lateral root specific length (LRSL), by 67.08%, and to decrease the root average diameter by 6.29%.

Regardless of the general treatment effect, there were significantly different responses among genotypes (Table 3 and Table S4). It is worth to mention the significant increase of percentage of root length devoted to fine lateral roots (PLFR) and lateral root specific length (LRSL) observed in mu_esp and sp_bola (Table 3), with mu_esp having the higher absolute values for these traits of the whole collection under NoP conditions. Another interesting response was presented by accession bol_144, which outperformed the other genotypes for reducing its roots average diameter (21.81%) and increase PLFR and LRSL under the NoP treatment.

3.4. Principal Components Analysis for Trial 1

Principal components analysis (PCA) was pursued in order to determine possible correlations between the response of the different measured traits to different inputs of P (% of increase or decrease, as in Table 3), trying to demonstrate how accessions differed in terms of response to different P levels. The first two principal components (PC) explained in combination 59.79% of the total variability (Figure 3A). Response in terms of total biomass dry weight (BW), physiological P use efficiency (PPUE), fruit total P content (FruitP), total shoot dry weight (SW), plant total P content (PTP), P efficiency ratio (PER) and root hairs dry weight (RHW), and values for P uptake efficiency (PUpE) and P utilization efficiency (PUtE) were the traits that contributed the most to the positive component of PC1, which explained 36.96% of the total variation. Response of lateral root average diameter (LRAD) and root total P content (RootP) were negatively correlated to PC1 (Figure 3A). Therefore, accessions plotted in the extreme right of the graph (Figure 3B), such as usa_chi, mu_esp, mex_pas and sp_piq, have in common that they have a great reduction in biomass when passing from control to NoP, and have good PUpE and PUtE. In other words, those are accessions that react very positively to any P addition to the soil but probably will not be appropriate to cultivate on poor soils (Figure 3B). At the same time, accession plotted at the upper most left part of the graph (Figure 3B), such as bol_144 and eq_973, are grouped for having high reductions in the total amount of P in the roots with a high reduction in the diameter of the roots (LRAD) as adaptation to low P, while having little difference in biomass under the two assayed conditions.

In addition, PC2 explained 22.83% of variability with the response of lateral root dry weight (LRW), lateral root average diameter (LRAD), root dry weight (RW) and P utilization efficiency (PUtE) being the traits that contributed the most to it. Conversely, shoot P concentration ([P]$_{Shoot}$), fruit P concentration ([P]$_{Fruit}$) and shoot total P content (ShootP) were negatively correlated with PC2 (Figure 3A). Hence, accession located in the top part of the graph, such as mex_pas and sp_bola (Figure 3B), change the allocation of root resources, reducing the lateral root weight and diameter in situation of P restriction, maintaining the P level status of the shoots. On the contrary, the accessions located on the lower part of the graph such as mex_scm stand out by changing the level of P of the shoots and fruits ([P]$_{Shoot}$, [P]$_{Fruit}$, and ShootP) as a strategy to adapt to the low P conditions without modifying the lateral root morphology or size (Figure 3A). Interestingly, there was a cluster of parameters, such as the response in terms of [P]$_{Root}$, R/S and LRSL, indicating correlations among them.

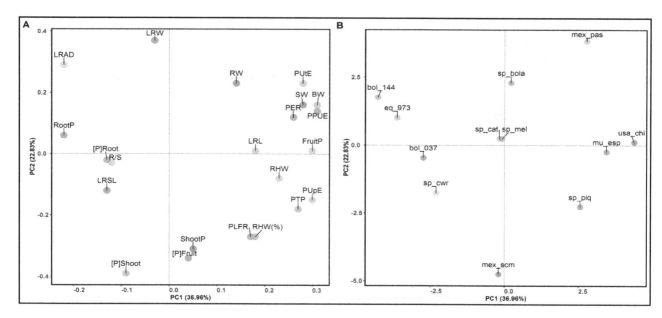

Figure 3. Principal Component Analysis (PCA) for the first two components based on trait differences between treatments for Trial 1. (**A**) Correlation between traits and the first two principal components. (**B**) Distribution of accessions based on studied traits. P tissue concentration traits [P]$_{Tissue}$, P tissue total content traits RootP, ShootP, FruitP, plant total P content (PTP) trait, efficiency parameters PPUE (physiological P use efficiency), PER (P efficiency ratio), PUpE (P uptake efficiency) and PUtE (P utilization efficiency) and morphometric traits RW (root dry weight), LRW (lateral root dry weight), RHW (root hairs dry weight), SW (shoot dry weight), BW (total biomass dry weight), R/S (root to shoot ratio), LRL (lateral root length), LRAD (lateral root average diameter), RHW% (root hairs dry weight %), PLFR (proportion of length dedicated to fine roots) and LRSL (lateral root specific length) were considered.

Bearing these results, the second trial was designed. In it, five accessions from Trial 1 (mu_esp, mex_pas, sp_bola, sp_piq and mex_scm) were re-trialed and used as a comparison standard against 13 new *C. annuum* accessions. These genotypes were selected based on their above average P uptake efficiency (PUpE) and P utilization efficiency (PUtE) scores and differential behavior against the lack of P. Note that an insufficient number of seeds to re-trial sp_cat and sp_mel, and the poor germination of usa_chi dictated their exclusion of Trial 2. The second trial was focused on checking the diversity within the germplasm belonging to *Capsicum annuum* species; for that reason, bol_144 and eq_973 were not selected, despite their interesting features. In this case, only P from the shoots was analyzed by a colorimetric protocol as a faster general measure of the P status of the plant, instead of a multi-elemental analysis by tissue. In addition, root hairs weight clustered together with P efficiency parameters was analyzed, and it was demonstrated that the lateral roots increase their length and reduce their diameter; thus, it was decided to analyze the root hairs' behavior as well. Both lateral and hair roots were scanned and analyzed (Figure 1).

3.5. Treatment Effect on P Accumulation and Efficiency Parameters for Trial 2 by Accessions

As in Trial 1, ANOVA showed that accession and treatment effects significantly affected P-related traits (Table S5). Interestingly, for physiological P use efficiency (PPUE) the accession effect was more important than treatment (Table S5). Remarkably, accession per treatment interaction was significant for a plant's total P content (PTP), PPUE and P efficiency ratio (PER) (Table S5). Accessions' individual variation between treatments are shown in Table 4 as $\left(\frac{\mu_{NoP} - \mu_{Control}}{\mu_{Control}}\right) \times 100\%$ negative then indicating lower values under NoPtraits. To consult the accessions' mean values per treatment, please refer to Table S6.

In Trial 2 ($n = 6$), all accessions but two showed significant differences between treatments for shoot P concentration ([P]$_{\text{Shoot}}$), plant total P content (PTP) and P efficiency ratio (PER) showing an average reduction of -31.5% and -66.17%, and an increase of 49.26%, respectively (Table 4). Accessions mex_096D and sp_piq stood out for their substantial [P]$_{\text{Shoot}}$ reduction and high PER value. In addition, accession sp_piq showed a significant reduction of its PTP level (-86.93%), which, along with mex_scm (-84.16%) and usa_sandia (-87.88%), represented the highest reductions of the whole collection (Table 4). Contrarily, sp_lam and sp_lobo showed no differences between treatments regarding PTP (Table 4 and Table S6). In the case of PPUE, NoP treatment presented an average reduction of 24.98%; however, significant differences were only detected for six accessions and with extremely erratic behavior within the collection; some accessions showed a reduction up to 72.75% (mex_scm), while others showed increases up to 45.04% (usa_jap).

Regarding P uptake efficiency (PUpE), average value was 298 mg P, ranging from 63 mg P (sp_lobo) to 796 mg P (usa_sandia). Accessions presenting above the average mean values were mex_096D, mex_103B, mex_ng, usa_conq and the re-trialed mex_scm and sp_piq (Figure 4). Contrarily to what happened in Trial 1, mu_esp was above the average for PUpE in this trial. Finally, average P utilization efficiency (PUtE) was 110 g DW g^{-1} P, while the minimum observed value was 43 g DW g^{-1} P (sp_lam) and the maximum was 183 g DW g^{-1} P (mex_pas). Like in Trial 1, accessions mex_pas (183 g DW g^{-1} P) and sp_bola (147 g DW g^{-1} P) presented the best performance for this parameter (Figure 4).

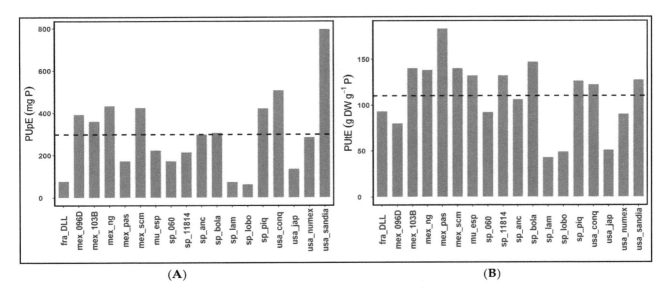

(A)   **(B)**

Figure 4. (**A**) Average P uptake efficiency (PUpE) and P utilization efficiency (PUtE) values for the 12 accessions studied in Trial 2 ($n = 6$). (**B**) Black dashed line represents average value for the whole collection for both PUpE (298 mg P) and PUtE (110 g DW g^{-1} P) parameters.

3.6. Treatment Effect on Root and Shoot Biomass and Morphometrics for Trial 2 by Accessions

In trial 2 ($n = 6$), multi-factorial ANOVA detected significant accession and treatment effects as well as the accession per treatment interaction for all biomass traits except root to shoot ratio (R/S) (Table S5). As expected, the effect of the NoP treatment led to lower dry weight accumulation of all sampled organs. This time, the most affected organs were the roots (RW, -52.96%) and root hairs (RHW, -59.10%) (Table 4). The genotypes usa_sandia, mex_scm and sp_piq showed the highest biomass reduction when passing from control to NoP. On the contrary, fra_DLL, sp_lam, sp_lobo and usa_jap showed no statistical differences between treatments, although it is important to note that fra_DLL, sp_lobo, and sp_lam displayed the smallest plants within the collection for both treatments, which could explain their results. Accession usa_jap, on the other hand, showed medium-sized plants (Table S6).

Table 4. Accession behavior given by differences (%) between the control and NoP conditions for Trial 2. Twenty different P accumulation and efficiency, biomass and root traits ($n = 6$) and parameters were considered.

Accession	P Accumulation and Efficiency Traits				Biomass Traits						Root Traits				
	[P] Shoot	PTP	PPUE	PER	RW	LRW	RHW	SW	BW	R/S	RTL	TAD	RHW%	PFR	RSL
fra_DLL	-31.92 *	-42.13 *	25.76	48.75 *	-31.04	-20.87	-34.20	-14.49	-15.61	-20.51	6.27	4.23	-6.84	-0.31	58.00
mex_096D	-44.10 *	-74.54 *	8.80	87.88 *	-42.62 *	-29.17 *	-50.76 *	-49.18 *	-48.86 *	27.27	-0.69	2.96	-19.79 *	-2.46	71.79 *
mex_103B	-25.45 *	-68.35 *	-42.41 *	33.91 *	-63.97 *	-57.22 *	-68.38 *	-56.97 *	-57.48 *	-4.26	-10.44	-7.53 *	-10.29	5.91 *	131.10 *
mex_ng	-38.75 *	-80.58 *	-53.53	66.18 *	-70.25 *	-68.60 *	-71.38 *	-69.72 *	-69.75 *	7.95	-0.06	-13.30 *	-7.35	9.79 *	229.58 *
mex_pas	-11.63	-54.71 *	-39.68 *	14.69	-45.43 *	-37.30 *	-52.56 *	-48.13 *	-47.97 *	13.16	6.22	-0.80	-11.40	1.83	81.25 *
mex_scm	-24.46 *	-84.16 *	-72.75 *	30.99 *	-71.09 *	-53.17 *	-79.62 *	-79.40 *	-78.97 *	60.71	-15.45	-0.72	-21.79	2.51	96.42
mu_esp	-28.78 *	-76.55 *	-45.93	41.15 *	-58.48 *	-36.78 *	-66.55 *	-64.64 *	-64.02 *	33.93	83.39	-8.37 *	-16.18	6.53 *	293.57 *
sp_060	-23.21 *	-57.00 *	-19.84	30.93 *	-58.36 *	-32.94 *	-64.94 *	-41.59 *	-41.30 *	-24.24 *	17.90	-1.52	-18.90	5.90	181.31 *
sp_11814	-27.72 *	-66.80 *	-40.61	34.34 *	-54.26 *	-9.84 *	-65.36 *	-55.36 *	-54.65 *	-6.46	110.07	-0.85	-21.27 *	3.37	282.31 *
sp_anc	-35.55 *	-66.88 *	-20.37	53.60 *	-47.93 *	-37.11 *	-53.30 *	-48.66 *	-48.08 *	2.94	-9.82	-5.51	-7.61	6.16	44.62
sp_bola	-29.23 *	-78.82 *	-59.09 *	41.52 *	-63.73 *	-6.21 *	-73.15 *	-70.90 *	-70.35 *	23.08	52.67	-3.10	-26.14 *	4.62	261.22 *
sp_lam	-30.84 *	-36.71	29.12	42.54 *	-35.96	-24.57	-42.23	-7.24	-10.66	-30.56 *	-7.73	2.45	-8.40	0.22	38.68
sp_lobo	-24.83	-35.31	16.75	31.29	-13.58	-9.56	-15.06	-14.14	-10.95	6.90	20.75	-7.69	-0.93	8.97 *	27.40
sp_piq	-42.48 *	-86.93 *	-50.75 *	91.29 *	-72.33 *	-51.80 *	-79.12 *	-75.78 *	-75.48 *	8.33	4.58	3.29	-21.96 *	2.35	220.45 *
usa_conq	-38.95 *	-79.74 *	-47.09	61.46 *	-59.27 *	-41.08 *	-65.12 *	-67.50 *	-66.97 *	13.04	3.73	-8.02	-14.25	8.23	122.83 *
usa_jap	-39.15 *	-49.02 *	45.04	68.30 *	-48.83	-23.47	-60.76	-13.55	-15.53	-45.95 *	9.01	-0.44	-21.29	0.00	72.51 *
usa_numex	-33.47 *	-64.88 *	-17.21	50.09 *	-49.78 *	-38.25 *	-53.04 *	-45.86 *	-46.10 *	-9.52	-39.33 *	8.76	-7.89	-3.70	21.77
usa_sandia	-36.52 *	-87.88 *	-65.94 *	57.77 *	-66.39 *	-59.05 *	-68.20 *	-80.15 *	-79.31 *	66.67	10.55	1.80	-10.35	0.10	262.06
Global mean	-31.50 *	-66.17 *	-24.98 *	49.26 *	-52.96 *	-35.39 *	-59.10 *	-50.18 *	-50.11 *	6.80	13.42	-1.91	-14.03 *	3.33 *	138.71 *

Shoot P concentration ([P]$_{Shoot}$), plant total P content (PTP), physiological P use efficiency (PPUE), P efficiency ratio (PER), total root dry weight (RW), lateral root dry weight (LRW), root hairs dry weight (RHW), shoot dry weight (SW), total biomass dry weight (BW), root to shoot ratio (R/S), root total length (RTL), total root average diameter (TRAD), percentage of root dry weight devoted to root hairs (RHW%), proportion of root length devoted to fine roots (PFR) and root specific length (RSL). * Indicates significant differences between treatments for that accession and trait.

All root parameters were significantly affected by the accession effect, while only a percentage of root hairs (RHW%), proportion of root length devoted to fine roots (PFR) and root specific length (RSL) were significantly affected by the treatment. For root total length (RTL), accession usa_numex (−39.33%) was the only genotype that significantly reduced its root length, while the general population tendency was to increase it (Table 4). Accessions mex_103B (−7.53%), mex_ng (−13.30%) and mu_esp (−8.37%) significantly decreased their total average diameter under NoP conditions (Table 4) in accordance with the population general trend (−1.91%). Likewise, the percentage of root dry weight devoted to root hairs (RHW%) was 14.03% lower under P-stress conditions, with accessions mex_096D (−19.79%), sp_11814 (−21.27%), sp_bola (−26.14%) and sp_piq (−21.96%) being the significantly affected ones (Table 4). Contrarily, the proportion of root length devoted to fine roots (PFR) showed a slight increase under NoP (3.33%), compared to control conditions, although only four accessions were significantly affected. Thus, accessions mex_103B (5.91%), mex_ng (9.79%), mu_esp (6.53%) and sp_lobo (8.97%) significantly increased this parameter under NoP conditions (Table 4). Ultimately, root specific length (RSL) was, on average, 138.71% higher under NoP conditions. Most accessions were significantly affected by the treatment; mu_esp (293.57%) and sp_11814 (282.31%) were the accessions with a higher increase for root specific length (Table 4).

3.7. Principal Components Analysis for Trial 2

The first two PCs explained 63.84% of total variation for Trial 2 (Figure 5). PC1 explained 47.48% of the total variation and was defined by the response of total biomass dry weight (BW), shoot dry weight (SW), physiological P use efficiency (PPUE), plant total P content (PTP), root dry weight (RW), root hair dry weight (RHW) and absolute values for P uptake efficiency (PUpE), while the traits that most contributed negatively were the response of the root to shoot ratio (R/S) and root specific length (RSL) (Figure 5A). PC2 on the other hand explained 16.36% of total variability and was positively correlated with response of shoot P concentration ([P]$_{Shoot}$), proportion of root length devoted to fine lateral roots (PFR) and root total length (RTL), while being negatively correlated with the total root average diameter (TAD) and P efficiency ratio (PER) (Figure 5A).

Based on those results, accessions located at the right side of the graph (usa_sandia, mex_ng, sp_piq, usa_conq, mex_scm and mex_103B), presented an important biomass reduction under NoP conditions and, at the same time, interesting PUpE and PUtE values and an increase at the R/S and RSL level when in NoP, indicating that these are good candidates for high input conditions due to their excellent response to the addition of P through fertilization (Table 4 and Figure 5B). On that matter, usa_sandia stood out for its impressive PUpE values and high increase of R/S and high increase of RSL (Figure 4B). On the opposite side, sp_lam, fra_DLL and usa_jap accessions were located, with negative values of R/S and relative low increase of RSL (Figure 5B) and poor values for PUpE and PUtE. From this group, usa_jap and fra_DLL had good values of biomass under NoP (Table S6). Altogether, this indicates that they perform well under NoP conditions but do not improve with additional units of P. Furthermore, in the upper part of the graph, accessions usa_numex and mex_096D were characterized by decreasing the P concentration in the shoot ([P]$_{Shoot}$), and thus, increasing PER, and having higher root's total length (RTL), proportion of length dedicated to fine roots (PFR) and root diameter (TAD) in the control than in NoP conditions (Table 4 and Figure 5B). Finally, on the bottom part of the plot, accessions mu_esp, mex_pas, sp_11814 and sp_bola were positively correlated with changes in TAD and PER and high values of PUtE, indicating a tendency to reduce their roots' average diameter while maintaining the shoot [P] concentration (Table 4 and Figure 5B).

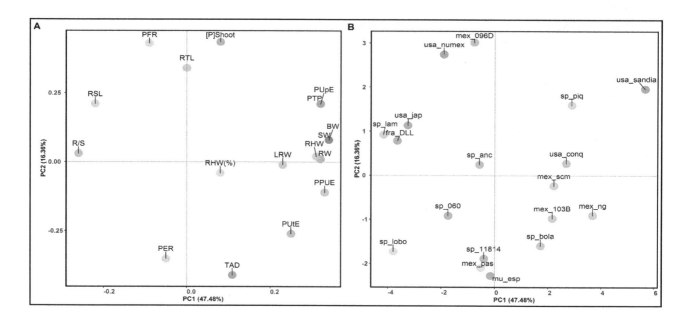

Figure 5. PCA based on trait increments between treatments for the Trial 2 experiment. (**A**) Correlation between traits and the first two principal components. (**B**) Distribution of accessions based on the studied traits. P tissue concentration traits $[P]_{Shoot}$, P total plant content (PTP) trait, efficiency parameters PPUE (physiological P use efficiency), PER (P efficiency ratio), PUpE (P uptake efficiency) and PUtE (P utilization efficiency) and morphometric traits RW (root dry weight), LRW (lateral root dry weight), RHW (root hairs dry weight), SW (shoot dry weight), BW (total biomass dry weight), R/S (root to shoot ratio), RTL (root total length), TAD (root total average diameter), RHW% (root hairs dry weight %), PFR (proportion of length dedicated to fine roots) and RSL (root specific length) were considered.

4. Discussion

4.1. Peppers Change Their Mineral Homeostasis and re-Allocate Their P Reserves to Adjust to Low-P Conditions

A comparison of P (root, shoot and fruit in Trial 1 and just shoot in Trial 2) concentrations provided relevant information on the impact of the different levels of P on pepper. There is evidence to suggest that pepper plant organs require P in different amounts, and the minimum levels are drastically different between tissues. Regarding that, roots presented the highest drop of P concentrations between treatments, indicating that they are able to mobilize P in order to benefit above-ground biomass. This response has been described in other crops, in which physiological and morphological changes, such as changes in root porosity and aerenchyma proportion, have been reported as mechanisms for reducing both the metabolic expenses and P requirements of the root system, while maintaining the foraging ability [15,22,39]. Interestingly, there were also differences among genotypes on P-tissue allocation, which opens the door to breeding materials with minimal P levels in the fruits and less need for fertilization without hampering production. For instance, some authors believe that we consume more P than required for a healthy diet, and often in the form of phytate, which is not fully absorbed by the human digestive system [40,41].

Homeostatic processes by which plants take up, transport and store nutrients are not independent, and therefore, the absence or excess of some elements can affect how the rest are processed, as was observed herein [6,38,39]. However, despite some significant differences between treatments for other macro minerals and tissue combinations, the values observed for this experiment are within the normal range, and therefore, no deficiency or excess was detected apart from P [6,42].

4.2. P Efficiency Parameters Measure Different Aspects of the Plant Response

The use of parameters to describe a plant's mineral uptake and use efficiency is a widely adopted practice in this scientific field [18,21]. Thus, physiological P use efficiency (PPUE) provides information on how productive a genotype may be, based on its tissues P concentration under a specific treatment; hence, high values indicate higher efficiency transforming absorbed P into biomass. Under these conditions, accessions mex_pas (control) and bol_144 (NoP) presented the highest PPUE for Trial 1, whereas in Trial 2, usa_sandia and fra_DLL presented the highest values for control and NoP, respectively. These results indicate a differential response, making these accessions interesting candidates for different P-fertilizers input conditions (e.g., high and low). Interestingly, the general response of increasing PPUE from control to NoP was not observed for trial 2. It is important to point out that although it is the same parameter, it was calculated in a different way depending on the trial. For Trial 1, concentration of P was a mean of all plant tissues whereas for Trial 2 it was extrapolated from shoot only, which may have caused a behavior distortion. P efficiency ratio (PER), on the other hand, relates the amount of yielded biomass with the amount of accumulated P in the plant; thus, high values indicate a higher ability to generate biomass with less P. Thus, bol_144 (Trial 1) and fra_DLL (Trial 2) are extremely efficient genotypes, especially under low-input conditions. These results indicate an interesting ability to use every unit of absorbed P and convert it into biomass and suggest that aptitude should be used in low-input systems.

Regarding P uptake efficiency (PUpE), accessions mex_scm and sp_piq showed an above average performance in both trails, although in the Trial 2 both usa_conq and usa_sandia showed higher values. This indicates that these accessions responded well to fertilization and were able to take up high amounts of P when it is present. In terms of P utilization efficiency (PUtE), accessions mex_pas and sp_bola showed the highest values in both trials, indicating that they are able to use the absorbed P into biomass generation more efficiently than the rest of the accessions. Furthermore, genetic variation regarding P acquisition and use efficiency has been widely reported in soybean, maize, sunflower, brassica and melon [15,18,21,43,44]. However, to our knowledge, this is the first work that provides such information for pepper germplasm. Herein, a wide range of variability is reported regarding P efficiency parameters, as well as several combinations among them, offering numerous possibilities for breeding for improved P uptake and P use efficiency parameters (PUE). Several authors have reported independence between uptake and use efficiency, which enables the improvement of both as well as selecting materials with different purposes (e.g., high- and low-input environments) [12,18,21,44]. These results seem to point towards that idea, since both parameters were located separately in both trials' PCA.

4.3. Modifications at Root Level

Many species promote root instead of aerial growth in order to enhance foraging capability [15,17,21]. In this experiment, a loss of root mass was observed under NoP conditions; however, this reduction was lower than that of the aerial part. This resulted in an increased root to shoot ratio under NoP conditions compared to control plants. The results indicate that, apart from lower biomass accumulation and redistribution of it, there are important modifications, particularly at root level, that help the plant to cope with P-stress. This was also observed in previous works with other crops for P-starvation conditions [20].

Morphological adaptations to low P concentrations in the soil aim at enhancing P acquisition by enabling exploitation of a greater soil volume, as well as enhancing P uptake without significantly increasing metabolic costs [17,18,45,46]. This is achieved mainly by the stimulation of root hairs [15,18,45], by halting secondary growth of the root and promoting primary growth and elongation [46] or increasing porosity and aerenchyma in roots [22]. Herein, lateral root length (LRL), but not total root length (RTL), increased under NoP conditions. It seems that lateral root elongation was a key response of the plant to reach possible P patches in the soil. This response has been described as an adaptive response to low P in Phaseolus vulgaris [46]. Other parameters, such as the lateral root specific length (LRSL), root specific length (RSL), percentage of fine roots (PFR) and percentage of fine lateral roots (PFLR), were higher under NoP, whereas the LRAD was lower. Therefore, pepper genotypes react to low P by producing thinner and lighter roots (with less carbon cost), which is in concordance with the literature [15,18,45,47]. On that regard, bol_144 (Trial 1) and mex_ng (Trial 2) stood out for their significant reduction of root average diameter while increasing the proportion of length dedicated to thinner roots under NoP conditions. In addition, accessions such as mu_esp and sp_bola showed a significant stimulation of their root specific length and proportion of length dedicated to thinner roots under the NoP conditions, despite presenting a lower root weight than under control conditions.

Although the percentage of root hair weight (RHW%) decreased in Trial 2, and was not significantly different in Trial 1, it must be pointed out that this measure includes fine roots and not specifically root hairs; therefore, it must be investigated if root hairs are modified in pepper under contrasting P conditions. Analyzing roots is a difficult task, and specific protocols must be set up to increase accuracy of root traits' study. The differences regarding root scanning and the analysis procedure between trials indicate that the first methodology (scanning just lateral roots) was more effective in finding root responses, since scanning all the fine roots has technical limitations.

4.4. A Wide Range of Responses to Breed Efficient Genotypes

Despite the general responses of pepper to low P described in the previous section, there was a wide range of responses depending on the accession studied. PCA's projection showed a widely differentiated behavior among accessions, creating several accession clusters depending on their overall response to NoP. For example, sp_piq, mex_scm, usa_sandia, usa_conquistador and mex_ng showed high PUpE values associated with increases in root to shoot ratio and root specific length. Sp_bola, mex_pas and mu_esp were associated with high PUtE values, no changes in their concentration of P in the shoots, reduction of the root diameter and an increase of percentage of fine roots and root total length. On the other hand, there were accessions that were poorly responsive to the changes in P levels, such as bol_37 and eq_973, or sp_lam, fra_DLL or usa_jap. Results indicated that some accessions were more suited to grow under low input conditions (bol_144, eq_973 and usa_jap) and others were highly responsive to increasing amounts of P available in the soil (sp_piq, sp_pas and mu_esp). It was also observed that P uptake efficiency and P utilization efficiency seem to be controlled independently, and here, this is demonstrated by the positioning of both parameters in opposite quadrants of the PCAs' second component, and accessions with contrasting levels.

On that matter, the availability of diversity is of paramount importance for crop breeding, enabling the combination of several favorable traits or behaviors in a single genotype, which in return can be a more effective solution than to have those traits in separate genotypes. For example, Miguel et al. [48] demonstrated, in common bean, that combining shallow basal roots and long root hairs yielded a larger effect regarding P acquisition than their additive effects separately. Breeding for efficient genotypes needs an accurate definition of the target to be improved; this is not the same as improving the ability to grow under low inputs than reacting favorably to P addition. Defining the best ideotype to each condition and the combination of different adaptation opens the possibility to breed towards different goals [12,18,21,44].

5. Conclusions

Herein, a diverse collection of 25 pepper accessions has been characterized for their behavior under P low-input conditions. A considerable amount of diversity has been reported for the response to phosphorus low-input conditions for several phosphorus uptake and use efficiency parameters, and root and biomass traits. Evidence suggests that P low-input conditions play an important role in the plant's tissues allocation for this mineral and that different organs show different critical levels of phosphorus. In addition, the responses of this collection indicate the existence of genetic diversity, which may be used in breeding programs to generate materials with different applications.

Accessions bol_144 and fra_DLL showed promising results for low-input conditions, whereas mex_scm, sp_piq, usa_conq and usa_sandia were on the opposite spectrum and are probably best used under high-input conditions due to their uptake efficiency. In addition, mex_pas and sp_bola showed the best results regarding P use efficiency. Finally, P low-input conditions proved to be an important factor controlling root morphology. Under these conditions, roots presented longer and thinner roots. These traits correlated to a higher efficiency and biomass accumulation under P-starving conditions. This work provides relevant first insights into pepper's response to phosphorus low-input conditions. More works are needed in order to dissect the mechanisms controlling the response, and consequently, to be introgressed into new materials.

Supplementary Materials
Table S1: Ion concentrations for irrigation water, Control and NoP solutions used in both Trial 1 and Trial 2. Table S2: Effect of Control and NoP treatments on P, K, Ca, Mg, Na and S (g/100g DW) concentrations for root, shoot and fruit tissues for Trial 1. Overall mean values, standard deviation and p-value for each plant tissue and treatment are provided. Table S3: Trial 1 multi-factor ANOVA's mean square values of accession and treatment effects, their interaction, and error for P concentration traits $[P]_{Root}$, $[P]_{Shoot}$, $[P]_{Fruit}$, P content traits RootP (g P), ShootP (g P), FruitP (g P), PTP (mg P), P efficiency parameters PPUE (g^2 DW g^{-1} P) and PER (g DW g^{-1} P) and for morphometric traits RW (g), LRW (g), RHW(g), SW (g), BW (g), R/S, LRL (m), LRAD (mm), RHW% (%), PLFR (%) and LRSL (m/g). Table S4: Trial 1 mean values and standard deviation for P accumulation and efficiency traits and parameters, and biomass and root traits and parameters. Table S5: Trial 2 multi-factor ANOVA's mean square values of accession (A) and treatment (T) effects, their interaction and error (E) for P concentration trait $[P]$Shoot, P content trait PTP (mg P), for efficiency parameters PPUE (g^2 DW g^{-1} P) and PER (g DW g^{-1} P) and for morphometric traits RW (g), LRW (g), RHW (g), SW (g), BW (g), R/S, RTL (m), TAD (mm), RHW%, PFR (%) and RSL (m/g). Table S6: Trial 2 mean values and standard deviation for P accumulation and efficiency traits and parameters, and biomass and root traits and parameters.

Author Contributions: Conceptualization, methodology and validation: A.F., Á.C. and A.R.-B.; Data curation: L.P.-D. and D.G.-V.; Formal analysis and investigation: L.P.-D., D.G.-V., V.C.-Z. and A.Q.; Resources, funding acquisition and project administration: A.F. and Á.C.; Writing–original draft: L.P.-D. and A.F.; Writing—review and editing: L.P.-D., D.G.-V., V.C.-Z., A.Q., Á.C., A.R.-B. and A.F. All authors have read and agreed to the published version of the manuscript.

Acknowledgments: Authors thank seed providers included in Table 1, such as P.W. Bosland, François Jourdan and the different Regulatory Boards of the PDOs and GPIs included in this study. Additionally, we want to thank Jose J. Luna for his advice on Mexican peppers.

References

1. Jaggard, K.W.; Qi, A.; Ober, S. Possible changes to arable crop yields by 2050. *Philos. Trans. R. Soc. B Biol. Sci.* **2010**, *365*, 2835–2851. [CrossRef] [PubMed]

2. Grafton, R.Q.; Daugbjerg, C.; Qureshi, M.E. Towards food security by 2050. *Food Secur.* **2015**, *7*, 179–183. [CrossRef]

3. Tilman, D.; Cassman, K.G.; Matson, P.A.; Naylor, R.; Polasky, S. Agricultural sustainability and intensive production practices. *Nature* **2002**, *418*, 671–677. [CrossRef] [PubMed]

4. Tsiafouli, M.A.; Thébault, E.; Sgardelis, S.P.; de Ruiter, P.C.; van der Putten, W.H.; Birkhofer, K.; Hemerik, L.; de Vries, F.T.; Bardgett, R.D.; Brady, M.V.; et al. Intensive agriculture reduces soil biodiversity across Europe. *Glob. Chang. Biol.* **2015**, *21*, 973–985. [CrossRef] [PubMed]

5. Raza, A.; Razzaq, A.; Mehmood, S.S.; Zou, X.; Zhang, X.; Lv, Y.; Xu, J. Impact of climate change on crops adaptation and strategies to tackle its outcome: A review. *Plants* **2019**, *8*, 34. [CrossRef] [PubMed]

6. Jones, J.B.J. *Plant Nutrition and Soil Fertility Manual*, 2nd ed.; Press, C., Ed.; Taylor & Francis Group: Boca Raton, FL, USA, 2012; ISBN 9781439816103.

7. Schnug, E.; Haneklaus, S.H. Assessing the plant phosphorus status. In *Phosphorus in Agriculture: 100% Zero*; Springer: Dordrecht, The Netherlands, 2016; pp. 95–125.

8. Vance, C.P.; Uhde-Stone, C.; Allan, D.L. Phosphorus acquisition and use: Critical adaptations by plants for securing a nonrenewable resource. *New Phytol.* **2003**, *157*, 423–447. [CrossRef]

9. Cordell, D.; Drangert, J.O.; White, S. The story of phosphorus: Global food security and food for thought. *Glob. Environ. Chang.* **2009**, *19*, 292–305. [CrossRef]

10. Mogollón, J.M.; Beusen, A.H.W.; van Grinsven, H.J.M.; Westhoek, H.; Bouwman, A.F. Future agricultural phosphorus demand according to the shared socioeconomic pathways. *Glob. Environ. Chang.* **2018**, *50*, 149–163. [CrossRef]

11. Schnug, E.; Haneklaus, S.H. The enigma of fertilizer phosphorus utilization. In *Phosphorus in Agriculture: 100% Zero*; Springer: Dordrecht, The Netherlands, 2016; pp. 7–26.

12. Lynch, J.P. Roots of the second green revolution. *Aust. J. Bot.* **2007**, *55*, 493–512. [CrossRef]

13. Fernández, J.M.; Selma, M.A.E. Estimación de la contaminación agrícola en el Mar Menor mediante un modelo de simulación dinámica. In *Proceedings of the El Agua y Sus Usos Agrarios*; Universidad de Zaragoza: Zaragoza, Spain, 1998; Volume 9, pp. 1–9.

14. Kauranne, L.-M.; Kemppainen, M. Urgent need for action in the Baltic sea area. In *Phosphorus in Agriculture: 100% Zero*; Springer: Dordrecht, The Netherlands, 2016; pp. 1–6.

15. Fernández, M.C.; Rubio, G. Root morphological traits related to phosphorus-uptake efficiency of soybean, sunflower, and maize. *J. Plant Nutr. Soil Sci.* **2015**, *178*, 807–815. [CrossRef]

16. Fita, A.; Bowen, H.C.; Hayden, R.M.; Nuez, F.; Picó, B.; Hammond, J.P. Diversity in expression of Phosphorus (P) responsive genes in *Cucumis melo* L. *PLoS ONE* **2012**, *7*, e35387. [CrossRef]

17. Li, J.; Xie, Y.; Dai, A.; Liu, L.; Li, Z. Root and shoot traits responses to phosphorus deficiency and QTL analysis at seedling stage using introgression lines of rice. *J. Genet. Genom.* **2009**, *36*, 173–183. [CrossRef]

18. Hammond, J.P.; Broadley, M.R.; White, P.J.; King, G.J.; Bowen, H.C.; Hayden, R.M.; Meacham, M.C.; Mead, A.; Overs, T.; Spracklen, W.P.; et al. Shoot yield drives phosphorus use efficiency in *Brassica oleracea* and correlates with root architecture traits. *J. Exp. Bot.* **2009**, *60*, 1953–1968. [CrossRef]

19. Lynch, J.P.; Brown, K.M. Topsoil foraging—An architectural adaptation of plants to low phosphorus availability. *Plant Soil* **2001**, *237*, 225–237. [CrossRef]

20. Niu, Y.F.; Chai, R.S.; Jin, G.L.; Wang, H.; Tang, C.X.; Zhang, Y.S. Responses of root architecture development to low phosphorus availability: A review. *Ann. Bot.* **2013**, *112*, 391–408. [CrossRef]

21. Fita, A.; Nuez, F.; Picó, B. Diversity in root architecture and response to P deficiency in seedlings of *Cucumis melo* L. *Euphytica* **2011**, *181*, 323–339. [CrossRef]

22. Fan, M.; Zhu, J.; Richards, C.; Brown, K.M.; Lynch, J.P. Physiological roles for aerenchyma in phosphorus-stressed roots. *Funct. Plant Biol.* **2003**, *30*, 493–506. [CrossRef]

23. Richardson, A.E.; Lynch, J.P.; Ryan, P.R.; Delhaize, E.; Smith, F.A.; Smith, S.E.; Harvey, P.R.; Ryan, M.H.; Veneklaas, E.J.; Lambers, H.; et al. Plant and microbial strategies to improve the phosphorus efficiency of agriculture. *Plant Soil* **2011**, *349*, 121–156. [CrossRef]

24. van de Wiel, C.C.M.; van der Linden, C.G.; Scholten, O.E. Improving phosphorus use efficiency in agriculture: Opportunities for breeding. *Euphytica* **2016**, *207*, 1–22. [CrossRef]

25. Bosland, P.W.; Votava, E.J. *Peppers: Vegetable and Spice Capsicums*; CABI Publishing: Wallingford, Oxon, UK, 2012; ISBN 178064020X.

26. Food and Agriculture Organization of the United Nations. *FAOSTAT Statistics Database*; FAO: Rome, Italy, 2019.

27. DeWitt, D.; Bosland, P.W. *Peppers of the World: An Identification Guide*; Ten Speed Press: Berkeley, CA, USA, 1996; ISBN 0898158400.

28. Sahitya, U.L.; Krishna, M.S.R.; Suneetha, P. Integrated approaches to study the drought tolerance mechanism in hot pepper (*Capsicum annuum* L.). *Physiol. Mol. Biol. Plants* **2019**, *25*, 637–647. [CrossRef]

29. Hwang, E.-W.; Kim, K.-A.; Park, S.-C.; Jeong, M.-J.; Byun, M.-O.; Kwon, H.-B. Expression profiles of hot pepper (*Capsicum annuum*) genes under cold stress conditions. *J. Biosci.* **2005**, *30*, 657–667. [CrossRef] [PubMed]

30. Jing, H.; Li, C.; Ma, F.; Ma, J.-H.; Khan, A.; Wang, X.; Zhao, L.-Y.; Gong, Z.-H.; Chen, R.-G. Genome-wide identification, expression diversication of dehydrin gene family and characterization of *CaDHN3* in pepper (*Capsicum annuum* L.). *PLoS ONE* **2016**, *11*, e0161073. [CrossRef] [PubMed]

31. Pereira-Dias, L.; Vilanova, S.; Fita, A.; Prohens, J.; Rodríguez-Burruezo, A. Genetic diversity, population structure, and relationships in a collection of pepper (*Capsicum* spp.) landraces from the Spanish centre of diversity revealed by genotyping-by-sequencing (GBS). *Hortic. Res.* **2019**, *6*, 54. [CrossRef]

32. Fita, A.; Alonso, J.; Martínez, I.; Avilés, J.; Mateu, M.; Rodríguez-Burruezo, A. Evaluating Capsicum spp. root architecture under field conditions. In Proceedings of the Breakthroughs in the Genetics and Breeding of Capsicum and Eggplant; Lanteri, S., Rotino, G.L., Eds.; Università degli Studi di Torino: Torino, Italy, 2013; pp. 373–376.

33. Fita, A.; Picó, B.; Roig, C.; Nuez, F. Performance of *Cucumis melo* ssp. agrestis as a rootstock for melon. *J. Hortic. Sci. Biotechnol.* **2007**, *82*, 184–190. [CrossRef]

34. Ministerio de Agricultura, Pesca y Alimentación. *Métodos Oficiales de Análisis*; MAPA: Madrid, Spain, 1994.

35. Hills, T.M.L.; Jackson, F. *Agricultural Experimentation: Design and Analysis*; Wiley: New York, NY, USA, 1978; ISBN 978-0-471-02352-4.

36. R Development Core Team. *R: A Language and Environment for Statistical Computing*; R Foundation for Statistical Computing: Vienna, Austria, 2009.

37. Wickham, H. *ggplot2: Elegant Graphics for Data Analysis*; Springer International Publishing: Basel, Switzerland, 2016; ISBN 978-3-319-24277-4.

38. Bouain, N.; Shahzad, Z.; Rouached, A.; Khan, G.A.; Berthomieu, P.; Abdelly, C.; Poirier, Y.; Rouached, H. Phosphate and zinc transport and signalling in plants: Toward a better understanding of their homeostasis interaction. *J. Exp. Bot.* **2014**, *65*, 5725–5741. [CrossRef] [PubMed]

39. Ham, B.-K.; Chen, J.; Yan, Y.; Lucas, W.J. Insights into plant phosphate sensing and signaling. *Curr. Opin. Biotechnol.* **2018**, *49*, 1–9. [CrossRef] [PubMed]

40. Rose, T.J.; Pariasca-Tanaka, J.; Rose, M.T.; Fukuta, Y.; Wissuwa, M. Genotypic variation in grain phosphorus concentration, and opportunities to improve P-use efficiency in rice. *Field Crop. Res.* **2010**, *119*, 154–160. [CrossRef]

41. Bryant, R.J.; Dorsch, J.A.; Peterson, K.L.; Rutger, J.N.; Raboy, V. Phosphorus and mineral concentrations in whole grain and milled low phytic acid (*lpa*) 1-1 rice. *Cereal Chem.* **2005**, *82*, 517–522. [CrossRef]

42. Russo, V.M. *Peppers: Botany, Production and Uses*; Russo, V.M., Ed.; CABI: Wallingford, UK, 2012; ISBN 9781845937676.

43. Akhtar, M.S.; Oki, Y.; Adachi, T. Genetic variability in phosphorus acquisition and utilization efficiency from sparingly soluble P-sources by *Brassica* cultivars under P-stress environment. *J. Agron. Crop Sci.* **2008**, *194*, 380–392. [CrossRef]

44. Hu, Y.; Ye, X.; Shi, L.; Duan, H.; Xu, F. Genotypic differences in root morphology and phosphorus uptake kinetics in *Brassica napus* under low phosphorus supply. *J. Plant Nutr.* **2010**, *33*, 889–901. [CrossRef]

45. Bates, T.R.; Lynch, J.P. Root hairs confer a competitive advantage under low phosphorus availability. *Plant Soil* **2001**, *236*, 243–250. [CrossRef]

46. Strock, C.F.; Morrow de la Riva, L.; Lynch, J.P. Reduction in root secondary growth as a strategy for phosphorus acquisition. *Plant Physiol.* **2018**, *176*, 691–703. [CrossRef] [PubMed]

47. López-Bucio, J.; Cruz-Ramírez, A.; Herrera-Estrella, L. The role of nutrient availability in regulating root architecture. *Curr. Opin. Plant Biol.* **2003**, *6*, 280–287. [CrossRef]

48. Miguel, M.A.; Postma, J.A.; Lynch, J.P. Phene synergism between root hair length and basal root growth angle for phosphorus acquisition. *Plant Physiol.* **2015**, *167*, 1430–1439. [CrossRef]

4

New Eco-Friendly Polymeric-Coated Urea Fertilizers Enhanced Crop Yield in Wheat

**Ricardo Gil-Ortiz [1,*], Miguel Ángel Naranjo [1,2], Antonio Ruiz-Navarro [3],
Marcos Caballero-Molada [1,2], Sergio Atares [2], Carlos García [3] and Oscar Vicente [4]**

[1] Institute for Plant Molecular and Cell Biology (UPV-CSIC), Universitat Politècnica de València,
   46022 Valencia, Spain; mnaranjo@ibmcp.upv.es (M.Á.N.); marcamo2@ibmcp.upv.es (M.C.-M.)
[2] Fertinagro Biotech S.L., Polígono de la Paz, C/ Berlín s/n, 44195 Teruel, Spain; sergio.atares@tervalis.com
[3] Centre for Soil and Applied Biology Science of Segura (CEBAS-CSIC), Espinardo University Campus,
   30100 Murcia, Spain; ruiznavarro@cebas.csic.es (A.R.-N.); cgarizq@cebas.csic.es (C.G.)
[4] Institute for the Preservation and Improvement of Valencian Agrodiversity (COMAV),
   Universitat Politècnica de València, 46022 Valencia, Spain; ovicente@upvnet.upv.es
*   Correspondence: rigilor@ibmcp.upv.es

Abstract: Presently, there is a growing interest in developing new controlled-release fertilizers based on ecological raw materials. The present study aims to compare the efficacy of two new ureic-based controlled-release fertilizers formulated with water-soluble polymeric coatings enriched with humic acids or seaweed extracts. To this end, an experimental approach was designed under controlled greenhouse conditions by carrying out its subsequent field scaling. Different physiological parameters and crop yield were measured by comparing the new fertilizers with another non polymeric-coated fertilizer, ammonium nitrate, and an untreated 'Control'. As a result, on the microscale the fertilizer enriched with humic acids favored a better global response in the photosynthetic parameters and nutritional status of wheat plants. A significant 1.2-fold increase in grain weight yield and grain number was obtained with the humic acid polymeric fertilizer versus that enriched with seaweed extracts; and also, in average, higher in respect to the uncoated one. At the field level, similar results were confirmed by lowering N doses by 20% when applying the humic acid polymeric-coated produce compared to ammonium nitrate. Our results showed that the new humic acid polymeric fertilizer facilitated crop management and reduced the environmental impact generated by N losses, which are usually produced by traditional fertilizers.

Keywords: coated-urea fertilizer; humic acid; lignosulfonate; natural polymers; seaweed extract; wheat

1. Introduction

According to the Food and Agriculture Organization of the United Nations (FAO), wheat is the world's largest cultivated crop per hectare and the third largest cereal to be produced [1]. In the European Union, 144.5 million tons were harvested in 2016 and production is estimated to increase by 3.5% each year. In fact, the world's production in 2017 was expected to come to 744.5 million tons, an increase that comes close to 1000% since 1990/1991. Current cereal production demand and gradual soil impoverishment mean that it is increasingly necessary to apply more fertilizers, mainly nitrogenous ones [2]. High quantities of nitrogen (N) per hectare need to be applied to soil to produce optimum wheat grain yields [3]. N-organic mineralization in soil is a slow process that requires the action of soil microorganisms and must also be given the necessary environmental conditions [4,5]. Plants absorb N in the form of exchangeable ammonium (NH_4^+) or nitrates (NO_3^-), which are highly mobile compounds in soil that can be easily lost by volatilization or leaching, which leads to environmental problems and/or toxicity for plants [6,7]. The legal restrictions associated with pollution limitations have been set to preserve the environment [8]. Traditional N fertilizers, such as sulfates, nitrates, or urea, are characterized by high constant N-release kinetics [9–11]. Fertilizer granules rapidly decompose and a strong N release occurs when applied to soil. Later emissions slow down and additional covert applications are usually necessary for crops to achieve expected yields [12]. This lack of efficiency implies figures like 90% for N loss of the total N applied [8,13,14]. N fertilization efficiency depends on different variables, such as environmental conditions, coupling soil/plant or management practices [9,15].

Fertilizer manufacturers have concentrated in recent decades to produce slow-release and controlled-release fertilizers (SRFs/CRFs) as enhanced-efficiency nitrogen fertilizers (EENF) [16]. SRFs are long-chain molecules of lower solubility than traditional fertilizers like formaldehyde, isobutylene diurea, or methylene urea, for which the biodegradability is proportional to the microbiological activity of the soil. CRFs action, on the other hand, depends mainly on diffusion through coatings and not directly on biodegradation, thus being more efficient in controlling release of nutrients [17,18]. The main advantages of these slow- or controlled-release fertilizer generations are summarized in numerous reviews [8,9,18]: (1) extending the durability of fertilizers by providing small amounts for a longer time; (2) lowering the number of fertilizer applications, generally to a single background application, by prolonging their time of action; (3) cutting costs by eliminating the typical covert applications of traditional fertilizers; (4) reducing environmental pollution by limiting the amount of fertilizer released being assimilated in soil/the plant system. Urea is the major N source used in plant nutrition [19]. Synthetic SRF urea formulations are, for example, urea formaldehyde, isobutylidene diurea, crotonylidene diurea, or sulfur-coated urea fertilizers [20]. Today, controlled-released coated urea (CRCU) is the most important technology being developed in the fertilizer industry [13,18]. To manufacture CRCU, urea is usually coated with polymers that control N release by diffusion based on the permeability of polymer coatings [13,18,21]. Release of N from polymeric CRCU is not significantly influenced by microorganisms of soils because nutrient release can be better controlled compared to sulfur-based coatings [18]. In fact, emissions are influenced mainly by environmental factors like temperature or humidity [17,22] and also by intrinsic factors of fertilizers, such as nutrient composition, coating thickness, granular shape, and diameter [17,18].

In recent decades, the use of synthetic polymers, like those based on sulfur, resins, or thermoplastic materials, has been hampered by legal restrictions that limit pollution due to these materials' difficult degradation [13]. Such products are used in many other industries to manufacture pesticides, herbicides, pheromones, fungicides or growth regulators [8,23,24]. Their marketed forms are encapsulations, reservoir laminate structures, or monolithic systems [8]. Carbohydrate and lignin-based polymer-coated urea has been indicated as an alternative to solve problems related to N emissions of traditional fertilizers and to avoid environmental problems concerning synthetic polymers [18,25,26]. In fact, they are ecologically friendly and easily available at cheap prices, which are the main restrictions of using CRFs. For example, coatings based on starch, ethyl cellulose, or lignin have been successfully employed to slow down N release from urea [13,26,27]. Including bio-inhibitors as urease or nitrification inhibitors in fertilizers is a commonplace practice to slow down N releases [2,9,28]. Biostimulants like amino acids, humic/fulvic acids or seaweed extracts offer beneficial properties for crops, such as biofortification and resistance to different abiotic stresses, e.g., drought or salinity [29–33].

At the beginning of their development, SCRFs were limited mainly to horticultural and ornamental crops, and actually are not well established in extensive cropping as more research is necessary to perform cheaper and more ecological fertilizers [34–36]. The objective of this research was to compare the efficacy of new ecological CRCU with traditional fertilizers in physiological terms, and also grain yield and quality in wheat. The novelty of this research lies in the combination of eco-friendly polymers as byproducts from the production of wood pulp, urease inhibitors, and natural biostimulants in the same fertilizer.

2. Materials and Methods

2.1. Experimental Design

A comparison of the effectiveness of the different polymeric-coated formulations was made under greenhouse conditions and then scaled to field essays. Experiment 1 (*Triticum aestivum*). Wheat was grown in a greenhouse at ambient temperature and humidity at the facilities of the Valencian Institute of Agrarian Research (IVIA) (Moncada, Spain) from autumn 2014 to spring 2015. Sowing was carried out in pots (22 cm high × 16 cm Ø) placed in watertight trays, sized 54 × 39 × 9 cm (6 pots/tray) with three repetitions per treatment on 7 November. The pots were filled with a non-fertilized soil from a fallow area close to the greenhouse in IVIA's grounds. Culture was developed with extra lighting for a 12:12 photoperiod and irrigated with distilled water. High temperatures were reduced and partly controlled by means of automatic systems for shading, ventilation, and water cooling. Temperature and humidity were measured every hour with a digital thermo-hygrometer. The average temperatures during the whole culture cycle, from November to May were $18.8 \pm 7.2\,°C$ ($39\,°C$ max. and $4.9\,°C$ min.), and average relative humidity $55.3 \pm 12.8\%$ (88.0% max. and 20.4% min.). Experiment 2 (*Triticum spelta*). Three separated grids of $400\,m^2$, divided into 16 individual surfaces of $25\,m^2$ each, were designed at the field level in a plot located in Teruel (Spain) at GPS coordinates 40°22′17.4″ N, 1°05′58.9″ W. Four treatments, including an untreated Control, were placed by quadruplicate. Sowing was carried out mechanically in the winter at a dose of 250,000 seeds/ha. Culture was surface-irrigated on a bi-weekly basis with well water.

2.2. Fertilizer Treatments

Different N fertilizers developed by Fertinagro Biotech S.L. (Teruel, Spain) were tested and their efficacy was compared. As these fertilizers are patented, the exact composition is not herein presented. To study the influence of the different coating compositions and thicknesses, the following fertilizers were tested in Experiment 1: (1) DURAMON® (Fertinagro Biotech S.L., Teruel, Spain), composed of urea, including a urease inhibitor (monocarbamida dihidrogenosulfate—MCDHS) with no coating (ES 2 204 307 patent); (2) a new controlled-released urea fertilizer based on DURAMON® technology, but also 3% lignosulfonate-coated with humic acids (hereafter CRF_A); (3) the same as (2), but 5% lignosulfonate-coated with seaweed extracts (CRF_B). The three formulates had the same N composition (24–0–0), but a different coverture percentage. Fertilizers were applied in wheat at doses of 150 kg ha^{-1} (nitrogen fertilizer units—NFU) as a basal dressing for maximum yields based on theoretical extraction by crops. Experiment 2: Based on the physiological responses observed in the greenhouse experiment, the best CRF was selected and applied to the field at 100% doses and with a reduction to 80%. Both doses were compared with ammonium nitrosulfate (NSA, 26–0–0) (Fertinagro Biotech S.L., Teruel, Spain) as the traditional fertilizer. Maximum doses of 80 kg ha^{-1} were applied in the phenological state of tillering (27 April), based on the historical average yields obtained in the area. In both experiments, the same repetitions with untreated plants were included (CONTROL).

2.3. Soil Fertility Characterization

Several soil properties were measured to characterize soil fertility in both experiments. pH and EC were determined in a 1/5 (*w/v*) aqueous soil extract by shaking for 2 h, followed by centrifugation at 26916 g for 15 min and filtration. pH was measured by a pH meter (Crison mod.2001, Barcelona, Spain) and EC with a Conductivity meter (Crison micro CM2200, Barcelona, Spain). Total and organic soil C (SOC) and total N (N) were determined by combustion gas chromatography in a Flash EA 1112 Thermo Finnigan (Franklin, MA, USA) elemental analyzer after eliminating carbonate by acid digestion with HCl. The total nutrient contents (P, K, Ca, Mg, Cu, Fe, K, Mg, Mn, and Zn) were extracted by aqua regia digestion (3:1, *v/v*, HCl/HNO$_3$) and determined by ICP-AES (Thermo Elemental Iris Intrepid II XDL, Franklin, MA, USA). Analysis showed that both cultures grew on N-poor soil (Table 1).

Table 1. Fertility of the soil used in the experimental analysis from the first 15 cm of soil surface. Data on total nitrogen (N), total carbon (C) and organic carbon (CO) and other macro- and micronutrients are shown. Values are means ± SD ($n = 5$) at the beginning of the experiment.

Parameters	Mean ± SD (%)	
	Microscale	Field
Total nitrogen (g 100 g^{-1})	0.09 ± 0.02	0.19 ± 0.03
Total carbon (g 100 g^{-1})	2.06 ± 0.28	6.76 ± 0.34
Organic carbon (g 100 g^{-1})	0.66 ± 0.07	1.84 ± 0.28
pH	8.75 ± 0.095	8.26 ± 0.11
EC (µS cm^{-1})	120.7± 27.99	109.12 ± 47.40
P (g 100 g^{-1})	0.064 ± 0.001	0.03 ± 0.01
K (g 100 g^{-1})	0.339 ± 0.14	0.74 ± 0.08
Mg (g 100 g^{-1})	0.285 ± 0.04	0.26 ± 0.04
Ca (g 100 g^{-1})	3.561 ± 0.42	1.35 ± 0.61
Fe (g 100 g^{-1})	11.113 ± 1.77	15.94 ± 6.27
Cu (mg kg^{-1})	15.929 ± 1.45	7.46 ± 1.81
Mn (mg kg^{-1})	191.99 ± 11.68	216.75 ± 73.57
Zn (mg kg^{-1})	26.951 ± 2.50	24.66 ± 4.61

2.4. Plant Growth

Photosynthetically active flag leaves (PAFL) were characterized in the phenological state of panicles swelling (booting stage) by fresh weight (g) and foliar surface (cm^2) using a LI-3100C area meter (LI-COR®, Lincoln, Nebraska, USA). Some plant material was weighed before being dried at 65 °C until a constant mass was obtained to determine dry weight (g). Differences in the total dry weight, length (cm), primary stem length (cm), and tillers number were determined at the end of the culture.

2.5. Leaf Greenness and Effective Quantum Yield of Photosystem II

Leaf greenness was measured in the booting stage using an SPAD-502 Chlorophyll meter (Konica-Minolta, Osaka, Japan). The effective quantum yield of photosystem II electron transport (ΦPSII), which represents the electron transport efficiency between photosystems within light-adapted leaves, was checked with a leaf fluorometer (Fluorpen FP100, Photos System Instrument, Drásov, Czech Republic). Both parameters were measured in a minimum of 25 PAFL.

2.6. Gas Exchange Analysis

Gas exchange measurements were taken at noon in five plants per treatment using a portable infrared gas analyzer LCpro-SD, equipped with a PLU5 LED light unit (ADC BioScientific Ltd., Hoddesdon, UK). The selected flag leaves in wheat (booting stage) of Experiment 1 were analyzed in a leaf chamber (6.25 cm^2) to determine the following parameters: stomatal conductance (gs) (expressed as mmol m^{-2} s^{-1}), net photosynthetic rate (A) (µmol m^{-2} s^{-1}), transpiration (E) (mol m^{-2} s^{-1}), and intercellular CO_2 concentration (Ci) (µmol mol^{-1}) under ambient CO_2, temperature, and relative humidity conditions. They were recorded by programming increasing photosynthetically active radiations (PAR) of 348, 522, 696, 870, 1218, and 1566 µmol m^{-2} s^{-1}. Water-use efficiency (WUE) and intrinsic WUE were calculated as the ratio between A/gs and A/E, expressed as µmol (CO_2 assimilated) mol^{-1} (H_2O transpired).

2.7. Foliar Nutrient Analysis

Foliar analyses were performed from the fresh samples collected in the phenological state of panicles swelling (booting stage) 70 days after plants emerged. Samples were composed of a pool with a minimum of four flag leaves taken from different plants in the same treatment. Four replicates per treatment and culture were collected and kept at $-80\ °C$ until they were biochemically analyzed. The compositions in macro- (N, P, K, Ca, Mg and S) and micronutrients (Fe, Cu, Mn, Zn, B, and Mo) were determined by Inductively Coupled Plasma Optical Emission Spectrometry (ICP-OES). N content was estimated by an N-Pen N 100 apparatus (Photon System Instruments, Drásov, Czech Republic).

2.8. Growth, Yield, and Cereal Grain Composition

Once grain ripening had been completed at 138 days in greenhouses after emergence, the remaining plants per culture were harvested and characterized in growth and grain yield terms. The growth parameters of the total dry weight of aerial parts, primary stem length, tillers number, ears number, and ear length and weight were measured. Yield was determined by measuring the total dry grain weight, one-hundred grain weight and grain number. Nitrogen Use Efficiency (NUE) for each fertilizer treatment was calculated as agronomic efficiency according to [37]:

$$\text{NUE}\left(\text{kg kg}^{-1}\right) = \frac{\text{Grain yield of the fertilized area} - \text{Grain yield of the unfertilized area}}{\text{Quantity of N applied as N fertilizer}} \quad (1)$$

At the field level and 90 days after applying fertilizers, the biomass of aerial parts, ears weight, and grain weight was studied. The Harvest Index (HI) was calculated as the grain weight/biomass of aerial parts. Different quality parameters were also measured in the grain. A representative composite sample was prepared separately for each treatment, pooling fractions of plant material for each replication. Subsequently, each composite mixture was ground and analyzed based on food quality analysis methods (Comission Regulation EC N° 152/2009 of 27 January): humidity (gravimetric by drying in an oven at 130 °C), ashes (gravimetric by incineration at 550 °C), lipids (extraction without hydrolisis in Soxtec Avanti—Foss), protein (Kjeldahl method using Foss automatic distillation equipment, Foss, Hillerød, Denmark), crude fiber (gravimetric), and total carbohydrates (volumetric using Luff Schoorl reagent). Analysis were carried out by the Valencia's Agrifood Laboratory (Burjassot, Spain).

2.9. Statistics

The differences between fertilizers treatments were tested by analysis of variance (ANOVA) at 95% confidence. Prior to the ANOVA, the data requirements of normality and homogeneity of variances were checked according to Levene's and Shapiro–Wilk tests. When the null ANOVA hypothesis was rejected, post hoc comparisons were made to establish the possible statistical differences among the different treatments applied using the Fisher's LSD test. The statistical Statgraphics Centurion XV, version 15.2.05 software program (Statpoint Technologies, Inc., Warrenton, VA, USA) was used to perform the analysis.

3. Results

3.1. Plant Growth, Leaf Greenness and Effective Quantum Yield of Photosystem II

No significant differences were observed for the various treatments performed for chlorophyll content and ΦPSII, measured by nondestructive techniques in the phenological state of panicles swelling (booting stage), although they were significant compared to the CONTROL (Table 2). Similar results were obtained when studying PAFL fresh weight content, dry weight and area. The total fresh weight of aerial parts was 1.4- and 1.7-fold significantly higher for the plants fertilized with DURAMON® compared to CRF_A and CRF_B, but no differences were observed for dry weight. The responses of fertilizer treatments on plant growth, foliar area, and root development are shown in Figure 1.

Table 2. Effects of fertilizer treatments CRF$_A$, CRF$_B$, and DURAMON® on photosynthetic parameters (effective quantum yield of photosystem II—ΦPSII, leaf greenness, nitrogen content, total fresh weight of aerial parts (g), dry weight of aerial parts (%), photosynthetically active flag leaves—PAFL fresh weight, PAFL area, and leaf area index—LAI) in *Triticum aestivum* leaves compared to the Control in the phenological state of panicles swelling (booting state). Values are means ± SD (n = 18 for ΦPSII, leaf greenness, and N content; n = 8 for the other growth parameters).

Parameters	CRF$_A$	CRF$_B$	DURAMON®	CONTROL
ΦPSII	0.69 ± 0.03 b	0.69 ± 0.03 b	0.68 ± 0.03 b	0.58 ± 0.07 a
Leaf greenness content (SPAD units)	54.5 ± 2.3 b	52.5 ± 2.3 b	54 ± 1.5 b	40.2 ± 6.8 a
N content (%)	5.5 ± 0.4 b	5 ± 0.3 ab	5.3 ± 0.4 ab	3.5 ± 0.9 a
Total fresh weight (aerial part) (g)	58.9 ± 9.3 b	46.9 ± 20.9 ab	79.6 ± 15.9 c	38.5 ± 10.5 a
Dry weight (aerial part) (%)	29 ± 8 a	33.3 ± 7.6 ab	35.4 ± 3.2 ab	32.9 ± 3.5 b
PAFL fresh weight (g)	24.5 ± 12.7 b	26.6 ± 25.5 b	27.2 ± 9.2 b	2.8 ± 3.8 a
PAFL dry weight (%)	6.13 ± 3.02 b	6.67 ± 5.81 b	7.23 ± 2.09 b	0.89 ± 1.13 a
PALF area (cm^2)	200.2 ± 69.8 b	156.9 ± 96.7 b	229.5 ± 72.7 b	67.6 ± 43.1 a
LAI	1 ± 0.3 b	0.8 ± 0.5 b	1.1 ± 0.4 b	0.3 ± 0.2 a

[1] Different letters in the same row indicate significant statistical differences (Fisher's LSD test, P < 0.05).

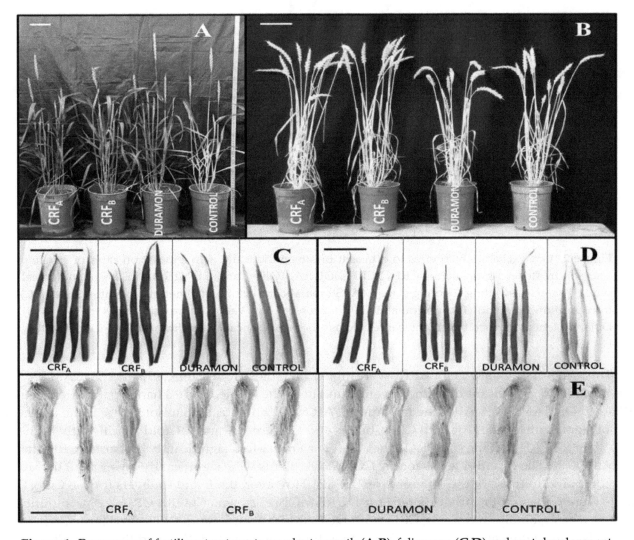

Figure 1. Responses of fertilizer treatments on plant growth (**A,B**), foliar area (**C,D**) and root development (**E**) of *Triticum aestivum* in the phenological state of panicles swelling (**A,C,D,E**) and at the end of the experiment (**B**). Treatments from left to right: CRF$_A$, CRF$_B$, DURAMON®, and CONTROL. Bars correspond to 10 cm.

3.2. Gas Exchange Analysis

Significant increases in *gs*, *A*, and *E* were noted in the plants treated with the different fertilizers as increasing PAR levels were applied from 348 to 1566 μmol m^{-2} s^{-1} (Figure 2A–C). Conversely, *Ci* showed a decreasing tendency in all the applied treatments (Figure 2D).

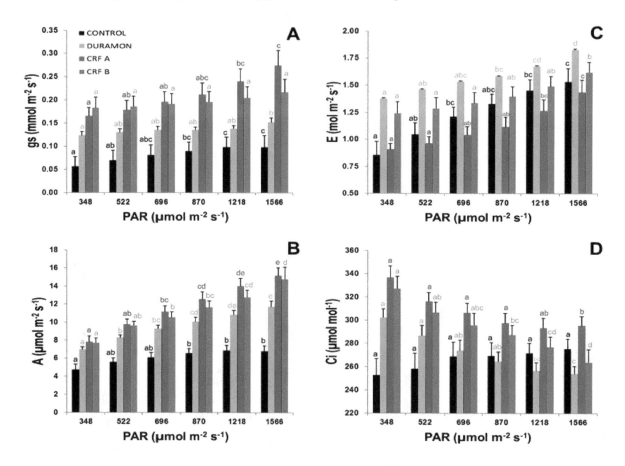

Figure 2. Gas exchange responses to different photosynthetically active radiation rates of *Triticum aestivum* treated with fertilizers CRF$_A$, CRF$_B$, DURAMON®, and CONTROL in the phenological state of panicles swelling (booting stage). (**A**) Stomatal conductance, (**B**) net photosynthetic rate, (**C**) transpiration rate, and (**D**) substomatal CO$_2$ concentration. Values represent means ± SE (*n* = 6). Different letters for each treatment (same color) indicate statistically significant differences (ANOVA, *P* < 0.05).

The levels of *gs*, *A* and *Ci* for CRF$_A$ were higher than those found for CRF$_B$ and DURAMON® but were lower for *E*. Significant differences were found in all the studied gas exchange parameters between CRFs and DURAMON®. After the maximum PAR application of 1566 μmol m^{-2} s^{-1}, the levels of *A* did not significantly differ for both CRFs, but *gs* and *Ci* were 1.3- and 1.1-fold significantly higher for CRF$_A$ than for CRF$_B$. At the same PAR, the *A* levels were 1.3-fold significantly higher for CRFs than for DURAMON®. The *gs* and *A* levels for the CONTROL plants were significantly lower at all the studied PAR compared to those of the different treatments. However, the *E* and *Ci* levels for the CONTROL plants were only significantly different for DURAMON®, as was *Ci* with CRF$_A$. *A/gs* significantly differed when globally comparing CRFs with DURAMON® and the CONTROL (Figure 3A), but the *A/E* levels were significantly higher for CRF$_A$ than for CRF$_B$, DURAMON® and the CONTROL, which also significantly differed from one another. The maximum *A/E* levels were produced within the 870 to 1218 μmol m^{-2} s^{-1} range (Figure 3B). The *A/E* levels in the CRF$_A$ leaves were 1.3- and 1.8-fold significantly higher than CRF$_B$ and DURAMON® at a PAR of 870 μmol m^{-2} s^{-1}.

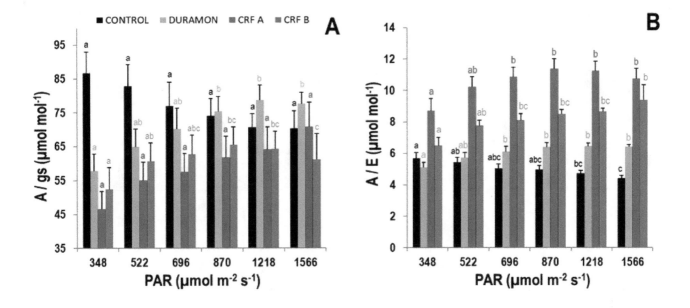

Figure 3. Water-use efficiency responses to different photosynthetically active radiation rates of *Triticum aestivum* treated with fertilizers CRF_A, CRF_B, DURAMON®, and the CONTROL in the phenological state of panicles swelling (booting stage). (**A**) Water-use efficiency and (**B**) intrinsic water-use efficiency. Values represent means ± SE ($n = 6$). Different letters for each treatment (same color) indicate statistically significant differences (ANOVA, $P < 0.05$).

3.3. Foliar Nutrient Content

No significant differences appeared in the PAFL macronutrient concentrations of N, K, Ca, Mg, and S at the beginning of ears formation (Figure 4A). The plants fertilized with CRF_A presented 1.1- and 1.2-fold higher foliar N average levels than CRF_B and DURAMON®. The foliar P concentrations in the plants fertilized with CRF_A were 1.2-fold significantly higher than for CRF_B and DURAMON®. On average, the plants fertilized with CRF_B presented slightly higher contents of K, Ca, Mg, and S. Regarding micronutrient foliar contents, the plants fertilized with CRF_A presented 1.7- and 1.5-fold higher Fe levels than CRF_B and DURAMON®, respectively (Figure 4B). Cu and Zn contents were 1.4- and 1.2-fold significantly higher in the plants fertilized with CRFs than in those fertilized with DURAMON®, respectively. Foliar Mn concentrations did not differ significantly among the distinct fertilizer treatments and the CONTROL levels came close to the critical thresholds. The foliar B levels were significantly higher in the plants treated with CRF_B compared to those treated with DURAMON®, but the quantitative CONTROL levels came close to CRF_B. Mo in PAFL content was < 2 mg kg^{-1} DW (dry weight) in all the treatments and the CONTROL.

No clear correspondence was obtained for the macro- and micronutrient concentrations quantified at the foliar level compared to those found in roots (Figure 4C, 4D). Quantitatively, the root N concentrations were around half of those obtained at the foliar level. The remaining macronutrient contents were slightly lower in roots, except for Ca. Micronutrients almost doubled in roots compared to foliar content but were 30-fold higher in Fe content. No significant differences were obtained for the three compared treatments in the macro- and micronutrient contents at the root level, except for Ca and Mg, which were higher in the plants treated with CRFs compared to DURAMON®.

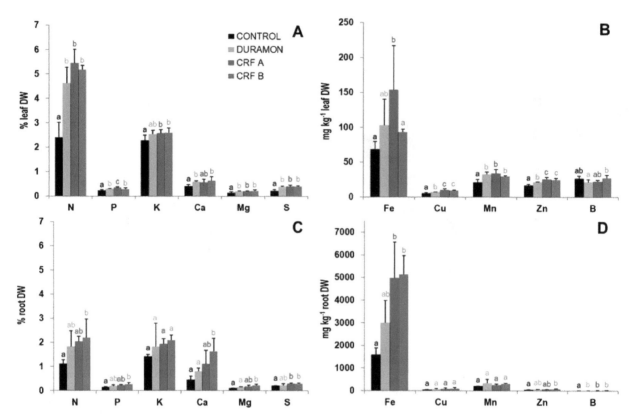

Figure 4. Macro- (**A,C**) and micronutrient (**B,D**) contents in leaves (**A,B**) and roots (**C,D**), for the different treatments with the polymeric-coated fertilizers (CRF$_A$, CRF$_B$), DURAMON®, and the CONTROL in the phenological state (booting stage). Results of macronutrients (N, P, K, Ca, Mg, and S) are expressed in percentage of DW, and micronutrients (Fe, Cu, Mn, Zn, B) in mg kg^{-1} DW. Values are means ± SD (n = 4). Different letters for a specific macro- or micronutrient in each panel indicate statistically significant differences between treatments (ANOVA, $P < 0.05$).

3.4. Growth, Yield, and Cereal Grain Composition

The results about the measured growth and yield parameters in greenhouses are shown in Table 3. No significant differences were found in the different treatments for the measured growth parameters. Significant differences were observed in the CONTROL with the CRF$_A$- and DURAMON®-treated plants for the total dry weight of aerial parts, and with DURAMON® for ear weight and number. Regarding the yield parameters, total dry grain weight was significantly higher for CRF$_A$ than for CRF$_B$. No significant differences were observed in grain weight among treatments, but differences were significant compared to the CONTROL. Total grain number was 1.2-fold significantly higher for CRF$_A$ than for CRF$_B$ but was not significant compared to DURAMON®. At the field level, no significant differences were found in the various applied treatments when comparing growth and grain yield parameters (Table 4). However, NUE was 28 and 38% higher for CRF$_A$ 80%, compared to CRF$_A$ 100% and NSA; these differences were statistically significant. No significant differences were observed among treatments and the CONTROL for the analyzed grain parameters. As a result, on average the values were 1.6% CRF$_A$ ash, 12.9% humidity, 1.7% lipids, 11.4% protein, 2.5% crude fiber, and 65.8% total carbohydrates. No significant differences were obtained comparing the studied quality parameters between CRF$_A$ and the rest of treatments and CONTROL.

Table 3. Comparison of growth and grain yield parameters among the applied fertilizer treatments CRF_A, CRF_B, and DURAMON® compared to the Control in *Triticum aestivum*. Values are means ± SD ($n = 10$) at the end of the culture—138 days after plant emergence.

Parameters	CRF_A	CRF_B	DURAMON®	CONTROL
Total dry weight (aerial part) (g)	40.2 ± 3.4 b	36.7 ± 6.0 ab	41.4 ± 7.7 b	31.7 ± 3.9 a
Primary stem length (cm)	60.8 ± 2.1 a	61.1 ± 3.0 a	53.3 ± 7.9 a	53.6 ± 10.0 a
Tillers number	10.0 ± 0.9 a	10.2 ± 1.6 a	9.8 ± 2.8 a	11.5 ± 2.4 a
Ears number	10.7 ± 1.3 a	9.8 ± 1.1 a	9.7 ± 2.3 a	9.2 ± 2.1 a
Ear weight (g)	2.7 ± 0.2 ab	2.4 ± 0.2 ab	3.0 ± 0.3 b	2.0 ± 0.2 a
Ear length (cm)	13.4 ± 0.4 ab	13.2 ± 0.9 ab	13.9 ± 0.3 b	12.9 ± 0.9 a
Total dry grain weight (g)	22.6 ± 2.0 c	18.6 ± 2.7 b	20.4 ± 3.2 bc	15.9 ± 3.0 a
Grain weight ($n = 100$)	4.8 ± 0.2 b	4.8 ± 0.4 b	4.8 ± 0.3 b	4.4 ± 0.3 a
Total grain number	473.7 ± 46.3 c	392.7 ± 67.4 ab	428.4 ± 79.8 bc	364.7 ± 52.8 a

[1] Different letters in the same row indicate statistically significant differences (Fisher's LSD test, $P < 0.05$).

Table 4. Field harvest comparison of the growth parameters and grain yield of *Triticum spelta* among the different applied fertilizer treatments CRF_A 100%, CRF_A 80%, and NSA compared to the Control. Values are means ± SD ($n = 10$) at the end of the culture—90 days after applying fertilizers.

Parameters	CRF_A 100%	CRF_A 80%	NSA	CONTROL
Biomass of the aerial part (t ha^{-1})	9.32 ± 1.34 b	9.61 ± 1.86 b	8.78 ± 1.41 b	4.91 ± 1.82 a
Ear weight (t ha^{-1})	3.58 ± 0.54 b	3.67 ± 0.54 b	3.76 ± 0.67 b	1.91 ± 0.46 a
Grain weight (t ha^{-1})	2.35 ± 0.45 b	2.49 ± 0.12 b	2.19 ± 0.37 b	1.16 ± 0.43 a
Nitrogen Use Efficiency (kg kg^{-1} N)	14.93 ± 5.58 a	20.76 ± 1.88 b	12.89 ± 4.66 a	-
Harvest Index	0.31 ± 0.05 a	0.33 ± 0.08 a	0.3 ± 0.01 a	0.29 ± 0.02 a

[1] Different letters in the same row indicate statistically significant differences (Fisher's LSD test, $P < 0.05$).

4. Discussion

Nutrient release contained in fertilizers depends on many factors, including environmental conditions, crop management, and the chemical composition of fertilizers [2,38,39]. Nowadays, the slowing down of nutrient emissions in fertilization is a challenge that has already been overcome [11,18,24]. However, high production costs and contamination linked to synthetic fertilizers and waste materials, together with increasingly restrictive environmental policies, have forced new ecological materials to be sought to allow sustainable fertilization [40,41]. The use of water-soluble synthetic products or natural polymers based on lignin has been indicated as an alternative to these problems because they can be obtained in large quantities and at cheap prices from the waste generated in the paper industry, from wood and other sources [42,43].

Research conducted with CRFs has shown that their efficiency is generally higher to that of traditional fertilizers and SRFs [9]. In fact, SRFs are more sensitive to high temperature and sandy soils [44]. Nevertheless, most research works conducted to date with CRFs have focused mainly on crops with a high added value, such as horticultural, ornamental or wood products, and have obtained different results [45–52]. It is important to point out that the main challenge of CRFs application to crops is to successfully provide the amount of nutrients that plants need and in a fractional manner. Moreover, there are also the important advantages that CRFs offer 'per se' in both crop management and the environment. In fact, CRF applications are usually unique, which means savings in crop handling from avoiding successive top-dressing fertilizer applications. Finally, nutrient doses are usually lower in CRFs than those applied with traditional fertilizers, and N losses by evaporation or leaching consistently lower.

In two experiments, this research compares the effectiveness of two lignosulfonate-based polymer-coated urea fertilizers: an analogous non-coated urea, and ammonium nitrosulfate as a traditional fertilizer. Based on the experimental design, the CRF with the best behavior was selected based on the responses of wheat to different physiological and yield parameters on the microscale. In a second stage, the selected CRF was compared with ammonium nitrosulfate in the field. In physiological terms, significant differences were found in growth, chlorophyll content and $\Phi PSII$ among CRF_A, CRF_B and DURAMON® compared with the untreated plants in early crop development stages. Lower values of E and higher of Ci were detected in plants treated with DURAMON®, as compared to the CRFs treatments. Intrinsic efficiency in the water use of CRF_A was significantly higher compared to CRF_B and DURAMON®. To explain these results, the enhanced effects of CRFs were produced by a combination of the individual effects of lignosulfonates and biostimulants on nutrient supply. In fact, lignosulfonates or sulfonated lignin have a variety of functional groups that provide unique colloidal properties and act as chelating agents [53]. The humic substances contained in CRF_A may promote plant development by stimulating root and shoot growth as they can enhance nutrient use efficiency by facilitating the assimilation of macro- and microelements [54,55]. Seaweed extracts can enhance chlorophyll and carotenoid contents in plant shoots, root thickness, and biomass [56]. However, the effectiveness of marketed biostimulants depends very much on their origin because their composition and proportions usually change [57].

Better physiological responses in the state of panicles formation suggest that plants would increase yield and biomass parameters at the end of the crop. A significant correlation was also found between higher levels of photosynthesis during grain formation and increased crop yields obtained in wheat [58,59]. In fact, on a microscale, the total yield expressed in dry grain weight was significantly higher in CRF_A than in CRF_B; and was, on average, also higher than DURAMON®. The observed differences were due to a large number of grains harvested by the production of 1.1-fold more spikes in the plants fertilized with CRF_A compared to CRF_B and DURAMON®. On average the plants fertilized with DURAMON® produced larger sized and heavier spikes with more grains. This could be explained by the faster N-release kinetics of DURAMON® compared to CRFs, as lacking lignosulfonate-polymeric coverage. A bigger N supply in the first crop stages could explain why the plants fertilized with DURAMON® seemed to be slightly more advanced in their phenological status compared to CRFs. Physiological requirements of wheat are established as 3 kg N Qm^{-1} grain; therefore, the theoretical yield that should have been obtained at a dose of 150 kg N ha^{-1} was 5 t ha^{-1}. Our results showed exceeded yields of 6.4 grain t ha^{-1} with CRF_A, 5.3 grain t ha^{-1} with CRF_B and 5.8 grain t ha^{-1} with DURAMON®. Despite DURAMON® not being a CRF because it lacks polymer-coating, it could be considered an SRF for being formulated with urease inhibitor MCDHS, which is also contained in CRFs. This would explain why the DURAMON®-treated plants gave yields close to CRFs. In fact, the minor differences between DURAMON® and CRFs might indicate that DURAMON® could be also used successfully to maintain N availability for plants over time in wheat. As examples, when using nitrification and urease inhibitors in wheat, maize and barley, it was obtained better crop yields and N_2O mitigation than SRCFs [2]. Further, better performance for CRFs and those formulated with nitrification inhibitors compared to traditional ones in maize with a reduction in N_2O emissions up to 21% that did not affect yields [39]. The lower yields obtained with CRF_B could be explained by excessive N emission slowdown by having formulated with a 2% thicker polymeric coverage. The best results obtained with CRF_A were confirmed at field level when comparing the NUE between the applied fertilizer treatments. It was possible to obtain yields close to those observed with CRF_A 100% and NSA, by applying CRF_A with a 20% less N content, but significantly increasing the NUE by more than four times. Even though it has been reported that NUE can vary depending on factors like the doses applied or climatic conditions [2], no reductions in grain yield were observed when applying different CRF formulations by reducing N content in a similar proportion [60,61].

The macronutrients analysis showed that wheat plants had a good NPK nutritional status in phenological state at the beginning of ear formation. The N concentrations in the treated plants fell within the N leaf DW 4–6% range, which is considered suitable for obtaining high yields [62], but no statistically significant differences appeared in the applied treatments. On the contrary, P concentrations were 1.2-fold significantly higher in CRF_A than in CRF_B and DURAMON®. In all cases, P levels fell within the range considered optimal for good plant development (0.2–0.5% leaf DW). Although the applied N fertilizer was not mixed with P, it is known that soil N applications can stimulate root growth and increase cation exchange capacity to favor Ca uptake, and P uptake indirectly [63]. In cereals, increased yields and improvements in the content of macro- and micronutrients of crops have been achieved in barley, maize, rice, or wheat using SCRFs [15,64–69]. No significant differences were found in the K, Ca, Mg, and S contents in leaves among treatments, and their levels were medium to high. Regarding micronutrients, the plants fertilized with CRF_A presented 1.7- and 1.5-fold higher Fe levels than CRF_B and DURAMON®, respectively. However, the Fe levels were optimum for maximum yields (21–200 mg kg^{-1} leaf DW). Cu and Zn contents were 1.4- and 1.2-fold significantly higher in the plants fertilized with CRFs compared to those fertilized with DURAMON®, but concentrations were at the lowest levels within the range considered normal for Cu and Zn (5–50 and 20–70 mg kg^{-1} foliar DW, respectively). The Mn content did not differ significantly for the different fertilizer treatments and presented lower levels (16–200 mg kg^{-1} leaf DW). Fertilizer treatments did not significantly affect B content as the CONTROL plants had similar levels. The Mo levels were very low and were lower than 2 mg kg^{-1} leaf DW in all the treatments.

5. Conclusions

In the present study, we have carried out a comparison of two lignin-coated controlled release fertilizers enriched with humic substances (CRF_A) or seaweed extracts (CRF_B) with a similar non polymeric-coated fertilizer (DURAMON®) and with an ammonium nitrosulfate one (NSA). Our results showed that plants performed better when they were fertilized with CRFs coated with humic substances, although the improvement in crop yield was not excessive compared to the seaweed-coated one and that the uncoated one. However, a significant improvement in crop yield and the measured physiological parameters of wheat plants was achieved with Fertinagro's controlled release fertilizers compared to the traditional NSA. Fertilization with these new technified CRFs greatly favored the wheat crop management by making it possible to carry out one single application as a basal dressing, which simplified crop handling. Smaller amounts of N in formulations gave important advantages, such as reduced costs and minimized N losses, which thus avoids the common contamination problems that usually occur when applying traditional fertilizers. We conclude that it is possible to use this technology in extensive cropping, as lignin-based polymers are economically feasible and environmentally friendly.

Author Contributions: Conceptualization, R.G.-O.; data curation, R.G.-O., and A.R.-N.; formal analysis, R.G.-O. and A.R.-N.; funding acquisition, M.Á.N., O.V. and S.A.; investigation, R.G.-O., A.R.-N., and M.C.-M.; methodology, R.G.-O., M.Á.N., A.R.-N., C.G., and O.V.; project administration, M.Á.N.; supervision, C.G. and O.V.; validation, O.V., A.R.-N.; visualization, M.Á.N., M.C.-M., and S.A.; writing–original draft, R.G.-O.; writing–review and editing, R.G.-O., O.V., and A.R.-N. All authors have read and agreed to the published version of the manuscript.

Acknowledgments: The authors are grateful to Manuel Talón (Centre of Genomics–IVIA) to provide the facilities to develop the greenhouse experiments and Ángel Boix for his support in the plants maintenance.

References

1. FAOSTAT. Available online: http://www.fao.org/faostat/en/#data/QC (accessed on 23 February 2020).
2. Feng, J.F.; Li, F.B.; Deng, A.X.; Feng, X.M.; Fang, F.P.; Zhang, W.J. Integrated assessment of the impact of enhanced-efficiency nitrogen fertilizer on N_2O emission and crop yield. *Agric. Ecosyst. Environ.* **2016**, *231*, 218–228. [CrossRef]
3. Zuk-Golaszewska, K.; Zeranskal, A.; Krukowska, A.; Bojarczuk, J. Biofortification of the nutritional value of foods from the grain of *Triticum durum* desf. *by an agrotechnical method: A scientific review. J. Elem.* **2016**, *21*, 963–975.
4. Barakat, M.; Cheviron, B.; Angulo-Jaramillo, R. Influence of the irrigation technique and strategies on the nitrogen cycle and budget: A review. *Agric. Water Manage.* **2016**, *178*, 225–238. [CrossRef]
5. Zak, D.R.; Holmes, W.E.; MacDonald, N.W.; Pregitzer, K.S. Soil temperature, matric potential, and the kinetics of microbial respiration and nitrogen mineralization. *Soil Sci. Soc. Am. J.* **1999**, *63*, 575–584. [CrossRef]

6.   Achat, D.L.; Augusto, L.; Gallet-Budynek, A.; Loustau, D. Future challenges in coupled C–N–P cycle models for terrestrial ecosystems under global change: A review. *Biogeochemistry* **2016**, *131*, 173–202. [CrossRef]

7.   Di, H.J.; Cameron, K.C. Inhibition of nitrification to mitigate nitrate leaching and nitrous oxide emissions in grazed grassland: A review. *J. Soils Sediments* **2016**, *16*, 1401–1420. [CrossRef]

8.   Akelah, A. Novel utilizations of conventional agrochemicals by controlled release formulations. *Mater. Sci. Eng. C-Biomimetic Mater. Sens. Syst.* **1996**, *4*, 83–98. [CrossRef]

9.   Shaviv, A.; Mikkelsen, R.L. Controlled-release fertilizers to increase efficiency of nutrient use and minimize environmental degradation - A review. *Fertil. Res.* **1993**, *35*, 1–12. [CrossRef]

10.   Ni, X.Y.; Wu, Y.J.; Wu, Z.Y.; Wu, L.; Qiu, G.N.; Yu, L.X. A novel slow-release urea fertiliser: Physical and chemical analysis of its structure and study of its release mechanism. *Biosyst. Eng.* **2013**, *115*, 274–282.

11.   Prasad R, R.G.B.; Lakhdive, B.A. Nitrification retarders and slow-release nitrogen fertilizers. *Adv. Agron.* **1971**, *23*, 337–383.

12.   Wang, Z.H.; Miao, Y.F.; Li, S.X. Wheat responses to ammonium and nitrate N applied at different sown and input times. *Field Crops Res.* **2016**, *199*, 10–20. [CrossRef]

13.   Naz, M.Y.; Sulaiman, S.A. Slow release coating remedy for nitrogen loss from conventional urea: A review. *J. Control. Release* **2016**, *225*, 109–120. [CrossRef] [PubMed]

14.   Chien, S.H.; Prochnow, L.I.; Cantarella, H. Recent developments of fertilizer production and use to improve nutrient efficiency and minimize environmental impacts. In *Advances in Agronomy*; Sparks, D.L., Ed.; Elsevier Academic Press Inc: San Diego, CA, USA, 2009; Volume 102, pp. 267–322.

15.   Diez, J.A.; Caballero, R.; Bustos, A.; Roman, R.; Cartagena, M.C.; Vallejo, A. Control of nitrate pollution by application of controlled release fertilizer (CRF), compost and an optimized irrigation system. *Fertil. Res.* **1996**, *43*, 191–195. [CrossRef]

16.   Halvorson, A.D.; Snyder, C.S.; Blaylock, A.D.; Del Grosso, S.J. Enhanced-Efficiency Nitrogen Fertilizers: Potential role in nitrous oxide emission mitigation. *Agron. J.* **2014**, *106*, 715–722. [CrossRef]

17.   Carson, L.C.; Ozores-Hampton, M. Factors affecting nutrient availability, placement, rate, and application timing of controlled-release fertilizers for Florida vegetable production using seepage irrigation. *Horttechnology* **2013**, *23*, 553–562. [CrossRef]

18.   Azeem, B.; KuShaari, K.; Man, Z.B.; Basit, A.; Thanh, T.H. Review on materials & methods to produce controlled release coated urea fertilizer. *J. Control. Release* **2014**, *181*, 11–21. [PubMed]

19.   Herrera, J.M.; Rubio, G.; Haner, L.L.; Delgado, J.A.; Lucho-Constantino, C.A.; Islas-Valdez, S.; Pellet, D. Emerging and Established Technologies to Increase Nitrogen Use Efficiency of Cereals. *Agronomy-Basel* **2016**, *6*, 25. [CrossRef]

20.   Dou, H.; Alva, A.K. Nitrogen uptake and growth of two citrus rootstock seedlings in a sandy soil receiving different controlled-release fertilizer sources. *Biol. Fert. Soils* **1998**, *26*, 169–172. [CrossRef]

21.   Razali, R.; Daud, H.; Nor, S.M. Modelling and Simulation of Nutrient Dispersion from Coated Fertilizer Granules. In Proceedings of the 3rd International Conference on Fundamental and Applied Sciences, Kuala Lumpur, Malaysia, 3–5 June 2014; pp. 442–448.

22.   Feng, C.; Lu, S.Y.; Gao, C.M.; Wang, X.G.; Xu, X.B.; Bai, X.; Gao, N.N.; Liu, M.Z.; Wu, L. "Smart" fertilizer with temperature- and pH-responsive behavior via surface-initiated polymerization for controlled release of nutrients. *ACS Sustain. Chem. Eng.* **2015**, *3*, 3157–3166. [CrossRef]

23.   Kenawy, E.R. Recent advances in controlled release of argochemicals. *J. Macromol. Sci.-Rev. Macromol. Chem. Phys.* **1998**, *C38*, 365–390. [CrossRef]

24.   Dubey, S.; Jhelum, V.; Patanjali, P.K. Controlled release agrochemicals formulations: A review. *J. Sci. Ind. Res. India* **2011**, *70*, 105–112.

25.   Majeed, Z.; Ramli, N.K.; Mansor, N.; Man, Z. A comprehensive review on biodegradable polymers and their blends used in controlled-release fertilizer processes. *Rev. Chem. Eng.* **2015**, *31*, 69–96. [CrossRef]

26.   Chowdhury, M.A. The controlled release of bioactive compounds from lignin and lignin-based biopolymer matrices. *Int. J. Biol. Macromol.* **2014**, *65*, 136–147. [CrossRef] [PubMed]

27.   Fernandez-Perez, M.; Garrido-Herrera, F.J.; Gonzalez-Pradas, E.; Villafranca-Sanchez, M.; Flores-Cespedes, F. Lignin and ethylcellulose as polymers in controlled release formulations of urea. *J. Appl. Polym. Sci.* **2008**, *108*, 3796–3803. [CrossRef]

28.   Abalos, D.; Jeffery, S.; Sanz-Cobena, A.; Guardia, G.; Vallejo, A. Meta-analysis of the effect of urease and nitrification inhibitors on crop productivity and nitrogen use efficiency. *Agric. Ecosyst. Environ.* **2014**, *189*, 136–144. [CrossRef]

29. Battacharyya, D.; Babgohari, M.Z.; Rathor, P.; Prithiviraj, B. Seaweed extracts as biostimulants in horticulture. *Sci. Hortic.* **2015**, *196*, 39–48. [CrossRef]
30. Calvo, P.; Nelson, L.; Kloepper, J.W. Agricultural uses of plant biostimulants. *Plant Soil* **2014**, *383*, 3–41. [CrossRef]
31. Canellas, L.P.; Olivares, F.L.; Aguiar, N.O.; Jones, D.L.; Nebbioso, A.; Mazzei, P.; Piccolo, A. Humic and fulvic acids as biostimulants in horticulture. *Sci. Hortic.* **2015**, *196*, 15–27. [CrossRef]
32. Colla, G.; Nardi, S.; Cardarelli, M.; Ertani, A.; Lucini, L.; Canaguier, R.; Rouphael, Y. Protein hydrolysates as biostimulants in horticulture. *Sci. Hortic.* **2015**, *196*, 28–38. [CrossRef]
33. du Jardin, P. Plant biostimulants: Definition, concept, main categories and regulation. *Sci. Hortic.* **2015**, *196*, 3–14. [CrossRef]
34. Birrenkott, B.A.; Craig, J.L.; McVey, G.R. A leach collection system to track the release of nitrogen from controlled-release fertilizers in container ornamentals. *Hortscience* **2005**, *40*, 1887–1891. [CrossRef]
35. Clark, M.J.; Zheng, Y.B. Species-specific fertilization can benefit container nursery crop production. *Can. J. Plant Sci.* **2015**, *95*, 251–262. [CrossRef]
36. Cox, D.A. Reducing nitrogen leaching-losses from containerized plants - the effectiveness of controlled-release fertilizers. *J. Plant Nutr.* **1993**, *16*, 533–545. [CrossRef]
37. Agegnehu, G.; Nelson, P.N.; Bird, M.I. The effects of biochar, compost and their mixture and nitrogen fertilizer on yield and nitrogen use efficiency of barley grown on a Nitisol in the highlands of Ethiopia. *Sci. Total Environ.* **2016**, *569–570*, 869–879. [CrossRef] [PubMed]
38. Huett, D.O.; Gogel, B.J. Longevities and nitrogen, phosphorus, and potassium release patterns of polymer-coated controlled-release fertilizers at 30 degrees C and 40 degrees C. *Commun. Soil Sci. Plan.* **2000**, *31*, 959–973. [CrossRef]
39. Yang, L.; Wang, L.G.; Li, H.; Qiu, J.J.; Liu, H.Y. Impacts of fertilization alternatives and crop straw incorporation on N_2O emissions from a spring maize field in Northeastern China. *J. Integr. Agric.* **2014**, *13*, 881–892. [CrossRef]
40. Harrison, R.; Webb, J. A review of the effect of N fertilizer type on gaseous emissions. In *Advances in Agronomy*; Sparks, D.L., Ed.; Elsevier Academic Press Inc: San Diego, CA, USA, 2001; Volume 73, pp. 65–108.
41. Obreza, T.A.; Rouse, R.E.; Sherrod, J.B. Economics of controlled-release fertilizer use on young citrus trees. *J. Prod. Agric.* **1999**, *12*, 69–73. [CrossRef]
42. Garcia, C.; Vallejo, A.; Diez, J.A.; Garcia, L.; Cartagena, M.C. Nitrogen use efficiency with the application of controlled release fertilizers coated with kraft pine lignin. *Soil Sci. Plant Nutr.* **1997**, *43*, 443–449. [CrossRef]
43. Treinyte, J.; Grazuleviciene, V.; Ostrauskaite, J. Biodegradable polymer composites with nitrogen- and phosphorus-containing waste materials as the fillers. *Ecol. Chem. Eng. S.* **2014**, *21*, 515–528.
44. Medina, L.C.; Sartain, J.B.; Obreza, T.A.; Hall, W.L.; Thiex, N.J. Evaluation of a soil incubation method to characterize nitrogen release patterns of slow- and controlled-release fertilizers. *J. AOAC Int.* **2014**, *97*, 643–660. [CrossRef]
45. Gasparin, E.; Araujo, M.M.; Saldanha, C.W.; Tolfo, C.V. Controlled release fertilizer and container volumes in the production of Parapiptadenia rigida (Benth.) Brenan seedlings. *Acta Sci.-Agron.* **2015**, *37*, 473–481. [CrossRef]
46. Haver, D.L.; Schuch, U.K. Production and postproduction performance of two New Guinea Impatiens cultivars grown with controlled-release fertilizer and no leaching. *J. Am. Soc. Hortic. Sci.* **1996**, *121*, 820–825. [CrossRef]
47. Jacobs, D.F.; Salifu, K.F.; Seifert, J.R. Growth and nutritional response of hardwood seedlings to controlled-release fertilization at outplanting. *For. Ecol. Manage.* **2005**, *214*, 28–39. [CrossRef]
48. Kaplan, L.; Tlustos, P.; Szakova, J.; Najmanova, J. The influence of slow-release fertilizers on potted chrysanthemum growth and nutrient consumption. *Plant Soil Environ.* **2013**, *59*, 385–391. [CrossRef]
49. Kinoshita, T.; Yano, T.; Sugiura, M.; Nagasaki, Y. Effects of controlled-release fertilizer on leaf area index and fruit yield in high-density soilless tomato culture using low node-order pinching. *PLoS ONE* **2014**, *9*, 10. [CrossRef]
50. Kinoshita, T.; Yamazaki, H.; Inamoto, K.; Yamazaki, H. Analysis of yield components and dry matter production in a simplified soilless tomato culture system by using controlled-release fertilizers during summer-winter greenhouse production. *Sci. Hortic.* **2016**, *202*, 17–24. [CrossRef]
51. Oliet, J.; Planelles, R.; Segura, M.L.; Artero, F.; Jacobs, D.F. Mineral nutrition and growth of containerized *Pinus halepensis* seedlings under controlled-release fertilizer. *Sci. Hortic.* **2004**, *103*, 113–129. [CrossRef]
52. Pack, J.E.; Hutchinson, C.M.; Simonne, E.H. Evaluation of controlled-release fertilizers for northeast Florida chip potato production. *J. Plant Nutr.* **2006**, *29*, 1301–1313. [CrossRef]

53. Vishtal, A.; Kraslawski, A. Challenges in industrial applications of technical lignins. *BioResources* **2011**, *6*, 3547–3568.

54. Cacco, G.; Attina, E.; Gelsomino, A.; Sidari, M. Effect of nitrate and humic substances of different molecular size on kinetic parameters of nitrate uptake in wheat seedlings. *J. Plant Nutr. Soil Sci.* **2000**, *163*, 313–320. [CrossRef]

55. Nardi, S.; Ertani, A.; Francioso, O. Soil-root cross-talking: The role of humic substances. *J. Plant Nutr. Soil Sci.* **2017**, *180*, 5–13. [CrossRef]

56. Michalak, I.; Gorka, B.; Wieczorek, P.P.; Roj, E.; Lipok, J.; Leska, B.; Messyasz, B.; Wilk, R.; Schroeder, G.; Dobrzynska-Inger, A.; et al. Supercritical fluid extraction of algae enhances levels of biologically active compounds promoting plant growth. *Eur. J. Phycol.* **2016**, *51*, 243–252. [CrossRef]

57. Lotze, E.; Hoffman, E.W. Nutrient composition and content of various biological active compounds of three South African-based commercial seaweed biostimulants. *J. Appl. Phycol.* **2016**, *28*, 1379–1386. [CrossRef]

58. Gaju, O.; DeSilva, J.; Carvalho, P.; Hawkesford, M.J.; Griffiths, S.; Greenland, A.; Foulkes, M.J. Leaf photosynthesis and associations with grain yield, biomass and nitrogen-use efficiency in landraces, synthetic-derived lines and cultivars in wheat. *Field Crops Res.* **2016**, *193*, 1–15. [CrossRef]

59. Richards, R.A. Selectable traits to increase crop photosynthesis and yield of grain crops. *J. Exp. Bot.* **2000**, *51*, 447–458. [CrossRef] [PubMed]

60. Ji, Y.; Liu, G.; Ma, J.; Xu, H.; Yagi, K. Effect of controlled-release fertilizer on nitrous oxide emission from a winter wheat field. *Nutr. Cycl. Agroecosyst.* **2012**, *94*, 111–122. [CrossRef]

61. Zhang, J.S.; Wang, C.Q.; Li, B.; Liang, J.Y.; He, J.; Xiang, H.; Yin, B.; Luo, J. Effects of controlled release blend bulk urea on soil nitrogen and soil enzyme activity in wheat and rice fields. *Chin. J. Appl. Ecol.* **2017**, *28*, 1899–1908.

62. Jones, J.B. Plant tissue analysis in micronutrients. In *Micronutrients in Agriculture*, 2nd ed.; Mortvedt, J.J., Ed.; Soil Science Society of America, Inc.: Madison, WI, USA, 1991; pp. 477–521.

63. Grunes, D.L. Effect of nitrogen on the availability of soil and fertilizer phosphorus to plants. *Adv. Agron.* **1959**, *11*, 369–396.

64. Zhao, B.; Dong, S.T.; Zhang, J.W.; Liu, P. Effects of controlled-release fertiliser on nitrogen use efficiency in summer maize. *PLoS ONE* **2013**, *8*, 8. [CrossRef]

65. Dong, Y.J.; He, M.R.; Wang, Z.L.; Chen, W.F.; Hou, J.; Qiu, X.K.; Zhang, J.W. Effects of new coated release fertilizer on the growth of maize. *J. Soil Sci. Plant Nutr.* **2016**, *16*, 637–649. [CrossRef]

66. Mi, W.H.; Yang, X.; Wu, L.H.; Ma, Q.X.; Liu, Y.L.; Zhang, X. Evaluation of nitrogen fertilizer and cultivation methods for agronomic performance of rice. *Agron. J.* **2016**, *108*, 1907–1916. [CrossRef]

67. Roshanravan, B.; Soltani, S.M.; Mahdavi, F.; Rashid, S.A.; Yusop, M.K. Preparation of encapsulated urea-kaolinite controlled release fertiliser and their effect on rice productivity. *Chem. Speciation Bioavailab.* **2014**, *26*, 249–256. [CrossRef]

68. Morikawa, C.K.; Saigusa, M.; Nakanishi, H.; Nishizawa, N.K.; Hasegawa, K.; Mori, S. Co-situs application of controlled-release fertilizers to alleviate iron chlorosis of Paddy rice grown in calcareous soil. *Soil Sci. Plant Nutr.* **2004**, *50*, 1013–1021. [CrossRef]

69. Morikawa, C.K.; Saigusa, M.; Nishizawa, N.K.; Mori, S. Importance of contact between rice roots and co-situs applied fertilizer granules on iron absorption by paddy rice in a calcareous paddy soil. *Soil Sci. Plant Nutr.* **2008**, *54*, 467–472. [CrossRef]

5

Adaptation of Plants to Salt Stress: Characterization of Na$^+$ and K$^+$ Transporters and Role of CBL Gene Family in Regulating Salt Stress Response

Toi Ketehouli, Kue Foka Idrice Carther, Muhammad Noman, Fa-Wei Wang, Xiao-Wei Li * and Hai-Yan Li *

College of Life Sciences, Engineering Research Center of the Chinese Ministry of Education for Bioreactor and Pharmaceutical Development, Jilin Agricultural University, Changchun 130118, China; stanislasketehouli@yahoo.com (T.K.); kuefokaidricecarther@yahoo.com (K.F.I.C.); mohmmdnoman@gmail.com (M.N.); fw-1980@163.com (F.-W.W.)
* Correspondence: xiaoweili1206@163.com (X.-W.L.); hyli99@163.com (H.-Y.L.)

Abstract: Salinity is one of the most serious factors limiting the productivity of agricultural crops, with adverse effects on germination, plant vigor, and crop yield. This salinity may be natural or induced by agricultural activities such as irrigation or the use of certain types of fertilizer. The most detrimental effect of salinity stress is the accumulation of Na$^+$ and Cl$^-$ ions in tissues of plants exposed to soils with high NaCl concentrations. The entry of both Na$^+$ and Cl$^-$ into the cells causes severe ion imbalance, and excess uptake might cause significant physiological disorder(s). High Na$^+$ concentration inhibits the uptake of K$^+$, which is an element for plant growth and development that results in lower productivity and may even lead to death. The genetic analyses revealed K$^+$ and Na$^+$ transport systems such as SOS1, which belong to the CBL gene family and play a key role in the transport of Na$^+$ from the roots to the aerial parts in the *Arabidopsis* plant. In this review, we mainly discuss the roles of alkaline cations K$^+$ and Na$^+$, Ion homeostasis-transport determinants, and their regulation. Moreover, we tried to give a synthetic overview of soil salinity, its effects on plants, and tolerance mechanisms to withstand stress.

Keywords: salinity; sodium; potassium; ion homeostasis-transport determinants; CBL gene family

1. Introduction

The adverse effects of salinity on plant growth are generally associated with the low osmotic potential of the soil solution and the high level of toxicity of sodium (and chlorine for some species) that causes multiple disturbances to metabolism, growth, and plant development at the molecular, biochemical, and physiological levels [1,2]. In vitro experiments have shown that the enzymes extracted from the halophyte plants *Triplex spongeosa* or *Suaeda maritima (L.)* are sensitive to NaCl to the same degree as those extracted from the glycophyte plants [3,4]. These experiments suggest that tolerance to salinity is not limited to a metabolic response in tolerant plants. Generally, sodium begins to have an inhibitory effect on enzymatic activity from a concentration of 100 mmol/L. Thus, the ability of plants to reduce sodium levels in the cytoplasm appears to be one of the decisive factors in salinity tolerance [5,6]. However, although chloride ions are micro-elements necessary as co-factors, for enzymatic activity, photosynthesis, and the regulation of cell turgor, pH, and electrical membrane potential, they remain no less toxic than Na$^+$ ions if their concentration reaches the critical threshold tolerated by plants [7]. Ionic cellular homeostasis is an essential and vital phenomenon for all organisms. Most cells maintain a high level of potassium and a low level of sodium in the cytoplasm through the coordination and regulation of different transporters and channels. There are two main strategies that plants use to cope with salinity—The compartmentalization of toxic ions within the vacuole and their exclusion outside the cell [5,6]. On the other hand, plants modify the composition of their sap; they can accumulate Na^+ and Cl^- ions to adjust the water potential of tissues necessary to maintain growth [6]. This accumulation should be consistent with a metabolic tolerance of the resulting concentration or with compartmentalization between the various components of the cell or plant. It requires relatively little energy expenditure. If this accumulation does not take place, the plant synthesizes organic solutes to adjust its water potential. It will require a large amount of biomass to ensure the energy expenditure necessary for such a synthesis. Therefore, one adaptation strategy consists of synthesizing osmoprotective agents, mainly amino compounds and sugars, and accumulating them in the cytoplasm and organelles [8,9]. These osmolytes, usually of a hydrophilic nature, are slightly charged but polar and highly soluble molecules [10], suggesting that they can adhere to the surface of proteins and membranes to protect them from dehydration. Another function attributed to these osmolytes is protection against the action of oxygen radicals following salt stress [11]. Under high sodium concentration levels, whether the latter is compartmentalized within the vacuole or excluded from the cell, the osmotic potential of the cytoplasm must be balanced with that of the vacuole and the external environment in order to maintain the cell turgor and the water absorption necessary for cell growth. This requires an increase in osmolyte levels in the cytoplasm, either by the synthesis of solutes (compatible with cellular metabolism) or by their uptake of the soil solution [12,13]. Among these synthesized compounds are some polyols, sugars, amino acids, and betaines, which, energetically, are very expensive to be produced by the cell [14]. The main role of these solutes is to maintain a low water potential inside the cells to generate a suction force for water absorption. Furthermore, the involvement of solutes such as glycine betaine, sorbitol, mannitol, trehalose, and proline in improving tolerance to abiotic stress has been demonstrated by genetic engineering and plant transgenesis [6,14,15]. On the other hand, salt stress induces the production of active forms of oxygen following the alteration of metabolism in the mitochondria and chloroplasts. These active forms of oxygen cause oxidative stress whose adverse effects are reflected in various cellular components such as membrane lipids, proteins, and nucleic acids [16]. As a result, the reduction of these oxidative damages through the deployment of a range of antioxidants could contribute to improving plant tolerance to stress [17]. Early events in plant stress adaptation begin with mechanisms of perception and signaling via signal and messenger transduction to activate various physiological and metabolic responses, including the expression of stress response genes. The main pathways activated during the salt stress signaling include calcium, abscisic acid (ABA), mitogen-activated protein kinases (MAPKinases), salt overly sensitive proteins (SOS), and ethylene [12]. In this chapter, we mainly discuss roles of alkaline cations K^+ and Na^+, ion homeostasis-transport determinants, and their regulation. Furthermore, we tried to give a hypothetical overview of soil salinity, its effects on plants, and tolerance mechanisms that allow the plants to withstand stress. A fundamental biological understanding and knowledge of the effects of salt stress on plants is needed to provide additional information for the study of the plant response to salinity and try to find other way for improving the impact of salinity in plants and accordingly enhance crop yields to cope with the starvation that persists in some parts of the world

2. Roles of Alkaline Cations K^+ and Na^+ in Plants

Potassium (K) is the third of the three primary nutrients required by plants, along with nitrogen (N) and phosphorus (P). Potassium, with about 100 to 200 mM concentration in the cytosol, is the major inorganic cation of the cytoplasm in plant and animal cells. The reasons for its preferential accumulation compared to Na^+ is probably due to the fact that Na^+ is more "chaotropic" (because of its smaller size and stronger electric field on its surface) [18].

Na^+ is not an essential nutrient for higher plants. For a high concentration of Na^+ in the soil, this cation becomes even toxic to the plant. At lower concentrations, the plant can use it beneficially as a vacuolar osmoticum.

2.1. Physiological Roles

As most inorganic cations are abundant in the cytoplasm, the potassium is involved in critical cell functions. In addition to its role in the neutralization of the net electric charge of biomolecules, the potassium participates, for example, in membrane transport processes, enzyme activation, and osmotic potential. In plants, in conjunction with osmotic potential [19], K^+ is involved in the control of the turgor pressure [20] and related functions, cell elongation and cell movement. Finally, K^+ plays a direct or indirect role, in the regulation of enzyme activities, the protein synthesis, photosynthesis and homeostasis of the cytoplasmic pH.

These different roles at the cellular level involve potassium in essential functions at the level of the whole plant, for example gas exchange control via regulation of the opening and closing of the stoma, the xylem sap ascension by root thrust, installation of potential osmotic gradient carrying phloem sap flow from original organs to hole organs or even port of herbaceous species.

2.2. Effect of K^+ Deficiency on Plants Physiology

In K^+ deficiency, the sap flow is disturbed, with spontaneous reduction of the phloem sap velocity of circulation. The photoassimilates then accumulate inside of the leaves. Symptoms of chlorosis and necrosis from the photooxidation of the photosynthetic system are frequently observed. It is well settled that K^+ deficiency induces the acidification of the extracellular medium. Minjian et al. [21], showed that root K^+ absorption depends on the activity of the proton pumps (H^+-ATPases) and the occurrence of K^+ transporters on the cellular membrane. The level of H^+ expulsion can be used as a criterion of tolerance to K^+ deficiency. Chen and Gabelman [22] observed in tomato strains that K^+ uptake efficiency is associated with a high net K^+ influx coupled with low pH around root surfaces. The proton-electrochemical gradient may contribute to energizing K^+ uptake, and indeed it is used by some KT/ KUP/HAK transporters, which co-transport K^+ and H^+ [23].

2.3. Toxicity of Na^+ in the Cytoplasm

In plants, the concentration of Na^+ in the cytosol is maintained at a lower value than that of K^+ in animals. In animal cells, the concentration of Na^+ is closely regulated to 10^{-2}mol L^{-1} value [24]. In plant cells, the concentration of Na^+ does not seem to be subjected to narrow homeostasis. When the plant grows in salinity conditions, the accumulation of Na^+ in the cytoplasm beyond a certain threshold becomes toxic, but this threshold is not clearly determined.

The toxicity of Na^+ in the cytosol would result from its "chaotropic" character by comparison with K^+ [18]. The toxicity of Na^+ would also probably mean its ability to compete with K^+ during the process of fixing important proteins. More than 50 enzymes require K^+ to be active, and Na^+ would not provide the same function [25]. Therefore, a high concentration of Na^+ in the cytoplasm inhibits the activity of many enzymes and proteins, leading to cell dysfunctions. In addition, protein synthesis requires a high concentration of K^+ for tRNA attachment to ribosomes [26], so the translation would also be affected.

2.4. Na^+ Acts as Osmoticum

If the plant cell cannot substitute Na^+ to K^+ in its cytosol, it can do it so in the vacuoles and use Na^+ as osmoticum. Different studies have actually shown that moderate amounts of Na^+ can improve the growth of many plant species [27]. For example, a beneficial "nutritious" effect of Na^+ has been described in tomato [28,29].

It is likely that the beneficial effect of Na^+ can especially be observed in conditions of K^+ deficiency. In these circumstances, a controlled build-up of Na^+ probably helps to ensure the regulation of cell turgor pressure [30,31]. Similarly, a moderate absorption of Na$^+$ can be beneficial if it helps the plant, for example, to quickly adjust their osmotic potential from the beginning of salt stress.

Despite these physiological observations, the genetic determinants of improving the growth of plants by sodium and genes may be involved in these processes, however, they are still poorly characterized. Research on rice [32] concerning the function in planta of a transporter HKT family provided genetic proof on the fact that an accumulation of Na$^+$ in K$^+$ deficiency conditions can promote the growth of the plant.

3. Interaction between K$^+$ and Na$^+$ Transport and Adaptation to Salt Stress

The adaptation of the plant to the presence of salt in the soil and salt stress involves various processes, occurring at different levels, from the cell to the whole organism, such as a modification of the metabolic activity leading to the accumulation of organic osmolytes [33], or morphological and developmental changes of the leaves [34]. Within this very complex network of responses, the control of membrane transport activities occurring through a variety of mechanisms, a selective accumulation of K$^+$ and an exclusion of Na$^+$ [25,35], appear as a central process. Thus, in a large number of models, from isolated cell culture to the whole plant, adaptation to salt stress appears to be correlated with the ability to selectively remove K$^+$, to control the Na$^+$ entrance, and maintain the K$^+$/Na$^+$ ratio of the internal contents of these two cations at a high level. In this context, the molecular and functional characterization of membrane transport systems of K$^+$ and Na$^+$ is, therefore, a priority objective. It is probable that the capacity of the channels and transporters to discriminate K$^+$ from Na$^+$ is essentially based on the difference of intensity of the electric field at the surface of these two ions, which results from their difference in size and hydration energies. The crystallographic resolution of the bacterial potassium channel structure [36,37] provides an example for understanding how carbonyl groups of the polypeptide chain can be spatially distributed along the permeation pathway to substitute, without energy barrier to the hydration shell of the ion.

4. Physiology of K$^+$ and Na$^+$ Transport in Plants

4.1. Structure–Function Relationship of the Root

The movement of the mineral elements by the roots and their transfer to the aerial parts involves at least two membrane steps—Ions *sensu stricto* absorption from the soil solution by the epidermal cells, cortical, and contingently endodermic, and the secretion inside the vessels at the level of xylem parenchyma cells. The ions radial movement from the orbital cells of the root to the stela can, in theory, take three paths [38]—The apoplastic pathway (through cell wall), the symplastic pathway (through cytoplasm), or a mixed path passing the ions alternately from apoplastic compartment to the symplastic compartment (Figure 1). Above the cell differentiation zone, the apoplastic path is interrupted by the endoderm of the root. The walls of these cells are impregnated with lignin and suberin. This deposition of hydrophobic compounds forms the framework of Caspary and constitutes a barrier that blocks water and solutes movement. The very close association of the endodermal cell membrane with the Caspary framework forces the ions and water to undergo membrane control to pass the endodermal barrier and migrate into the stele. However, at several levels in the root, the ions can take a direct apoplastic path from the external environment to the xylem: at the apex, where the endodermis is not yet suberized, at the level of endoderm discontinuity, induced by the appearance of the secondary roots [39], and in some species, at the level of some non-suberized endodermal cells, called passage cells which are thought to serve as cellular gatekeepers, controlling access to the root interior [40].

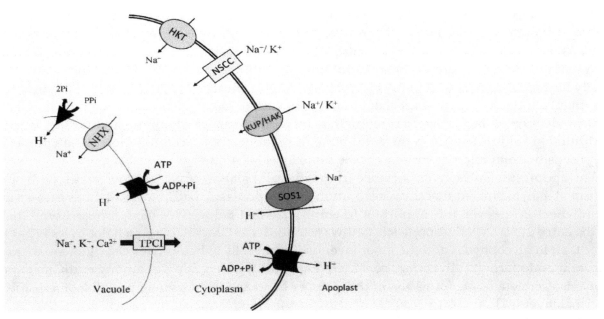

Figure 1. Sodium transport at the cellular level. Schematic representation of transport systems involved in Na$^+$ transport at the plants through the plasma membrane or the tonoplast. Primary transport systems consisting of proton pump ATPases on the plasmalemma and the tonoplast and a pyrophosphatase on the tonoplast create a pH gradient and a potential difference electric on both sides of the membranes (cytosolic side more alkaline and charged more negatively). Proton concentration gradients allow Na$^+$ excretion of cytoplasm towards the outside environment or the vacuole via the operation of antiports Na$^+$/H$^+$ (appealed SOS1 (Salt Overly Sensitive protein 1) on the plasmalemma or NHX1 (K$^+$, Na$^+$/H$^+$ antiporter), on the vacuole). Potential gradients electric created by the pumps cause the entry of Na$^+$ in the cytoplasm of the cell since the external environment or the vacuole via non-selective cationic channels (NSCC) (CNGC (Cyclic Nucleotide Gated Channels) on the plasma membrane? TPC1 (Two-Pore Channel 1) on the tonoplast) or possibly carriers of the HKT (High-Affinity K$^+$ Transporters type in some species. At high external concentration, Na$^+$ could also enter the cell by borrowing K$^+$ carriers KUP/HAK (K$^+$ uptake/High-Affinity K$^{+\cdot}$) type.

In the mature root areas of the majority of plants, a second concentric barrier to that formed by the endodermis is formed at the root periphery on the exoderm, subepidermal cell layer. The suberization of the exoderm would occur later during root development than that of the endoderm and would be accelerated in case of drought [40] or salt stress.

4.2. Structure–Function Relationship of Root and Salt Stress

The current data about root structure and function, as discussed above, indicate that sodium ions can take a direct apoplastic path from the outer medium to the xylem at several levels of the root because endodermal suberization is not yet in place in the young roots area, and leaks remain in secondary roots appearance, which induces a brief discontinuity of the endoderm [41]. The relative contributions of the apoplastic and symplastic pathways of Na$^+$ transport is therefore largely conditioned by root anatomy and are likely to alter according to plant species and soil salinity. The apoplastic pathway (also called apoplastic leak) could be predominant in Na$^+$ transport under salt stress conditions.

Studies carried out on rice have shown that there is a strong correlation between sodium transport and the apoplastic tracer. In two different lines of rice, one more tolerant to salt than the other, a significant difference between the proportions of sodium amount and accumulated PTS in their aerial parts was observed [42,43]. This phenomenon results from the fact that the Na$^+$ entrance into the rice is essentially by free migration in the apoplast up to the stele in spots where the endoplasmic barrier is not functional. This apoplastic leak could occur at the lateral root connection points, at root's apex before complete differentiation of rhizodermis and endodermis, and even in mature areas with differentiated endoderm because of the inherent permeability of the parietal broad outline [44].

It has been shown in rice that the control of apoplastic leakage of Na^+ into the roots is a critical determinant of salinity tolerance. The addition in the culture medium of silicon in sodium silicate partially blocked the apoplastic pathway and considerably improved the growth and photosynthesis of rice plants under salt stress, especially in the GR4 variety [44,45]. This improvement is correlated with the reduction of the Na^+ concentration in plant aerial parts. Furthermore, the authors found that the addition of sodium silicate in the culture medium reduced the accumulation of Na^+ in the aerial parts of sensitive and tolerant varieties at the same level [44,45].

The apoplastic pathway importance in the overall balance of Na^+ inflow varies with species. Garcia et al. [46] estimated that the contribution of the apoplastic pathway is 10 times greater in rice than in wheat. Moreover, it is important to emphasize that halophytes have root anatomy that can limit the entry of Na^+ via the apoplastic pathway. Indeed, the Caspary band in halophytes is 2–3 times thicker than in glycophytes, and the inner layer of cortical cells in halophytes can differentiate to form the second endoderm [2]. In cotton, considered as salinity-tolerant plant among cultivated species, salinity also accelerates the formation of the Caspary band and induces the formation of an additional exodermal layer [47].

All these findings show that there is a correlation between plant tolerance to salinity and the ability to control the apoplastic influx of Na^+ into the roots. It is, therefore, possible to postulate that reducing apoplastic leakage in sensitive species such as rice is a strategy for increasing plant tolerance to salinity. In this perspective, it is important to write down that complete blockage of apoplastic leakage is not likely to significantly affect water inflow and nutrient ion uptake because this leakage contributes little (less than 6%) in rice) to the incoming flows in the roots [46,48]. Some authors have estimated that the apoplastic flow contributes to the xylem flow feeding in a proportion that cannot exceed 1 to 5% [49]. This means that, concerning K^+, the symplasmic transport ensures the essential translocation of this ion from soil solution to the xylem vessels of the stele.

5. Potassium Availability in the Soil and Its Absorption by Plants

K^+ is an important cofactor in many biosynthetic processes, and in the vacuole, it plays key roles in cell volume regulation [50].

The concentration of K+ in the soil solution is generally between a few tens of $\mu mol. L^{-1}$ and a few $mmol. L^{-1}$ (i.e., approximately 10 to 10^3 times lower than that of the cell). The roots are thus confronted with a wide concentration range and the plants possess transport systems allowing them to grow over concentration ranges of K^+, ranging from 10^{-6} to 10^{-1} $mol. L^{-1}$ [51].

An enhancement of the absorption capacity of K^+ by the root is observed when the availability of this ion in the soil is limited [52]. In wheat, K^+ deprivation increases the high-affinity transport efficiency, without altering the characteristics of low-affinity transport. This type of response has also been observed in barley and ryegrass [53]. This reaction is not general, but there are many proteins involved in high-affinity potassium transport. However, in Arabidopsis, two proteins have been identified as the most important transporters in this process. Interestingly, one of these transporters, AtHAK5, is a carrier protein and is thought to mediate active transport of potassium into plant roots, whereas the other protein, AKT1, is a channel protein and likely mediates a passive transport mechanism with an increased affinity for K^+ under conditions of potassium limitation [54,55].

Several different natural phenomena could be involved in root absorption capacity enhancement observed in response to K^+ deficiency in the soil. An initial model to account for this response proposes an allosteric regulation of the absorption capacity in terms of the cytosolic concentration of K^+, resulting in an inhibition by "feedback" of the transporters when the availability of this ion in the area is high, leading to an increase in its concentration in the cytoplasm [56]. Under this model, the K^+ availability diminution in the area leads to a decrease of K^+ concentration in the cytoplasm, which would lift the allosteric inhibition of transport, thus causing absorption capacity augmentation. Another hypothesis, non-exclusive of the previous one, is based on the observation of modifications of the membrane polypeptide equipment when the plants are cultivated in a weakly concentrated potassium area, confirming the installation of new transport systems in barley [57], especially high-affinity transporters in barley [58], wheat [59], and *Arabidopsis thaliana*, [55,60]. In *Arabidopsis*, studies using the patch-clamp technical revealed that K^+ deficiency increases the activity of IRK-type channels (inward rectifying K^+ channel). This augmentation may reflect a corresponding gene(s) expression enhancement or the existence of a post-translational regulation mechanism (e.g., by dephosphorylation). However, the physiological meaning of the channels activity stimulation—And thus of passive transport systems in response to K^+ concentration diminution in the area—Is unclear, even though it is possible that channels may participate in the absorption function from a relatively low external K^+ concentration. Membrane potentials have indeed been found to be negative enough to be able to involve channels in the influx of potassium from an external solution of which K^+ concentration is less than 10 μM [61].

6. Long-range Transport in Xylem and Phloem

6.1. Transport into the Xylem

The Na^+ content of the roots appears to be relatively constant during salt stress. This steady-state probably results in part from root cells' ability to discharge Na^+ in the external area. It also results from Na^+ translocation in the stele and xylem vessels to the aerial parts. The sodium levels of the xylem and phloem may alter during the flow of plant sap. An increase of Na^+ concentration in xylem sap has been described in an "includer" type plant (definition below) *Plantago maritima* [48].

In opposition to this, a decrease of Na^+ concentration in xylem sap has been reported in "excluder" plants type—The sodium contained in the xylem is reabsorbed by roots during the ascent of the sap, and re-excreted toward the outside environment [48]. The amount of sodium that reaches the leaves via xylem sap can be controlled during transport in xylem vessels.

Unfortunately, there is a lack of knowledge about the mechanisms of Na^+ transport in the xylem. However, in *Arabidopsis* under moderate salt stress conditions (40 mM NaCl), *Sos1* mutants (having lost an H^+/Na^+ antiport system) accumulate fewer Na^+ in the aerial parts than wild-type plants [34,62]. This suggests that SOS1 plays a role in the transport of Na^+ from the roots to the aerial parts. However, the use of a reporter gene reveals that in the roots, SOS1 is expressed preferentially in the parenchymal cells around the xylem vessels [62]. Together, these data suggest that SOS1 has been involved in Na^+ secretion in xylem sap from stele parenchymal cells under moderate salt stress conditions.

In some plants, there is a reduction of Na^+ accumulation in the aerial parts. This reduction could be explained by sodium removal from the xylem before it reaches the foliar system. The existence of this strategy in plants has been clearly demonstrated by the research work of Adem et al. [63]. The authors have shown that in barley, the Na^+ concentration of the xylem sap varies together with the stem height (10 mM at the base of the stem and only 2 mM at the 8th leaf). This difference of concentration is important particularly for maintaining the photosynthetic activity of young leaves, which in return allows the formation and growth of new leaves. Molecular mechanisms of Na^+ removal from xylem sap ("desalting" of xylem sap) are beginning to be documented. In particular, the genetic analyses revealed that two transporters of the HKT family, *AtHKT1* in *Arabidopsis* and *OsHKT8* in rice, are involved in this desalting process.

The majority of plants maintain a high K^+/Na^+ ratio in their aerial parts, so it appears that the selectivity to the benefit of K^+ is ensured during the secretion. The ions are excreted in the xylem bundles via xylem transfer cells that can promote, or delay, the efflux of Na^+ in this vessel. The control of the Na^+ concentration in the xylem can also be carried out all along the stem by reabsorption of the sodium in exchange of potassium in the raw sap by the parenchyma cells [3]. H^+-ATPases of the plasma membrane would ensure the energization of the various transports resulting in the exchange of Na$^+$ against K$^+$. The H$^+$ gradient created by these pumps would allow the secretion of K$^+$ via an antiport H$^+$/K$^+$, and a uniporter of Na$^+$ would ensure sodium reabsorption.

Concerning potassium, the ions absorbed at the level of the plasma membrane of the root superficial cells (epidermal and cortical) are transported towards the tissues of the stele by diffusion from one cell to another through plasmodesmata (symplastic pathway). After migration beyond the endodermal barrier, the ions leave the symplasm crossing a second plasma membrane at the level of the last living cells that border the vessels (xylem parenchyma). Once in the apoplast stellar, the ions are driven by the centripetal flow of water to the vessels, and the convection flow of the raw sap (water and mineral units) carried by transpiration and/or root thrust then exports them to the aerial parts [64].

The inner position of xylem parenchyma cells in the root makes the electrophysiology analyses using microelectrodes difficult. As a result, the mechanisms of secretion of ions in the xylem have been less studied than the mechanisms of absorption. It has been acknowledged that the stela's tissues are not able to accumulate ions and that these ions, inflated at the entrance of the symplasm, passively diffuse to the vessels. This passive diffusion was thought to be the consequence of an oxygen deficiency in the central tissues of the root that results in cell depolarization [65]. The stele cells in hypoxic conditions were then unable to retain the ions. However, Zhu et al. have shown that aeration of root pivotal tissues allows cells sufficient oxygenation. CCCP instantly blocks efflux in the xylem of the ^{36}Cl$^-$ accumulated in advance but not efflux to the area through the epidermis [66]. These results show that the CCCP affects the existing system at the level of the stele and not the one located in the cells of the epidermis. Since the 1970s, it has been clearly established that the ions efflux in the stellar apoplast depends on specific transporters located on the plasma membrane of xylem parenchyma cells. Several experimental data indicate that absorption and secretion are controlled separately.

In general, the secretion of nutrient ions in the stellar apoplast could in many cases be a passive phenomenon, catalyzed by channels. For example, in *Arabidopsis*, the SKOR potassium channel of the Shaker family plays an important role in K$^+$ secretion in xylem sap [67]. The knowledge at the molecular level on the mechanisms of secretion of nutrient ions in the xylem sap is, however, still rather small.

6.2. Transport into the Phloem

The growth and development of the plant require distribution of photosynthesis products. These molecules synthesized in the so-called "source" organs (mature leaves) must then be relocated to the growing organs and non-photosynthetic plant tissues (organs called "wells," young leaves, flowers, seeds, fruits, roots). This relocation requires selective long-distance transport, which is provided by the phloem system.

Data obtained from barley show that the sodium contents of xylem and phloem sap are altered throughout transport in the vessels of the aerial parts [68]. The sodium contained in xylem would be absorbed and stored into leaf cells during its movement, and there would also be a translocation of a part of the sodium from xylem to phloem in the leaf, so that the sodium concentration in phloem sap has increased, as it moves from the top of the leaf to its base. Foliar anatomy, particularly in the area of young veins, suggests that such a transfer could occur either directly from apoplast to symplasm of phloem cells, or by symplasmic transport from parenchymal cells [68]. This recirculation of ions from xylem to phloem thus makes it possible to significantly reduce the salt content of the leaves. This has also been observed in some species such as Lupin [69], pepper [70], corn [71], and barley [13].

Perez-Alfocea et al. [72] have found that Na$^+$ translocation in the phloem of *Lycopersicon pennellii*, a wild type tomato that is tolerant to salinity, is more important than that observed in domesticated tomatoes. This suggests that the translocation of Na$^+$ into the phloem would be an adaptation strategy in plants. However, the Na$^+$ translocation direction and the conditions under which it occurs are probably critical. Indeed, it seems crucial that translocation by phloem does not transport Na$^+$ to the young tissues—Otherwise, it would completely inhibit their growth. In other words, translocation by the phloem should essentially redirect Na$^+$ to the roots. In the pepper plant, it has been shown that the translocation of Na$^+$ from the aerial parts to the roots only occurs when Na$^+$ is removed from the nutrient solution, i.e. when there is a favorable gradient between phloem and roots [70]. In *Arabidopsis*, it has been shown that the sodium transporter *AtHKT1*, expressed in phloem tissues, assure Na$^+$ recirculation from the leaves to the roots through phloem by removing Na$^+$ from the rising stream of raw sap at the aerial parts. This system thus plays the role of controlling the Na$^+$ accumulation in the leaves and plant resistance to salt stress [73].

With regard to potassium, the phloem loading and its discharge contribute to the establishment of the osmotic potential gradients (and therefore hydric) created between the source organs (high concentration of sugars and ions in the phloem sap) and the well organs (lower concentrations). The osmotic gradient is initiated at the level of the source organs by the creation of an electrochemical potential due to the H$^+$-ATPases activity of the fellow cells that are in direct electrical contact with the cells of the screened canals (making the phloem vessels) via plasmodesmata. This energization of the membrane allows the influx of sugars (essentially sucrose) and potassium into the cells. In summary, the available data indicate that control of K$^+$ transport in phloem tissues of source and well organs contribute to three main functions: (i) the phloem cells membrane potential regulation, tending to bring its value closer to that of equilibrium potential of K$^+$ (E_K), (ii) the installation of the osmotic gradient responsible for the sap flow between the source and well organs, and (iii) well organs (including seeds and fruits) potassium supply.

The electrophysiological characterization of the potassium conductance of phloem cells is still poorly advanced because of the difficulty to obtain protoplasts. This difficulty is less with corn roots, whose stele is easy to separate from the cortex. Phloem cells can then be obtained by dissection in which potassium conductance has been identified. They are close to the IRKs in their selectivity and responses to inhibitors, but they show a small correlation. It means that they allow an entrance or output of potassium according to the membrane potential value. In *Arabidopsis*, the AKT2 gene from the Shaker potassium channel family could code this type of conductance [74,75].

In *Arabidopsis thaliana*, the use of plants expressing the GFP gene under the AtSUC2 gene promoter (active specifically in phloem fellow cells) made it possible to isolate protoplasts from these cells and to identify two potassium conductance—An outgoing conductance of the ORK type and an incoming conductance of the IRK type. However, conductance that may be specific to cells located either in the source regions or in the well regions has not yet been demonstrated.

7. Adaptive Strategies of Plants to Na$^+$: Exclusion and Inclusion

The ability of a plant to compartmentalize Na$^+$ at the cellular level induces a difference of Na$^+$ management in the whole plant. We can distinguish two ways of plants responses to salt (exclusion and inclusion). "These strategies characterize behavior patterns that are not mutually exclusive" (Levigneron et al. reviewed in [76]). Excluder type plants are generally salinity-sensitive and are unable to control the level of cytoplasmic Na$^+$. This ion is transported in the xylem, conveyed to the leaves by transpiration stream, and then partly "re-circulated" by the phloem to be brought back to the roots. These sensitive species, therefore, contain little Na$^+$ in the leaves and an excess in the roots. Includer plants, which are resistant to NaCl, accumulate Na$^+$ in the leaves where it is sequestered (in the vacuole, foliar epidermis, and old limbs). However, excluder plants also accumulate Na$^+$ in the vacuole of root and stem cells. Of course, these two types of behavior are extreme, and some species can incorporate behaviors characteristic of both types of strategy.

7.1. K$^+$ and Na$^+$ Transport Systems in Plants

The kinetic characteristics of K$^+$ transport systems were studied (since 1950) using the ^{42}K and ^{86}Rb tracers, in particular by Epstein et al. The incorporation rate analysis of the tracer into the excised barley roots, in terms of the external concentration, reveals complex kinetics that presents two phases [77]. This kinetics, which can be analyzed according to the Michaelis-Menten formalism, suggests the existence of two absorption mechanisms. The first mechanism corresponds to a high-affinity saturable system (Km $\approx$ 20 µM), which allows the influx of K$^+$ from low concentrations in the area (less than

1 mM). The second mechanism corresponds to a low-affinity system (Km $\approx$10 mM) responsible for ions absorption from high concentrations. The second absorption mechanism differs from the first by the fact that it is not selective for K^+ (vis-a-vis of Na^+), and its ability to transport K^+ depends on the nature of the accompanying anion [78]. Electrophysiological data obtained on roots suggest that H^+-K^+ symports are responsible for high-affinity K^+ transport [79]. Low-affinity absorption is passive and involves channels.

For sodium, it is established that its initial entrance from the external environment into the cytoplasm of the roots cortical cells is passive [48], either via non-selective voltage-dependent cation channels (NSCCs) [2] or probably via some family members of sodium transporter [80,81] (Figure 2).

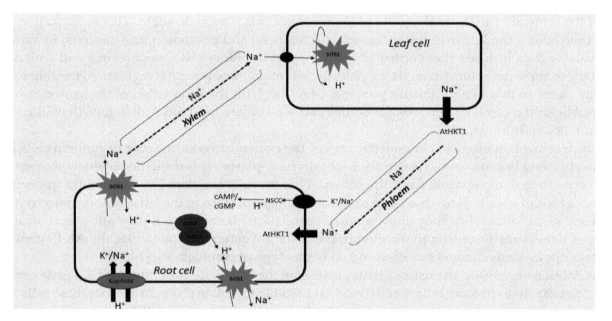

Figure 2. Na^+ transport at the level of the whole plant. Sodium ions can enter the cells of the root through non-selective channels (NSCCs) not formally identified at the molecular level, some of which appear to be inactivated by cyclic nucleotides (cAMP and cGMP; Maathuis and Sanders 2001), transporters HKT and high concentration of Na, KUP/HAK carriers. Na excretion cell roots to the soil solution or to the vessels of the xylem involve the antiport H^+/Na^+ SOS1, whose activity is regulated by the SOS3 CBL protein associated with the SOS2 kinase of the HKT conveyors allow desalinization of xylem sap and phloem loading in Na^+ at the leaf level.

Several families of channels and transporters involved in K^+ and Na^+ transport have been identified at the molecular level in plants.

7.2. Channels

7.2.1. Shaker Channels

These channels exist in plants, fungi, bacteria, and animals. The first members of this family were identified in animals.

These channels are formed with four subunits, which are organized around a central pore. The hydrophobic region of each subunit includes six transmembrane segments (TMS). A membrane loop (called P, for pore) between the fifth and sixth TMS participates in wall constitution of the central pore. Subunits can gather into homotetramers or heterotetramers. These channels are all voltage-regulated and active on the plasma membrane. They are very selective of K^+ beside Na^+. In higher plants, several Shaker channels have been cloned and characterized. There are nine members in *Arabidopsis*, with different functional properties, expression patterns, and localization [82]. The first two Shaker channels identified in plants are AKT1 and KAT1, cloned in 1992 in *Arabidopsis* [83].

In a very interesting way, the characterization of these functional systems has shown that they act as inward rectifying channels [55,84], despite strong homology with voltage-dependent, highly selective K^+ channels, which act as outgoing channels. This observation has generated a lot of interest and triggered numerous studies on the structure–function relationship of these channels, with the main objectives of understanding the mechanisms of opening-closing of the pore and the regulation by the voltage.

There are three main functional types of Shaker channels—incoming rectification channels (KAT, AKT1, and ATKC1 families), outgoing rectification channels (SKOR family), and low rectification channels (AKT2 family). The fourth TMS, carrying positively charged residues (R or K), is the cause of the channel sensitivity to the voltage. The pore loop (P) studies by controlled mutagenesis have identified a motif (TxGYG) involved in ionic selectivity.

The role in the plant of several *Arabidopsis* Shakers was analyzed by reverse genetics. In a general way, these channels allow the massive exchanges of K^+ (influx or efflux), between the symplast and the apoplast (K^+ entrance of the cell for the incoming channels, exit for the outgoing channels), entry, and exit by low rectification channels. They play a role in the removal of K^+ from the soil solution (AKT1 and AtKC1 channels in Arabidopsis), long-distance K^+ transport in the xylem and phloem (SKOR and AKT2 channels), or in the transport of K^+ in the guard cells at the origin of stomatal movements (GORK, KAT1, and KAT2 channels) [85,86]. The shaker channels, outstanding the potassium conductance of the plasmalemma, participate in parallel to the regulation of cellular potassium concentration, to the control of the membrane potential, and to the osmotic potential regulation.

7.2.2. KCO Channels

KCO (or TPK) is the second family of specific K^+ channels identified in plants. These channels are probably composed of either two subunits (family KCO-2P) or four subunits (family KCO-1P), which are organized around a central pore associating four domains (P for pore). In *Arabidopsis*, the KCO-2P family (two P domains per subunit) has five members, and the KCO-1P family (one pore domain per subunit) has only one member [85]. The first member of the KCO-2P family, KCO1, was discovered in silico via the use of the highly conserved GYGD motif in Shaker channels [87]. It has been expressed in insect cells, where it acts as a selective channel of K^+. At the subcellular level, *AtKCO1* has been localized at the level of the tonoplast [61,88], suggesting that it plays a different role from that of the Shaker channels in transporting K^+ through intracellular membranes. Electrophysiological analysis of vacuolar currents on invalidated mutant kco1 suggests that KCO1 contributes to SV type currents, which are outgoing and slow vacuolar currents [61].

7.2.3. Non-Selective Cationic Channels (NSCCs)

These channels, less selective of K^+ than Shaker, have been characterized in different cell types [89] NSCCs include CNGCs and GLRs, which are still poorly characterized. Obviously, all transporters have not significant role in potassium or sodium uptake, thus recent studies on GLRs showed that their expression throughout the plant, open up the possibility that GLR receptors could have a pervasive role in plants as non-specific amino acid sensors in diverse biological processes [90]. There has been no progress in elucidating their role in potassium and sodium uptake for the last two decades.

An indication of the CNGCs involvement in Na^+ influx is that the addition of similar cyclic nucleotides in the environment inhibits Na^+ influx and non-selective cation channel activity [91]. In animals, CNGCs are non-selective cationic channels involved in signal transduction in response to different stimuli. They are permeable to Ca^{2+}, Na^+, and K^+ [92]. Activation of these channels leads to membrane depolarization and cytoplasmic calcium concentration enhancement, thereby activating the signaling pathways dependent on this ion. These channels have a similar structure to the Shaker-type voltage-dependent potassium channels, a hydrophobic domain formed by 6 TMSs (named S1 to S6), and a P domain forming the pore between the fifth and the sixth TMS. In their hydrophilic N- and

C-terminal ends, they have respectively a calmodulin binding domain (CaMBD) and a cyclic nucleotide binding domain (CNBD).

In plants, a family of ion channels homologous to CNGCs animal channels was identified in the late 1990s. The first cDNA encoding a channel belonging to this family was cloned in barley by screening an expression library by searching for calmodulin-interacting proteins and was named HvCBT1 [93]. The second member of the family was isolated from tobacco by the same approach [94]. This cDNA, named NtCBP4, has 61.2% identity with HvCBT1. In Arabidopsis, 20 family members, named CNGC-1 to 20, have been identified in silico by sequence analogy [95].

CNGCs are a class of nonselective cation channels that are permeable to monovalent and divalent cations such as Na^+, K^+, and Ca^{2+} [89,96,97]. Although their down-regulation can prevent Na^+ uptake, it can potentially be concomitantly harmful to the plants, as the uptake of other elements will be compromised. However, in rice root, the downregulation of the rice (Oryza sativa) OsCNGC1 contributed to the superior tolerance of the cultivar FL478 to salt stress [25], as it could avert toxic Na^+ influx, in contrast to the sensitive cultivar, in which the gene was up-regulated by salinity stress. Also, Arabidopsis thaliana null mutants, Atcngc10, were found to have enhanced growth under salt stress compared to wild-type plants [98]. Furthermore, Atcngc3 T-DNA insertion mutants showed an increase in tolerance to high levels of NaCl and KCl [99]. With regard to the correlation between CNGC down-regulation and stress tolerance, Mekawy et al. (2015) evaluated the relative tolerance of two rice cultivars, Egyptian Yasmine and Sakha 102. They observed that the greater tolerance of Egyptian Yasmine was partially attributable to the down-regulation of OsCNGC1, with the concomitant up-regulation of plasma membrane protein 3 (PMP3), a plasma membrane protein involved in the inhibition of excess Na^+ uptake at the level of the root [100].

Also, some observations show that, in Arabidopsis, the AtCNGC1 and AtCNGC2 channels introduced into yeast expression plasmids appear to complement a defective yeast mutant for K^+ transport [95]. In tobacco, over-expression of NtCBP4 confers transgenic plants nickel tolerance and tin hypersensitivity that decrease Ni^{2+} accumulation and increase Pb^{2+} accumulation [94]. Subsequently, it has been shown that NtCBP4 is expressed on the plasma membrane of tobacco cells [94]. The hypothesis is that NtCBP4 would be a transport system (perhaps permeable to Ca^{2+}) allowing Pb^{2+} entry into the cell.

The data in planta on the function of a CNGC were obtained indirectly following genetic analysis on an altered Arabidopsis mutant in response to a pathogen [101]. This study has made it possible, for the first time, to highlight the involvement of a CNGC ion channel in a signaling pathway. In general, CNGCs are probably involved, like their homologs in animal cell signaling [89,102]. They would be permeable to monovalent and/or Ca^{2+} cations and regulated by cyclic nucleotides and calmodulin. In plant CNGCs, the cyclic nucleotide-binding domain and the calmodulin-binding domain are both located in the C-terminal cytoplasmic region, where they overlap slightly [102].

7.3. Transporters

The KUP/HAK/KT family. A transporter belonging to a new family of K^+ transport systems has been identified in Escherichia coli (KUP1) [103] and in yeast Schwanniomyces occidentalis (SoHAK1) [104]. The SoHAK1 expression in a mutated strain of S. cerevisiae for K^+ uptake systems restored growth onto a low K^+ concentration environment [104], SoHAK1 seems to be a high-affinity K^+ transporter. The homologs in plants, named KUP, HAK, or KT (for "K^+ uptake," "High-Affinity K^+ transporter," and K^+ Transporter, respectively), form a large family containing at least 17 members in rice [105]. The structure of these transporters is poorly known. The hydrophobicity profiles suggest that they have 12 TMSs and a long cytoplasmic loop between the second and third segments.

In plants, the first gene of the HAK/KT/KUP family, named HvHAK, was cloned in barley by qRT-PCR, with corresponding primers to conserved regions of E. coli KUP1 transporters and SoHAK1 [58]. In Arabidopsis, the first members identified in the HAK/KT/KUP family were cloned by complementation of a yeast mutant [106] or by the search for homologous sequences to KUP1 and HvHAK in the data banks [60]. Overexpression of AtKUP1 and AtKUP2 cDNAs induces an $^{86}Rb^+$

influx in yeast or in *Arabidopsis* growth cells [60,106]. For AtKUP1, the absorption kinetics in terms of concentration shows a Michaellian style in the low concentration range (less than 100 μmol. L^{-1}), raising the kinetics associated with the mechanism I in roots [60,106]. This similarity suggested that KUP-type systems are responsible for active K^+ transport with high affinity in plant cells. However, the analysis of absorption kinetics by the AtKUP1 system as a function of K^+ concentration also reveals low-affinity transport activity [60]. In other words, the AtKUP1 system alone can generate biphasic absorption kinetics, which evokes the kinetics observed in the roots (mechanism I plus mechanism II). The duality of transport kinetics by AtKUP1 could reflect two different modes of operation for this system. No current was detected by heterologous expression of AtKUP1 in the Xenopus oocyte, and the transport mechanisms unable to be determined [60]. However, the K^+ influx generated by *HvHAK1* and *AtKUP1* proteins in yeast is inhibited by the presence of Na^+ in the environment [58,106]. The localization of *AtKUP1* gene expression analyzed by northern blot, has led to variable results in which, the mRNA is undetectable in the roots but present in the aerial parts [60], mainly expressed in roots [58,106] or undetectable throughout the plant [107]. These variations could be associated with differences in plant growth conditions. This would mean that the accumulation of *AtKUP1 mRNA* is highly dependent on environmental conditions.

By a classic genetic approach based on the search for altered mutants in absorbent hairs growth, the authors of reference [108] have isolated another family member, named *TRH1* or *AtKUP4*. The trh1 mutant shows a decrease in $^{86}Rb^+$ uptake. The phenotype of absorbent hair growth of the mutant plants is not restored when they are grown in an environment containing 50 mM of K^+. The high-affinity K^+ transporter function of *TRH1* has been demonstrated by the complementation of yeast mutant *trk1*. TRH1 is expressed in the roots and in the aerial parts. It could be involved in the absorbent hair formation by allowing the influx of K^+ necessary for the growth and the elongation of these cells.

In general, all these HAK/KT/KUP transporters are not sufficiently characterized at the functional level, because of difficulties in expressing them in a heterologous system (a few rare members, however, express themselves in the yeast *S. cerevisiae* and/or in *E. coli* bacteria). In plants, they are present in many cell types and seem to be found on both the plasma membrane and the vacuolar membrane [105].

7.3.1. HKT Transporters

HKT transporters have homologs in fungi (TRK) and bacteria. Their predicted global structure, based on sequence analyses, is similar to that of potassium channels (at 2 TMS) that exist for example in bacteria. The hydrophobic region of the HKT polypeptides comprises four repeats of the (1 TMS/1 P/1 TMS) module. In the functional protein, the four loops are arranged to form a central pore [109].

All HKT transporters characterized so far in plants are permeable to Na^+, and some are also permeable to K^+. The role of these transporters in planta of K^+ transport has not yet been clarified. Several studies have demonstrated the role of these systems in planta in the transport of Na^+ and revealed that HKTs are involved in the tolerance of plants to salinity.

The protein sequence of *TaHKT1* has about 20% homology with the TRK systems identified in yeast and its structure would integrate 10 to 12 hydrophobic regions likely to correspond to TMS.

The *TaHKT1* expression in the Xenopus oocyte causes an activated current by the addition of K^+ or other cations to the external medium. The intensity of this current increases when the pH of the external medium is lowered. However, the analysis of transgenic plants overexpressing *TaHKT1* did not make it possible to highlight a contribution of this system to the absorption function of K^+ by the root [110]. Subsequent analyses revealed a sensitivity of the transport to the presence of Na^+ in the area. These data suggested that *TaHKT1* would rather function as a high-affinity Na^+: K^+ symport for K^+ (ca = 10 μM), energized by the electrochemical gradient of Na^+ across the membrane [111], which is completely unexpected energy coupling mechanism in plants. Moreover, this type of operation is limited to conditions of low external concentration of Na^+.

When the Na^+ concentration is higher, the transport of K^+ by *TaHKT1* would be blocked and this system would function as a low-affinity Na^+ transporter (*Km* close to 5 mM) [111]. The physiological significance of this result remains unclear since in vivo K^+-transport analyses in higher plants have never revealed Na^+-K^+ symport activity (e.g., the addition of Na^+ in the medium does not stimulate K^+ uptake).

The only member of the HKT family in *Arabidopsis*, orthologue of the wheat *TaHKT1* gene, has been identified and designated as *AtHKT1*. The expression of this gene in yeast strains lacking the Na^+ efflux system aggravates their sensitivity to Na^+, but it does not suppress K^+ transport deficiency in *trk1* and *trk2* mutants that have difficulty to absorb potassium [112]

When expressed in the *Xenopus* oocyte, *AtHKT1* exhibits strictly selective Na^+ transport activity, without any permeability to K^+. Similarly, *AtHKT1*expression does not complement a type of *E. coli* mutant unable to absorb K^+, which helps to show that *AtHKT1* carries only Na^+.

AtHKT1 is expressed in the vascular tissues of the root and the aerial parts, at the level of the phloem and the xylem parenchyma [113].

While the *AtHKT1* gene is unique in *Arabidopsis*, it is interesting to note that the HKT family in rice has 7–9 members, depending on the cultivars [114]. The analysis of the polypeptide sequences of the transporters encoded by these genes shows a rather significant difference between the members—apart from two pairs of highly homologous transporters (OsHKT3/OsHKT9 and OsHKT1/OsHKT2, 93 and 91% identity, respectively), the percentage of identity between the different transporters is between 40 and 50%. Nipponbare (japonica), *Ni-OsHKT2*, and *Ni-OsHKT5* probably do not encode functional transporters due to large deletions or the presence of "stop codons" in the reading frame. However, *OsHKT2* is identified in another cultivar (*indica*) and codes for a functional transporter, Po-OsHKT2 [115].

Localization studies by analysis of transformed plants with a promoter (GUS fusion) has shown that these two HKTs are expressed at the vascular tissue level. Specifically, all of the available data (including in situ hybridization analyses) reveal that *OsHKT1* is localized in foliar vascular tissue but also in the root cortex and endoderm [32], whereas *OsHKT8* is mainly localized at the level of the xylem parenchyma, in the roots and in the leaves [116].

The most detailed data at the functional level concerns OsHKT1. This system is one of the closest counterparts in rice of the first HKT characterized, TaHKT1 (wheat), which is a transporter of K^+ and Na^+ (*OsHKT1* and *TaHKT1* have 67% identity). *OsHKT1* has been characterized by three different teams, leading to conflicting results. Expressed in the Xenopus oocyte, OsHKT1 is described as a cationic transport system, with little discrimination with respect to the different alkaline cations [117], or as a very selective transporter of Na^+ [115]. Expressed in yeast, it is described either as a K^+ permeable transport system [117] or as a Na^+ transport system blocked by K^+ [114]. OsHKT1 expression in *S. cerevisiae* yeast mutants deficient for K^+ transport did not allow growth on medium poor in K^+ (0.1 mM KCl). The growth inhibition test on *S. cerevisiae* G19 yeast strains, highly sensitive to Na^+ following the disruption of ENA genes (which code for Na^+ excretory ATPases), revealed that the cells expressing OsHKT1 exhibited more sensitivity to Na^+ than those expressing TaHKT1 in the presence of 50 and 100 mM NaCl.

7.3.2. CHX Transporters (Monovalent Cation H^+ Exchanger)

These transport systems have been identified in plants on the basis of their homology with systems previously characterized in other organisms, such as bacteria, yeasts or algae. Only transporters involved in sodium compartmentalization in the plant vacuole are now relatively well known.

As in unicellular organisms, transports through the tonoplast is activated by an H^+-ATPase pump that establishes a proton gradient [118]. The operation of the CHXs is electron-based and thus does not disturb the potential difference across the membrane. These systems are probably involved in both monovalent cation homeostasis and cytoplasmic and/or vacuolar pH regulation [119].

From a biochemical point of view, tonoplast antiport Na^+/H^+ activity, which may be involved in sodium vacuolar compartmentalization, was initially demonstrated by the Blumwald group in several species [120]. This Na^+/H^+ antiport activity was associated with a 170 kDa vacuolar protein identified in *Beta vulgaris*, whose accumulation is increased by NaCl treatments [121]. Antibodies planned against this protein inhibited the Na^+/H^+ antiport activity. This protein was, therefore, a good candidate for the antiport activity detected on the tonoplast but its coding gene remains unknown.

From the molecular point of view, an *Arabidopsis* cDNA, named *AtNHX1*, related to the yeast ScNHX1 protein, constituted the first characterized system. Only this tonoplast antiport Na^+/H^+ of Arabidopsis antigen has yet clearly been involved in sodium vacuolar compartmentalization [5,120,122,123]. The expression of this plant cDNA complements defective yeasts in the Na^+/H^+ transporter present in the vacuolar membrane [123]. In *Arabidopsis*, AtNHX1 overexpression confers to transgenic plant tolerance to external Na^+ concentrations above 200 mM [5]. *AtNHX1* is expressed in all plant tissues and is found on the internal system tonoplast and on the membranes (RER, Glogi). Systematic sequencing of the *Arabidopsis thaliana* genome has identified 35 genes that can encode proteins being similar to antiport Na^+/H^+. Constitutive overexpression of *AtNHX1* improves salinity tolerance also in tomato [124], *Brassica napus* [125], and soybean [126]. Fukuda et al. have identified an *AtNHX1* homologue in rice, *OsNHX1*. OsNHX1 expression is induced into the roots and into the aerial parts during salt stress. The authors found that OsNHX1overexpression enhances the salinity tolerance of transgenic cells and plants [127].

Within the *CHX* family, some members may be good candidates for K^+ transport. This is the case in *Arabidopsis* for AtKEAs that resemble the K^+/H^+ bacterial antigens KefB and KefC. However, no experimental data for these systems are available, except for expression data in *Arabidopsis* tissues. Of the 28 KEA genes in this plant, 18 are specifically expressed during the microgametogenesis phase or in sporophytic tissues, suggesting that CHXs are involved in the regulation of potassium homeostasis in the pollen growth phase and germination [128]. Two CHXs have been characterized in more detail. *AtCHX17* appears to be preferentially expressed in roots under stress conditions, such as high salt concentrations, low external pH, low external K^+ concentration, and/or basic acid treatment [125]. The analysis of the mutant *AND-T atnhk17* suggests that this gene has a function in potassium homeostasis since the mutant plants accumulate less potassium than the wild ones. When expressed in yeast, *AtNHX17* co-localizes with markers of the *Golgi apparatus* and complements the pH sensitivity of a *kha1* mutant yeast strain [129], suggesting a role in potassium homeostasis and pH regulation under stress conditions. Loss of function mutants of this gene showed alteration in the ultrastructure of the chloroplast with a sharp decrease in chlorophyll level in the leaves, and an increase in cytosolic pH in the guard cells. The growth of *atnhx23* mutants was enhanced by the addition of high concentrations of potassium in the environment but altered by the addition of NaCl [130]. All these data suggest that *AtNHX23* is an antiport K^+ (Na^+/H^+) active at the level of the chloroplast envelope and involved in potassium homeostasis and perhaps in regulating the pH of the stroma.

The Na^+/H^+ antiport systems of the plasma membrane are still poorly characterized. The only information relates to the SOS1 protein in *Arabidopsis*, which has a homologous sequence with antiport Na^+/H^+ and would be involved in sodium efflux at the plasmalemma level [62]. Evidence has been provided that SOS1 does have antiport Na^+/H^+ activity [62].

The sodium hypersensitive *Arabidopsis* mutant *sos1* exhibits, when cultivated in presence of moderate NaCl concentrations (40 mM), higher Na^+ content in its roots than those observed in the plant control of wild-type genotype. Moreover, using the reporter gene system, the authors have highlighted the localization of SOS1 in epidermal cells at the root end. These results suggest the involvement of *SOS1* in Na^+ efflux from the roots in the environment. In addition, it is interesting to note that SOS1 overexpression in *Arabidopsis* significantly improves plants tolerance to salinity. AtSOS1 is, therefore, an important determinant of salt sensitivity in plants. *AtSOS1* activity is controlled by *AtSOS2* and *AtSOS3*. AtSOS3 (a Ca^{2+} affine protein belonging to the CBL family) directly interacts with AtSOS2 which a serine/threonine protein kinase is [131]. The interaction of *AtSOS3 and AtSOS2* triggers *AtSOS2* protein kinase activity, which phosphorylates and activates SOS1. Moreover, CBL/CIPK perceive cytosolic Ca^{2+} signals resulting from salt stress and have important roles in regulating salt stress response and ion homeostasis [132].

7.4. Ion Transporters Mediating Role in Salt Tolerance

In Arabidopsis roots, *AtCNGC3* is thought to be involved in Na$^+$ fluxes. It has been reported that a null mutation in *AtCNGC3* would reduce the net Na$^+$ uptake during the early stages of NaCl exposure (40–80 mM). However, longer exposure of wild type (WT) and mutant seedlings to NaCl (80–120 mM), induces the accumulation of similar Na$^+$ concentrations in both plants [99].

These results indicate the involvement of *AtCNGC3* in Na$^+$ uptake during the early stages of salt stress. In salt-tolerant rice varieties, *OsCNGC1* is negatively more regulated than in salt-sensitive varieties subjected to salt stress conditions [133]. *Arabidopsis thaliana AtHKT1;1*, facilitates the influx of Na$^+$ into heterologous expression systems [134]. Apparently, there is a determinant of salt stress tolerance that controls the influx of Na$^+$ into the roots, resulting in lower accumulation of Na$^+$ in *athkt1* mutants than in WT plants [135]. Horie et al. demonstrated that *OsHKT2;1*, regulates the influx of Na$^+$ into root cells [32]. Plants lacking the *OsHKT2;1* gene, when exposed to 0.5 mM Na$^+$ in the absence of K$^+$, exhibit lower Na$^+$ accumulation and reduced growth [32]. *OsHKT2;2/1*, a new isoform of HKT isolated from the rice plant roots that is no more than an intermediate between *OsHKT2;1* and *OsHKT2;2*, was supposed to confer salt tolerance to the Nona Bokra rice cultivar by allowing the absorption of K$^+$ in roots under salt stress [136]. It has now been shown that OsHKT2;2/1 regulates the influx of Na$^+$ into the roots of plants exposed to salt stress [137]. Note that the constitutive overexpression of *AtNHX1* in Arabidopsis improves salt tolerance [138]. Besides, overexpression of NHX1 in various transgenic plants, such as Brassica [139], cotton [17], maize [140], rice [141], tobacco [142], tomato [143], and wheat [144], exposed to NaCl concentrations ranging from 100 mM to 200 mM improve their tolerance to salt stress. The induction of NHX1 and NHX2 in response to salt stress depends on ABA [145,146]. It is widely known that, during salt stress, NHX activity increases, which promotes salt stress tolerance in many plants [147]. AtCHX21, expressed in the endodermal cells of the roots and its mutants, subjected to salt stress, accumulate less Na$^+$ in their xylem and leaves sap, indicating that CHX21 could be involved in the transport of Na$^+$ through the endoderm in the stele [148]. Under moderately saline conditions, SOS1 most likely occurs in the xylem load of Na$^+$, due to the fact that Na$^+$ accumulates to a lesser extent in sos1 mutants [149]. In high salinity conditions, the xylem load of Na$^+$ is probably a passive process because a high concentration of cytosolic Na$^+$ in xylem parenchyma cells and a comparatively depolarized plasma membrane would favor the movement of Na$^+$ in the xylem [150]. Plants can recover xylem Na$^+$ from root cells to avoid high concentrations of Na$^+$ in aerial tissues [151]. This recovery has been observed in the basal regions of the roots and shoots of plants such as maize, beans, and soybeans [2,65]. In *Arabidopsis*, the *HKT1* mutation renders the mutants hypersensitive to salt stress and causes a greater accumulation of Na$^+$ in the leaves [152–154]. Inactivation lines have higher levels of Na$^+$ but low levels of K$^+$ in shoots. These results show that AtHKT1 is involved in the recovery of Na$^+$ from xylem while directly stimulating the load of K$^+$. This is one of the mechanisms to maintain a higher K$^+$/Na$^+$ ratio in shoots during salt stress in plants [155]. Synergistic effects of *SOS1*, *HKT1;5*, and *NHX1* have been proposed to regulate Na$^+$ homeostasis in *Puccinellia tenuiflora*, a halophytic plant [156]. The NaCl stress–induced vacuolar compartmentalization of its xylem load has been attributed to regulation by the differential expression of *NHX1* and *HKT1;5*. The NaCl stress-induced expression of SOS1 and NHX1 in the roots would also have been more effective in excluding Na$^+$ and Cl$^-$ in the intertidal population of *Suaeda salsa* [157]. The genetic or environmental variation of salt tolerance among halophyte populations is related to the differential expression of Na$^+$ efflux channels. Detailed structural analysis of HKT1;5 was performed in *Triticum sp.* [158]. Variations in its amino acid sequences result in a change in Na$^+$ affinity and a subsequent change in salt tolerance in two species of *Triticum*. Comparative analysis of antioxidant mechanisms in *Cynodon dactylon* (salt-tolerant grass) and *Oryza sativa* (salt-sensitive plant) was corroborated by the high expression levels of SOS 1 and NHX1 transporters in Cynodon [159]. Salt tolerance in barley has been attributed to the regulation of Na$^+$ loading in root xylem elements [160]. This is controlled by a cross between reactive oxygen species (ROS), nicotinamide adenine dinucleotide phosphate oxidase (NADPH oxidase), Ca^{2+}, and K$^+$.

8. Calcineurin B–Like Proteins (CBL) and CBL-Interacting Protein Kinases (CIPK) and Salt Tolerance in Plants

Calcium serves as a pivotal messenger in many adaptation and developmental processes. Cellular calcium signals are detected and transmitted by sensor molecules such as calcium-binding proteins. In plants, the calcineurin B-like protein (CBL) family seems to be a unique group of calcium sensors and plays a key role in decoding calcium transients by specifically interacting with and regulating a family of protein kinases (CIPKs) [161]. Several CBL proteins appear to be targeted to the plasma membrane by processes of dual lipid modification by myristoylation and S-acylation. Additionally, CBL/CIPK complexes have been identified in other cellular localizations, suggesting that this network may confer spatial specificity in Ca^{2+} signaling.

Molecular genetics analysis of loss-of-function mutants involves various CBL proteins and CIPKs as important components of abiotic stress responses, hormone reactions, and ion transport processes. The event of CBL and CIPK proteins appears not to be restricted to plants, raising the question about the function of these Ca^{2+} decoding components in non-plant species.

8.1. Organization of the CBL–CIPK Network

CBL proteins have been initially identified from *Arabidopsis thaliana* [162]. Bioinformatics and comparative genomic analysis in plants have provided details about the sequence specificity, conservation, function, and complexity, and ancestry of CBL and CIPK proteins families from lower plants to higher plants. Bioinformatics research reports showed that *Arabidopsis thaliana* has 10 CBLs and 26 CIPKs [163], while in other plants, *Populus trichocarpa* has 10 CBLs and 27 CIPKs [164], *Oryza sativa* has 10 CBLs and 31 CIPKs [165], *Zea mays* has 8 CBLs and 43 CIPKs [165], *Vitis vinifera* has eight CBLs and 21 CIPKs [166], *Sorghum bicolor* has 6 CBLs and 32 CIPKs [166], *Glycine max* has 52 CIPKs [62], and *Brassica rapa L.* (Chinese cabbage) has 17 CBL genes [167].

All CBL proteins share a rather conserved core region consisting of four EF-hand calcium-binding domains that are separated by spacing regions encompassing an absolutely conserved number of amino acids in all CBL Proteins [161].

In contrast to CNB from animals and fungi, CBLs do not interact with a PP2B-type phosphatase that appears to be absent in plants.

Instead, CBL proteins interact with a group of serine-threonine protein kinases that evolutionary belong to the superfamily of CaM-dependent kinases (CaMKs) and form a phylogenetically separate cluster within the group of SNF1 related kinases. Therefore, this group has also been indicated as *Snf1* related kinase group 3 (*SnRK3*; [168]. As in other kinases of the CaMK group, the kinase domain in CIPKs is segregated by a domain called "junction domain" from the less-conserved C-terminal regulatory domain. Within the regulatory region of CIPKs, a conserved NAF domain (designated according to the prominent amino acids N, A and F) mediates binding of CBL proteins and simultaneously functions as an auto-inhibitory domain [169]. Binding of CBLs to the NAF motif removes the auto-inhibitory domain from the kinase domain, thereby conferring auto-phosphorylation and activation of the kinase [170]. Additional phosphorylation of the activation loop within the kinase domain by a yet unknown kinase further contributes to the activation of CIPKs [171].

Kinases related to CIPKs, like the AMP-activated protein kinase (AMPK), are dephosphorylated by type 2C protein phosphatases (PP2C) [172]. Interestingly, CIPKs can associate with PP2Cs like ABI1 and ABI2 via a C-terminal protein-phosphatase interaction (PPI) domain [173]. Currently, it is not known if PP2Cs may dephosphorylate CIPKs or if phosphorylation of PP2Cs by CIPKs occurs in vivo. Alternatively, the generation of CIPK/PP2C complexes could serve the formation of signaling kinase/phosphatase modules allowing for rapid alternating phosphorylation/dephosphorylation of target proteins.

In this regard, crystallization studies of CBL4 in complex with the regulatory domain of CIPK24 suggest that either CBLs or PP2Cs may mutually exclusively interact with the regulatory domain of CIPKs, and that formation of a trimeric complex is unlikely [174]. Therefore, it is tempting to speculate that PP2C interaction with the PPI domain of CIPKs leads to competitive replacing of the CBL protein, which binds to the NAF and partly to the PPI domain. Dissociation of the CBL protein would release the otherwise masked auto-inhibitory domain of the CIPK resulting in inactivation of the kinase. Alternative Ca^{2+}-dependent binding of CBL proteins to the CIPK would favor phosphorylation of a given substrate by out-competing the PP2C from the complex. However, as the target stowage domains are still unknown for CIPKs and PP2Cs, such models can currently not consider the influence of substrate binding.

Interestingly, the PPI domain was shown to be structurally related to the kinase-associated domain1 (KA1) of the kinase KIN2/PAR-1/MARK subfamily [174,175]. Moreover, SnRK1, the SNF1 homologous in plants, also contains such a structural domain [175]. Although the function of this domain is not known, this finding may point to a mechanism of protein regulation that is conserved from animals to plants [174,175].

8.2. Mechanisms of CBL-CIPK Pathway

Structural characteristics of CBL and CIPK proteins provide the basis for their interaction. The crystal structure of the complex of Ca^{2+}-CBL4 with the C-terminal regulatory domain of CIPK24 was first resolved [174]. It reveals how the CBL-CIPK complex decodes intracellular Ca^{2+} signals provoked by extracellular stimulation [176]. The CBL protein harbors four elongation factor hands (EF-hands), and each EF-hand contains a conserved α-helix-loop-α-helix structure responsible for Ca^{2+} binding [163]. The EF-hands are organized in fixed spaces that are 22, 25, and 32 amino acids distant from EF1 to EF4 in turn [177,178]. The loop region is characterized by a consensus sequence of 12 residues DKDGDGKIDFEE [163]. Amino acids in positions 1 (X), 3 (Y), 5 (Z), 7 (−X), 9 (−Y), and 12 (−Z) are responsible for Ca^{2+} coordination [176]. EF1 contains an insertion of two amino acid residues between position X and position Y. Variation of amino acids in these positions causes the change of Ca^{2+}-binding affinity [163]. Amino acid residues of CBL4 at positions X, Y, Z, and −Z bind Ca^{2+} depending on side-chain donor oxygen, while backbone carbonyl oxygen atom and water facilitation are used at positions −Y and −X, respectively [176].

The CIPK protein consists of two domains, one is the conserved N-terminal kinase catalytic domain, which comprises a phosphorylation site-containing activation loop, and the other is the highly variant C-terminal regulatory domain harboring NAF/FISL motif and a phosphatase interaction motif (PPI) [170]. The NAF motif, named by its highly conserved amino acids Asn (N), Ala (A), Phe (F), Ile (I), Ser (S), and Leu (L), is necessary for binding CBL protein. This motif is necessary for sustaining the interaction between CIPK24 and CBL4 and is able to attach the C-terminal regulatory domain of CIPK24 to cover its activation loop for keeping the kinase in an auto-inhibited state (Figure 3) [179]. Attachment of Ca^{2+} by EF-hands leads to the modification of molecular surface properties of CBL4 [176] and supports CBL4 interact with CIPK24 via the NAF motif. The interaction triggers the conformational changes of CIPK24 and exposes its activation loop [180]. Once the activation loop is free, the auto-inhibited CIPK24 is phosphorylated by an unknown upstream kinase and activates CIPK24. Subsequently, the activated CIPK24 phosphorylates the Na^+/H^+ exchanger SOS1 on the PM to exclude the excess Na^+ from the cell (Figure 3a) [180]. Abscisic acid-insensitive 2 (ABI2), a member of protein phosphatase 2C (PP2C), was identified as a CIPK24-interacting phosphatase [179]. The salt-tolerant phenotype of abi2 indicated that ABI2 is a negative regulator of CIPK24 in the SOS pathway. Up to now, the blocking mechanism of ABI2 in the CBL4-CIPK24 pathway is not yet elucidated. It is assumed that ABI2 might function in the process of dephosphorylating SOS1 (Figure 3b) or CIPK24 (Figure 3c) [179].

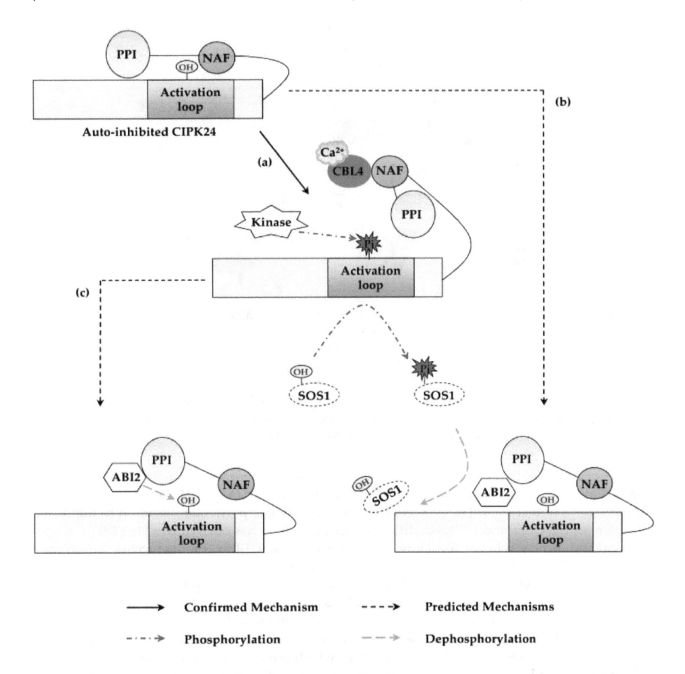

Figure 3. Mechanism of Calcineurin B-like protein 4 (CBL4)-CBL–interacting protein kinase (CIPK24) signaling pathway. **(a)** The Ca^{2+}-binding CBL4 interacts with the NAF motif of CIPK24 and changes the conformation of CIPK24. CIPK24 exposes its activation loop and then is phosphorylated by an unknown upstream protein kinase. Activated CIPK24 phosphorylates and stimulates salt overly sensitive 1 (SOS1), subsequently. **(b)** Abscisic acid-insensitive 2 (ABI2) binds to the phosphatase interaction (PPI) domain of CIPK24 and dephosphorylates SOS1 which was phosphorylated by CIPK24. **(c)** Activated CIPK24 is dephosphorylated by ABI2, and its activity is inhibited. (Adapted from Mao et al. (2016)).

8.3. Physiological Roles of CBLs and CIPKs in Plant Responses to Abiotic Stress Signals

The physiological roles of CBL and CIPK were firstly uncovered in salt overly sensitive (SOS) pathway (Figure 4) [180]. The *Arabidopsis* mutants *sos1*, *sos2*, and *sos3* produced the same salt-sensitive phenotype under high-salt stress [181]. SOS3 and SOS2, also known as CBL4 and CIPK24 respectively, were demonstrated to synergistically up-regulate the activity of plasma membrane (PM)-located Na^+/H^+ exchanger SOS1 in *Arabidopsis*, leading to the Na^+ efflux from cells in the high-salt environment [180].

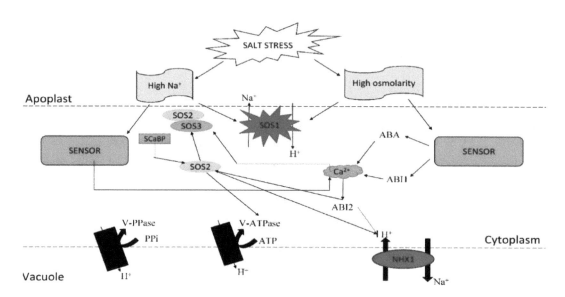

Figure 4. Signaling pathways responsible for Na$^+$ extrusion in Arabidopsis under salt stress. Excess Na$^+$ and high osmolarity are separately sensed by unknown sensors at the plasma membrane level, which then induce an increase in cytosolic Ca^{2+}. This increase is sensed by SOS3, which activates SOS2. The activated SOS3-SOS2 protein complex phosphorylates SOS1, the plasma membrane Na$^+$/H+ antiporter, resulting in the efflux of Na$^+$ ions. SOS2 can regulate NHX1 antiport activity and V-H+-ATPase activity independently of SOS3, possibly by SOS3-like Ca^{2+}-binding proteins (SCaBP) that target it to the tonoplast. Salt stress can also induce the accumulation of ABA, which, by means of ABI1 and ABI2, can negatively regulate SOS2 or SOS1 and NHX1.

It has been found that CBL-CIPK pathways work as regulators in nutrients transport systems, regulating sodium (Na$^+$) [180], potassium(K$^+$) [182], magnesium (Mg^{2+}) [183], nitrate (NO$_3$$^-$) [184], and proton (H$^+$) homeostasis [185]. Recently, in some reviews, particular attention to the possible involvement of the CBLs and CIPKs in different ions sensitivity has been drawn [186,187].

As calcium sensor relieves in plants, calcineurin B–like (CBL) proteins provide an important contribution to decoding Ca^{2+} signatures elicited by a variety of abiotic stresses. Currently, it is well known that CBLs perceive and transmit the Ca^{2+} signals mainly to a group of serine/threonine protein kinases called CBL-interacting protein kinases (CIPKs).

In the year 2016, Cho et al. reported that the CBL10 member of this family has a novel interaction partner besides the CIPK proteins. Yeast two-hybrid screening with CBL10 as bait identified an *Arabidopsis* cDNA clone encoding a TOC34 protein, which is a member of the translocon of the outer membrane of chloroplasts (TOC) complex and possesses the GTPase activity. Bimolecular fluorescence complementation (BiFC) analysis verified that the CBL10–TOC34 interaction takes place at the outer membrane of chloroplasts in vivo and thus decreases its GTPase activity in *Arabidopsis* [188].

These findings indicate that a member of the CBL family, CBL10, can modulate not only the CIPK members but also TOC34, allowing the CBL family to relay the Ca^{2+} signals in more diverse ways than currently known.

In tomato, the calcium sensor Cbl10 and its interacting protein kinase *Cipk6* define a signaling pathway in plant immunity [189]. Ca^{2+} signaling is an early and necessary event in plant immunity. The tomato (*Solanum lycopersicum*) kinase *Pto* triggers localized programmed cell death (PCD) upon recognition of *Pseudomonas syringae* effectors AvrPto *or* AvrPtoB. In a virus-induced gene silencing screen in *Nicotiana benthamiana*, Fernando and al. identified two components of a Ca^{2+}-signaling system, Cbl10 (for calcineurin B-like protein) and *Cipk6* (for calcineurin B-like interacting protein kinase), as their silencing inhibited Pto/AvrPto-elicited PCD. *N. benthamiana* Cbl10 and Cipk6 are also required for PCD triggered by other plant resistance genes and virus, oomycete, and nematode effectors and for host susceptibility to two *P. syringae* pathogens.

Tomato *Cipk6* interacts with Cbl10 and its in vitro kinase activity is enhanced in the presence of *Cbl10* and Ca^{2+}, suggesting that tomato Cbl10 and Cipk6 constitute a Ca^{2+}-regulated signaling module. Overexpression of tomato *Cipk6* in *N. benthamiana* leaves causes accumulation of reactive oxygen species (ROS), which requires the respiratory burst homolog *RbohB*. Tomato Cbl10 and Cipk6 interact with *RbohB* at the plasma membrane. Finally, Cbl10 and Cipk6 contribute to ROS generated during effector-triggered immunity in the interaction of *P. syringae* pv tomato DC3000 and *N. benthamiana*. The role of the Cbl/Cipk signaling module in PCD has been identified, establishing a mechanistic link between Ca^{2+} and ROS signaling in plant immunity [189].

Xu et al. showed that the protein kinase CIPK23, encoded by the Arabidopsis Low-K^+-sensitive 1 (*LKS1*) gene, regulates K^+ uptake under low K^+ conditions. Lesion of *LKS1* has reduced K^+ uptake and caused leaf chlorosis and growth inhibition, whereas overexpression of *LKS1* significantly enhanced K^+ uptake and tolerance to low K^+. They demonstrated that *CIPK23* directly phosphorylates the K^+ transporter *AKT1* and further found that CIPK23 is activated by the binding of two calcineurin B-like proteins, CBL1 and CBL9 [55]. Further research on protein kinase *CIPK23* in *Arabidopsis* has revealed that CIPK23 is expressed in a variety of cell types and tissues and regulates distinct physiological processes including the opening/closing of stomata in the leaves, and the potassium uptake in the roots [190]. In addition, the authors showed that CIPK23 kinase interacts and functions with both CBL1 and CBL9 calcium sensors, providing a molecular link between intracellular calcium fluctuations and the regulation of transpiration and nutrient uptake. CBL1 and CBL9 can both recruit CIPK23 on the plasma membrane, suggesting that CIPK23-CBL complexes associated with the plasma membrane modulate the membrane on which the target proteins are located, including the AKT1 potassium channel [190,191] by proteins phosphorylation. Cheong et al provided more information on the mechanistic aspects of calcium signaling by plants. According to their finds, the combination of CIPK23 with a specific set of other components in the guard cells results in the regulation of the stomatal response to ABA, while CIPK23 and another set of components in the root tissues participate in the regulation of potassium absorption. Since CIPK23 is also present in other tissues, such as vascular tissues of roots, stems, and leaves, the authors hypothesized that CIPK23 could also be associated with other components of these tissues, for example during long-distance transport and distribution of K^+ throughout the plant. They showed that the other components that interact with CIPK23 include the CBL1 and CBL9 calcium sensors that functionally overlap in regulating stomatal movement and K^+ uptake. It is possible that other CBLs may also interact with CIPK23 in regulating K^+ nutrition. Such selective and overlapping interactions can encode unique responses that are different from any CBL–CIPK interaction. Among the CBLs that regulate a specific CIPK in the same process, some may play a more dominant role than others. For example, the functions of CIPK23 in stomatal response and K^+ absorption appear to be primarily regulated by CBL1 and CBL9, each functioning in other processes by regulating other CIPKs [190,192].

Hashimoto et al. have identified a novel general regulatory mechanism of CBL-CIPK complexes in that CBL phosphorylation at their flexible C-terminus probably induces conformational changes that enhance specificity and activity of CBL-CIPK complexes toward their target proteins. The phosphorylation status of CBLs does not appear to influence the stability, localization, or CIPK interaction of these calcium sensor proteins in general. However, proper phosphorylation of CBL1 is absolutely required for the in vivo activation of the AKT1, K^+ channel by CBL1-CIPK23 and CBL9-CIPK23 complexes in oocytes [190,193]. Moreover, the authors have shown that, by combining CBL1, CIPK23, and AKT1, the reconstituted CBL-dependent enhancement of phosphorylation of target proteins by CIPKs in vitro. In addition, they reported that phosphorylation of CBL1 by CIPK23 is also required for the CBL1-dependent enhancement of CIPK23 activity toward its substrate.

Recent studies have uncovered the crucial functions of CBL-CIPK complexes in an increasing number of biological processes like salt tolerance, potassium transport, nitrate sensing, and stomatal regulation [194]. CBL proteins determine the cellular localization of their interacting protein kinases in vivo and are essential for the activity of the resulting CBL-CIPK complexes toward their target proteins [55,184]. Despite the established importance of CBL-CIPK complexes in regulating the activity of ion channels and transporters like SOS1, AKT1, AKT2, and NRT1 [195], only very few target phosphorylation sites of CIPKs have been clearly identified. The occurrence of phosphorylation of CBLs by CIPKs appears not to be restricted to the model organism *Arabidopsis*.

In 2017, it was reported that *BdCIPK31*, a CIPK gene from *Brachypodium distachyon*, functions positively to drought and salt stress through the ABA signaling pathway [196]. Overexpressing *BdCIPK31* functions in stomatal closure, ion homeostasis, ROS scavenging, osmolyte biosynthesis, and transcriptional regulation of stress-related genes. In fact, it appears that transgenic tobacco plants overexpressing *BdCIPK31* presented improved drought and salt tolerance and displayed hypersensitive response to exogenous ABA [196]. Further investigations revealed that *BdCIPK31* functioned positively in ABA-mediated stomatal closure, and transgenic tobacco exhibited reduced water loss under dehydration conditions compared with the controls. *BdCIPK31* also affected Na^+/K^+ homeostasis and root K^+ loss, which contributed to maintaining intracellular ion homeostasis under salt conditions. Moreover, the reactive oxygen species scavenging system and osmolyte accumulation were enhanced by *BdCIPK31* overexpression, which was conducive for alleviating oxidative and osmotic damages. Additionally, overexpression of *BdCIPK31* could elevate several stress-associated gene expressions under stress conditions [196].

In 2013, *TaCIPK14* and *TaCIPK29* were found to confer single or multiple stress tolerance in transgenic tobacco [197]. Transgenic tobaccos overexpressing TaCIPK14 exhibited higher contents of chlorophyll and sugar, higher catalase activity, while decreased amounts of H_2O_2 and malondialdehyde (MDA), and lesser ion leakage under cold and salt stresses. In addition, overexpression also enhanced the seed germination rate, root elongation and decreased Na^+ content in the transgenic lines under salt stress. Higher expression of stress-related genes was observed in lines overexpressing *TaCIPK14* compared to controls under stress conditions [197].

Under conditions of high salinity, *TaCIPK25* expression was markedly down-regulated in wheat roots [198]. Overexpression of *TaCIPK25* resulted in hypersensitivity to Na^+ and superfluous accumulation of Na^+ in transgenic wheat lines. The *TaCIPK25* expression did not decline in transgenic wheat and remained at an even higher level than that in wild-type wheat controls under high-salinity treatment. Furthermore, the transmembrane Na^+/H^+ exchange was impaired in the root cells of transgenic wheat. These results suggested that *TaCIPK25* negatively regulated salt response in wheat [198].

9. Conclusions and Perspectives

The data available on the CHX family in *Arabidopsis* and other plants clearly highlight a novel and original mechanism involved in plants' tolerance to the salinity. This mechanism, which was previously not demonstrated in plants, allows detoxification of Na^+ in leaves by recirculation of this ion to the roots via the phloem. Plants face a dilemma regarding the transport of sodium. Sodium absorption is useful for lowering osmotic potential, being able to absorb water and maintaining turgor, but excess sodium is toxic. Many studies have focused on the toxic role of Na^+ in the plant during salt stress and the elucidation of the mechanisms of tolerance to this stress.

The role of Na^+ at lower concentrations is not well known. The current consensus is that the energization of the cell membrane is based solely on a proton gradient. However, the available data for some *CHXs* encourage us to continue to imagine that Na^+ (at non-toxic concentrations) can lead to symport systems and energize active K^+ uptake. Several indices seem to support this hypothesis, for example, AtNHX23 an antiport K^+ (Na^+/H^+) active at the level of the chloroplast envelope and involved in potassium homeostasis and perhaps in regulating the pH of the stroma. However, the Na^+/H^+ antiport systems of the plasma membrane are still poorly characterized. The available information is only related to the SOS1 protein in *Arabidopsis*, which has a homologous sequence with antiport Na^+/H^+ and would be involved in sodium efflux at the plasmalemma level. *SOS1* overexpression in *Arabidopsis* significantly improves plants' tolerance to salinity. AtSOS1 is, therefore, an important determinant of salt sensitivity in plants. AtSOS1 activity is controlled by AtSOS2 and AtSOS3. AtSOS3 (a Ca^{2+} affine protein belonging to the CBL family) directly interacts with AtSOS2, which is a serine/threonine protein kinase.

Studies on CBLs and CIPKs over the past decade have greatly advanced our knowledge of the function of single proteins in distinct physiological processes. Major advances in understanding this signaling system were through the identification of an increasing number of targets regulated by the CBL-CIPK complexes. The progress of the research on the CBL and CIPK families in different plant species other than *Arabidopsis thaliana* is still at an infant stage; in most cases, it is limited to interaction studies and expression analysis of these families.

The CBL-CIPK signaling model emphasizes the importance of future research that focuses on the molecular mechanisms underlying the regulation of transporters that allow us to better understand plant's response to abiotic stress such as salt stress and also establish a proficient method of identifying molecular targets for genetically engineered resistant crops with enhanced tolerance to various environmental stresses. Therefore, the most important challenge for future research is not only functional characterization but also the elucidating of the details of synergistic functions in this interaction network and revealing the molecular mechanisms of the complexes regulating target proteins.

Author Contributions: Conceptualization, H.-Y.L. and T.K.; T.K. wrote this manuscript. X.-W.L., F.-W.W., K.F.I.C., and M.N. participated in the writing and modification of this manuscript. Validation, X.-W.L. All authors read and approved the final manuscript.

Acknowledgments: The authors are grateful to Prof. Li Haiyan for the critical discussion of this article.

References

1. Munns, R. Comparative physiology of salt and water stress. *Plant Cell Env.* **2002**, *25*, 239–250. [CrossRef]
2. Tester, M.; Davenport, R. Na$^+$ tolerance and Na$^+$ transport in higher plants. *Ann. Bot.* **2003**, *91*, 503–527. [CrossRef] [PubMed]
3. Flowers, T.J.; Colmer, T.D. Plant salt tolerance: Adaptations in halophytes. *Ann. Bot.* **2015**, *115*, 327–331. [CrossRef] [PubMed]
4. Ruan, C.J.; da Silva, J.A.T.; Mopper, S.; Qin, P.; Lutts, S. Halophyte Improvement for a Salinized World. *Crit. Rev. Plant Sci.* **2010**, *29*, 329–359. [CrossRef]
5. Apse, M.P.; Blumwald, E. Na$^+$ transport in plants. *Febs Lett.* **2007**, *581*, 2247–2254. [CrossRef]
6. Munns, R. Genes and salt tolerance: Bringing them together. *New Phytol.* **2005**, *167*, 645–663. [CrossRef]
7. Teakle, N.L.; Tyerman, S.D. Mechanisms of Cl-transport contributing to salt tolerance. *Plant Cell Environ.* **2010**, *33*, 566–589. [CrossRef]
8. Ashraf, M.; Foolad, M. Roles of glycine betaine and proline in improving plant abiotic stress resistance. *Environ. Exp. Bot.* **2007**, *59*, 206–216. [CrossRef]
9. Ksouri, R.; Falleh, H.; Megdiche, W.; Trabelsi, N.; Mhamdi, B.; Chaieb, K.; Bakrouf, A.; Magné, C.; Abdelly, C. Antioxidant and antimicrobial activities of the edible medicinal halophyte Tamarix gallica L. and related polyphenolic constituents. *Food Chem. Toxicol.* **2009**, *47*, 2083–2091. [CrossRef]
10. Sairam, R.; Tyagi, A. Physiology and molecular biology of salinity stress tolerance in plants. *Curr. Sci.* **2004**, *86*, 407–421.
11. Adler, G.; Blumwald, E.; Bar-Zvi, D. The sugar beet gene encoding the sodium/proton exchanger 1 (BvNHX1) is regulated by a MYB transcription factor. *Planta* **2010**, *232*, 187–195. [CrossRef] [PubMed]
12. Chinnusamy, V.; Jagendorf, A.; Zhu, J.-K. Understanding and improving salt tolerance in plants. *Crop Sci.* **2005**, *45*, 437–448. [CrossRef]
13. Shabala, S.; Shabala, S.; Cuin, T.A.; Pang, J.; Percey, W.; Chen, Z.; Conn, S.; Eing, C.; Wegner, L.H. Xylem ionic relations and salinity tolerance in barley. *Plant J.* **2010**, *61*, 839–853. [CrossRef] [PubMed]
14. Majumder, A.; Parida, A.; Sankar, K.; Qureshi, Q. Utilization of food plant species and abundance of hanuman langurs (*Semnopithecus entellus*) in Pench Tiger Reserve, Madhya Pradesh, India. *Taprobanica J. Asian Biodivers.* **2011**, *2*, 105–108. [CrossRef]

15. Rontein, D.; Dieuaide-Noubhani, M.; Dufourc, E.J.; Raymond, P.; Rolin, D. The metabolic architecture of plant cells stability of central metabolism and flexibility of anabolic pathways during the growth cycle of tomato cells. *J. Biol. Chem.* **2002**, *277*, 43948–43960. [CrossRef] [PubMed]

16. Cuin, T.A.; Tian, Y.; Betts, S.A.; Chalmandrier, R.; Shabala, S. Ionic relations and osmotic adjustment in durum and bread wheat under saline conditions. *Funct. Plant Biol.* **2009**, *36*, 1110–1119. [CrossRef]

17. He, C.; Yan, J.; Shen, G.; Fu, L.; Holaday, A.S.; Auld, D.; Blumwald, E.; Zhang, H. Expression of an Arabidopsis vacuolar sodium/proton antiporter gene in cotton improves photosynthetic performance under salt conditions and increases fiber yield in the field. *Plant Cell Physiol.* **2005**, *46*, 1848–1854. [CrossRef]

18. Clarkson, D.T.; Hanson, J.B. The mineral nutrition of higher plants. *Ann. Rev. Plant Physiol.* **1980**, *31*, 239–298. [CrossRef]

19. White, P.J. Ion Uptake Mechanisms of Individual Cells and Roots: Short-Distance Transport. In *Marschner's Mineral Nutrition of Higher Plants*; Elsevier: Amsterdam, The Netherlands, 2012; pp. 7–47.

20. Gierth, M.; Mäser, P. Potassium transporters in plants–involvement in K$^+$ acquisition, redistribution and homeostasis. *FEBS Lett.* **2007**, *581*, 2348–2356. [CrossRef]

21. Cao, M.; Yu, H.; Yan, H.; Jiang, C. Difference in tolerance to potassium deficiency between two maize inbred lines. *Plant Prod. Sci.* **2007**, *10*, 42–46.

22. Chen, J.; Gabelman, W.H. Morphological and physiological characteristics of tomato roots associated with potassium-acquisition efficiency. *Sci. Hortic.* **2000**, *83*, 213–225. [CrossRef]

23. Rodríguez-Navarro, A. Potassium transport in fungi and plants. *Biochim. Et Biophys. ActaRev. Biomembr.* **2000**, *1469*, 1–30. [CrossRef]

24. Thomson, S.J.; Hansen, A.; Sanguinetti, M.C. Identification of the intracellular Na$^+$ sensor in Slo2. 1 potassium channels. *J. Biol. Chem.* **2015**, *290*, 14528–14535. [CrossRef] [PubMed]

25. Assaha, D.V.; Ueda, A.; Saneoka, H.; Al-Yahyai, R.; Yaish, M.W. The role of Na$^+$ and K$^+$ transporters in salt stress adaptation in glycophytes. *Front. Physiol.* **2017**, *8*, 509. [CrossRef]

26. Blaha, G.; Stelzl, U.; Spahn, C.M.T.; Agrawal, R.K.; Frank, J.; Nierhaus, K.H. Preparation of functional ribosomal complexes and effect of buffer conditions on tRNA positions observed by cryoelectron microscopy. *Methods Enzymol.* **2000**, *397*, 292–306.

27. Maathuis, F.J. Sodium in plants: Perception, signalling, and regulation of sodium fluxes. *J. Exp. Bot.* **2013**, *65*, 849–858. [CrossRef]

28. Figdore, S.S.; Gabelman, W.; Gerloff, G. The Accumulation and Distribution of Sodium in Tomato Strains Differing in Potassium Efficiency When Grown under Low-K Stress. In *Genetic Aspects of Plant Mineral Nutrition*; Springer: Berlin, Germany, 1987; pp. 353–360.

29. Tahal, R.; Mills, D.; Heimer, Y.; Tal, M. The relation between low K$^+$/Na$^+$ ratio and salt-tolerance in the wild tomato species Lycopersicon pennellii. *J. Plant Physiol.* **2000**, *157*, 59–64. [CrossRef]

30. Marschner, H.; Kuiper, P.; Kylin, A. Genotypic differences in the response of sugar beet plants to replacement of potassium by sodium. *Physiol. Plant.* **1981**, *51*, 239–244. [CrossRef]

31. Subbarao, G.; Wheeler, R.M.; Stutte, G.W.; Levine, L.H. How far can sodium substitute for potassium in red beet? *J. Plant Nutr.* **1999**, *22*, 1745–1761. [CrossRef]

32. Horie, T.; Costa, A.; Kim, T.H.; Han, M.J.; Horie, R.; Leung, H.-Y.; Miyao, A.; Hirochika, H.; An, G.; Schroeder, J.I. Rice OsHKT2; 1 transporter mediates large Na$^+$ influx component into K$^+$-starved roots for growth. *EMBO J.* **2007**, *26*, 3003–3014. [CrossRef]

33. Ghars, M.A.; Parre, E.; Debez, A.; Bordenave, M.; Richard, L.; Leport, L.; Bouchereau, A.; Savouré, A.; Abdelly, C. Comparative salt tolerance analysis between Arabidopsis thaliana and Thellungiella halophila, with special emphasis on K$^+$/Na$^+$ selectivity and proline accumulation. *J. Plant Physiol.* **2008**, *165*, 588–599. [CrossRef] [PubMed]

34. Munns, R.; Tester, M. Mechanisms of salinity tolerance. *Annu. Rev. Plant Biol.* **2008**, *59*, 651–681. [CrossRef] [PubMed]

35. Brini, F.; Hanin, M.; Mezghani, I.; Berkowitz, G.A.; Masmoudi, K. Overexpression of wheat Na$^+$/H+ antiporter TNHX1 and H$^+$-pyrophosphatase TVP1 improve salt-and drought-stress tolerance in Arabidopsis thaliana plants. *J. Exp. Bot.* **2007**, *58*, 301–308. [CrossRef] [PubMed]

36. Demidchik, V.; Straltsova, D.; Medvedev, S.S.; Pozhvanov, G.A.; Sokolik, A.; Yurin, V. Stress-induced electrolyte leakage: The role of K$^+$-permeable channels and involvement in programmed cell death and metabolic adjustment. *J. Exp. Bot.* **2014**, *65*, 1259–1270. [CrossRef]

37. Qie, L.; Lewis, S.L.; Sullivan, M.J.P.; Lopez-Gonzalez, G.; Pickavance, G.C.; Sunderland, T.; Ashton, P.; Hubau, W.; Abu Salim, K.; Aiba, S.-I.; et al. Long-term carbon sink in Borneo's forests halted by drought and vulnerable to edge effects. *Nat. Commun.* **2017**, *8*, 1966. [CrossRef]

38. Clarkson, D.T. Roots and the delivery of solutes to the xylem. Philosophical Transactions of the Royal Society of London. *Ser. B Biol. Sci.* **1993**, *341*, 5–17.

39. Hossain, M.R.; Pritchard, J.; Ford-Lloyd, B.V. Qualitative and quantitative variation in the mechanisms of salinity tolerance determined by multivariate assessment of diverse rice (*Oryza sativa* L.) genotypes. *Plant Genet. Resour.* **2016**, *14*, 91–100. [CrossRef]

40. Robbins, N.E.; Trontin, C.; Duan, L.; Dinneny, J.R. Beyond the barrier: Communication in the root through the endodermis. *Plant Physiol.* **2014**, *166*, 551–559. [CrossRef]

41. Flowers, T.J.; Munns, R.; Colmer, T.D. Sodium chloride toxicity and the cellular basis of salt tolerance in halophytes. *Ann. Bot.* **2014**, *115*, 419–431. [CrossRef]

42. Anil, V.S.; Krishnamurthy, H.; Mathew, M. Limiting cytosolic Na$^+$ confers salt tolerance to rice cells in culture: A two-photon microscopy study of SBFI-loaded cells. *Physiol. Plant.* **2007**, *129*, 607–621. [CrossRef]

43. Flam-Shepherd, R.; Huynh, W.Q.; Coskun, D.; Hamam, A.M.; Britto, D.T.; Kronzucker, H.J. Membrane fluxes, bypass flows, and sodium stress in rice: The influence of silicon. *J. Exp. Bot.* **2018**, *69*, 1679–1692. [CrossRef] [PubMed]

44. Yeo, A.; Flowers, S.; Rao, G.; Welfare, K.; Senanayake, N.; Flowers, T. Silicon reduces sodium uptake in rice (*Oryza sativa* L.) in saline conditions and this is accounted for by a reduction in the transpirational bypass flow. *Plant Cell Environ.* **1999**, *22*, 559–565. [CrossRef]

45. Farooq, M.A.; Saqib, Z.A.; Akhtar, J. Silicon-mediated oxidative stress tolerance and genetic variability in rice (*Oryza sativa* L.) grown under combined stress of salinity and boron toxicity. *Turk. J. Agric. For.* **2015**, *39*, 718–729. [CrossRef]

46. Garcia, A.; Rizzo, C.; Ud-Din, J.; Bartos, S.; Senadhira, D.; Flowers, T.; Yeo, A. Sodium and potassium transport to the xylem are inherited independently in rice, and the mechanism of sodium: Potassium selectivity differs between rice and wheat. *Plant Cell Environ.* **1997**, *20*, 1167–1174. [CrossRef]

47. Davis, L.; Sumner, M.; Stasolla, C.; Renault, S. Salinity-induced changes in the root development of a northern woody species, Cornus sericea. *Botany* **2014**, *92*, 597–606. [CrossRef]

48. Keisham, M.; Mukherjee, S.; Bhatla, S. Mechanisms of sodium transport in plants—progresses and challenges. *Int. J. Mol. Sci.* **2018**, *19*, 647. [CrossRef]

49. Yeo, A.; Yeo, M.; Flowers, T. The contribution of an apoplastic pathway to sodium uptake by rice roots in saline conditions. *J. Exp. Bot.* **1987**, *38*, 1141–1153. [CrossRef]

50. Barragán, V.; Leidi, E.O.; Andrés, Z.; Rubio, L.; De Luca, A.; Fernández, J.A.; Cubero, B.; Pardo, J.M. Ion exchangers NHX$_1$ and NHX$_2$ mediate active potassium uptake into vacuoles to regulate cell turgor and stomatal function in Arabidopsis. *Plant Cell* **2012**, *24*, 1127–1142.

51. Wang, Y.; Wu, W.-H. Potassium transport and signaling in higher plants. *Annu. Rev. Plant Biol.* **2013**, *64*, 451–476. [CrossRef]

52. Glass, A.D.; Fernando, M. Homeostatic processes for the maintenance of the K$^+$ content of plant cells: A model. *Isr. J. Bot.* **1992**, *41*, 145–166.

53. Glass, A.D.; Dunlop, J. The influence of potassium content on the kinetics of potassium influx into excised ryegrass and barley roots. *Planta* **1978**, *141*, 117–119. [CrossRef] [PubMed]

54. Pyo, Y.J.; Gierth, M.; Schroeder, J.I.; Cho, M.H. High.-affinity K$^+$ transport in Arabidopsis: AtHAK5 and AKT1 are vital for seedling establishment and postgermination growth under low-potassium conditions. *Plant Physiol.* **2010**, *153*, 863–875. [CrossRef] [PubMed]

55. Xu, J.; Li, H.-D.; Chen, L.-Q.; Wang, Y.; Liu, L.-L.; He, L.; Wu, W.-H. A protein kinase, interacting with two calcineurin B-like proteins, regulates K$^+$ transporter AKT1 in Arabidopsis. *Cell* **2006**, *125*, 1347–1360. [CrossRef] [PubMed]

56. Coskun, D.; Britto, D.T.; Kronzucker, H.J. Regulation and mechanism of potassium release from barley roots: An in planta42K$^+$ analysis. *New Phytol.* **2010**, *188*, 1028–1038. [CrossRef]

57. Zeng, J.; Quan, X.; He, X.; Cai, S.; Ye, Z.; Chen, G.; Zhang, G. Root and leaf metabolite profiles analysis reveals the adaptive strategies to low potassium stress in barley. *BMC Plant Biol.* **2018**, *18*, 187. [CrossRef] [PubMed]

58. Santa-María, G.E.; Rubio, F.; Dubcovsky, J.; Rodríguez-Navarro, A. The HAK1 gene of barley is a member of a large gene family and encodes a high-affinity potassium transporter. *Plant Cell* **1997**, *9*, 2281–2289.

59. Wang, T.-B.; Gassmann, W.; Rubio, F.; Schroeder, J.I.; Glass, A.D. Rapid up-regulation of HKT1, a high-affinity potassium transporter gene, in roots of barley and wheat following withdrawal of potassium. *Plant Physiol.* **1998**, *118*, 651–659. [CrossRef]

60. Kim, E.J.; Kwak, J.M.; Uozumi, N.; Schroeder, J.I. AtKUP1: An Arabidopsis gene encoding high-affinity potassium transport activity. *Plant Cell* **1998**, *10*, 51–62. [CrossRef]

61. Sharma, T.; Dreyer, I.; Riedelsberger, J. The role of K^+ channels in uptake and redistribution of potassium in the model plant Arabidopsis thaliana. *Front. Plant Sci.* **2013**, *4*, 224. [CrossRef]

62. Zhu, K.; Chen, F.; Liu, J.; Chen, X.; Hewezi, T.; Cheng, Z.-M.M. Evolution of an intron-poor cluster of the CIPK gene family and expression in response to drought stress in soybean. *Sci. Rep.* **2016**, *6*, 28225. [CrossRef]

63. Adem, G.D.; Roy, S.J.; Zhou, M.; Bowman, J.P.; Shabala, S. Evaluating contribution of ionic, osmotic and oxidative stress components towards salinity tolerance in barley. *Bmc Plant Biol.* **2014**, *14*, 113. [CrossRef] [PubMed]

64. Nardini, A.; Salleo, S.; Jansen, S. More than just a vulnerable pipeline: Xylem physiology in the light of ion-mediated regulation of plant water transport. *J. Exp. Bot.* **2011**, *62*, 4701–4718. [CrossRef] [PubMed]

65. De Boer, A.; Volkov, V. Logistics of water and salt transport through the plant: Structure and functioning of the xylem. *Plant Cell Environ.* **2003**, *26*, 87–101. [CrossRef]

66. Zhu, J.; Liang, J.; Xu, Z.; Fan, X.; Zhou, Q.; Shen, Q.; Xu, G. Root aeration improves growth and nitrogen accumulation in rice seedlings under low nitrogen. *Aob Plants* **2015**, *7*, plv131. [CrossRef]

67. Kim, H.Y.; Choi, E.-H.; Min, M.K.; Hwang, H.; Moon, S.-J.; Yoon, I.; Byun, M.-O.; Kim, B.-G. Differential gene expression of two outward-rectifying shaker-like potassium channels OsSKOR and OsGORK in rice. *J. Plant Biol.* **2015**, *58*, 230–235. [CrossRef]

68. Shabala, S. Learning from halophytes: Physiological basis and strategies to improve abiotic stress tolerance in crops. *Ann. Bot.* **2013**, *112*, 1209–1221. [CrossRef]

69. Munns, R.; Lorraine Tonnet, M.; Shennan, C.; Anne Gardner, P. Effect of high external NaCl concentration on ion transport within the shoot of *Lupinus albus*. II. Ions in phloem sap. *PlantCell Environ.* **1988**, *11*, 291–300. [CrossRef]

70. Blom-Zandstra, M.; Vogelzang, S.A.; Veen, B.W. Sodium fluxes in sweet pepper exposed to varying sodium concentrations. *J. Exp. Bot.* **1998**, *49*, 1863–1868. [CrossRef]

71. Lohaus, G.; Hussmann, M.; Pennewiss, K.; Schneider, H.; Zhu, J.J.; Sattelmacher, B. Solute balance of a maize (*Zea mays* L.) source leaf as affected by salt treatment with special emphasis on phloem retranslocation and ion leaching. *J. Exp. Bot.* **2000**, *51*, 1721–1732. [CrossRef]

72. Alfocea, F.P.; Balibrea, M.E.; Alarcón, J.J.; Bolarín, M.C. Composition of xylem and phloem exudates in relation to the salt-tolerance of domestic and wild tomato species. *J. Plant Physiol.* **2000**, *156*, 367–374. [CrossRef]

73. An, D.; Chen, J.-G.; Gao, Y.-Q.; Li, X.; Chao, Z.-F.; Chen, Z.-R.; Li, Q.-Q.; Han, M.-L.; Wang, Y.-L.; Wang, Y.-F. AtHKT1 drives adaptation of Arabidopsis thaliana to salinity by reducing floral sodium content. *PLoS Genet.* **2017**, *13*, e1007086. [CrossRef] [PubMed]

74. Cuin, T.; Dreyer, I.; Michard, E. The role of potassium channels in Arabidopsis thaliana long distance electrical signalling: AKT2 modulates tissue excitability while GORK shapes action potentials. *Int. J. Mol. Sci.* **2018**, *19*, 926. [CrossRef] [PubMed]

75. Gajdanowicz, P.; Michard, E.; Sandmann, M.; Rocha, M.; Corrêa, L.G.G.; Ramírez-Aguilar, S.J.; Gomez-Porras, J.L.; González, W.; Thibaud, J.-B.; Van Dongen, J.T.; et al. Potassium (K^+) gradients serve as a mobile energy source in plant vascular tissues. *Proc. Natl. Acad. Sci. USA* **2011**, *108*, 864–869. [CrossRef] [PubMed]

76. Renault, S.; Lait, C.; Zwiazek, J.J.; MacKinnon, M. Effect of high salinity tailings waters produced from gypsum treatment of oil sands tailings on plants of the boreal forest. *Environ. Pollut.* **1998**, *102*, 177–184. [CrossRef]

77. Epstein, E.; Rains, D.; Elzam, O. Resolution of dual mechanisms of potassium absorption by barley roots. *Proc. Natl. Acad. Sci. USA* **1963**, *49*, 684. [CrossRef]

78. Hamouda, S.B.; Touati, K.; Amor, M.B. Donnan dialysis as membrane process for nitrate removal from drinking water: Membrane structure effect. *Arab. J. Chem.* **2017**, *10*, S287–S292. [CrossRef]

79. Nieves-Cordones, M.; Alemán, F.; Martínez, V.; Rubio, F. K^+ uptake in plant roots. The systems involved, their regulation and parallels in other organisms. *J. Plant Physiol.* **2014**, *171*, 688–695. [CrossRef]

80. Ali, A.; Maggio, A.; Bressan, R.A.; Yun, D.-J. Role and functional differences of HKT1-type transporters in plants under salt stress. *Int. J. Mol. Sci.* **2019**, *20*, 1059. [CrossRef]

81. Haro, R.; Bañuelos, M.A.; Senn, M.E.; Barrero-Gil, J.; Rodríguez-Navarro, A. HKT1 mediates sodium uniport in roots. Pitfalls in the expression of HKT1 in yeast. *Plant Physiol.* **2005**, *139*, 1495–1506. [CrossRef]

82. Pilot, G.; Pratelli, R.; Gaymard, F.; Meyer, Y.; Sentenac, H. Five-group distribution of the Shaker-like K$^+$ channel family in higher plants. *J. Mol. Evol.* **2003**, *56*, 418–434. [CrossRef]

83. Anderson, J.A.; Huprikar, S.S.; Kochian, L.V.; Lucas, W.J.; Gaber, R.F. Functional expression of a probable Arabidopsis thaliana potassium channel in Saccharomyces cerevisiae. *Proc. Natl. Acad. Sci. USA* **1992**, *89*, 3736–3740. [CrossRef] [PubMed]

84. Ahmad, I.; Mian, A.; Maathuis, F.J. Overexpression of the rice AKT1 potassium channel affects potassium nutrition and rice drought tolerance. *J. Exp. Bot.* **2016**, *67*, 2689–2698. [CrossRef] [PubMed]

85. Lebaudy, A.; Véry, A.-A.; Sentenac, H. K$^+$ channel activity in plants: Genes, regulations and functions. *Febs Lett.* **2007**, *581*, 2357–2366. [CrossRef] [PubMed]

86. Saponaro, A.; Porro, A.; Chaves-Sanjuan, A.; Nardini, M.; Rauh, O.; Thiel, G.; Moroni, A. Fusicoccin activates KAT1 channels by stabilizing their interaction with 14-3-3 proteins. *Plant Cell* **2017**, *29*, 2570–2580. [CrossRef]

87. Czempinski, K.; Zimmermann, S.; Ehrhardt, T.; Müller-Röber, B. New structure and function in plant K$^+$ channels: KCO1, an outward rectifier with a steep Ca^{2+} dependency. *Embo, J.* **1997**, *16*, 2565–2575. [CrossRef]

88. Czempinski, K.; Gaedeke, N.; Zimmermann, S.; Müller-Röber, B. Molecular mechanisms and regulation of plant ion channels. *J. Exp. Bot.* **1999**, *50*, 955–966. [CrossRef]

89. Demidchik, V.; Maathuis, F.J. Physiological roles of nonselective cation channels in plants: From salt stress to signalling and development. *New Phytol.* **2007**, *175*, 387–404. [CrossRef]

90. Forde, B.G.; Roberts, M.R. Glutamate receptor-like channels in plants: A role as amino acid sensors in plant defence? *F1000prime Rep.* **2014**, *6*, 37. [CrossRef]

91. Zhao, N.; Zhu, H.; Zhang, H.; Sun, J.; Zhou, J.; Deng, C.; Zhang, Y.; Zhao, R.; Zhou, X.; Lu, C. Hydrogen sulfide mediates K$^+$ and Na$^+$ homeostasis in the roots of salt-resistant and salt-sensitive poplar species subjected to NaCl stress. *Front. Plant Sci.* **2018**, *9*, 1366. [CrossRef]

92. Gamel, K.; Torre, V. The Interaction of Na$^+$ and K$^+$ in the Pore of CyclicNucleotide-Gated Channels. *Biophys. J.* **2000**, *79*, 2475–2493. [CrossRef]

93. Schuurink, R.C.; Shartzer, S.F.; Fath, A.; Jones, R.L. Characterization of a calmodulin-binding transporter from the plasma membrane of barley aleurone. *Proc. Natl. Acad. Sci. USA* **1998**, *95*, 1944–1949. [CrossRef] [PubMed]

94. Arazi, T.; Sunkar, R.; Kaplan, B.; Fromm, H. A tobacco plasma membrane calmodulin-binding transporter confers Ni^{2+} tolerance and Pb^{2+} hypersensitivity in transgenic plants. *Plant J.* **1999**, *20*, 171–182. [CrossRef] [PubMed]

95. Köhler, C.; Merkle, T.; Neuhaus, G. Characterisation of a novel gene family of putative cyclic nucleotide-and calmodulin-regulated ion channels in Arabidopsis thaliana. *Plant J.* **1999**, *18*, 97–104. [CrossRef] [PubMed]

96. Hanin, M.; Ebel, C.; Ngom, M.; Laplaze, L.; Masmoudi, K. New insights on plant salt tolerance mechanisms and their potential use for breeding. *Front. Plant Sci.* **2016**, *7*, 1787. [CrossRef] [PubMed]

97. Mian, A.A.; Senadheera, P.; Maathuis, F.J. Improving crop salt tolerance: Anion and cation transporters as genetic engineering targets. *Plant Stress* **2011**, *5*, 64–72.

98. Jin, Y.; Jing, W.; Zhang, Q.; Zhang, W. Cyclic nucleotide gated channel 10 negatively regulates salt tolerance by mediating Na$^+$ transport in Arabidopsis. *J. Plant Res.* **2015**, *128*, 211–220. [CrossRef] [PubMed]

99. Gobert, A.; Park, G.; Amtmann, A.; Sanders, D.; Maathuis, F.J. Arabidopsis thaliana cyclic nucleotide gated channel 3 forms a non-selective ion transporter involved in germination and cation transport. *J. Exp. Bot.* **2006**, *57*, 791–800. [CrossRef]

100. Mekawy, A.M.M.; Assaha, D.V.; Yahagi, H.; Tada, Y.; Ueda, A.; Saneoka, H. Growth, physiological adaptation, and gene expression analysis of two Egyptian rice cultivars under salt stress. *Plant Physiol. Biochem.* **2015**, *87*, 17–25. [CrossRef]

101. Moon, J.Y.; Belloeil, C.; Ianna, M.L.; Shin, R. Arabidopsis CNGC Family Members Contribute to Heavy Metal Ion Uptake in Plants. *Int. J. Mol. Sci.* **2019**, *20*, 413. [CrossRef]

102. Very, A.-A.; Sentenac, H. Molecular mechanisms and regulation of K$^+$ transport in higher plants. *Ann. Rev. Plant Biol.* **2003**, *54*, 575–603. [CrossRef]

103. Schleyer, M.; Bakker, E.P. Nucleotide sequence and 3'-end deletion studies indicate that the K(+)-uptake protein kup from Escherichia coli is composed of a hydrophobic core linked to a large and partially essential hydrophilic C terminus. *J. Bacteriol.* **1993**, *175*, 6925–6931. [CrossRef]

104. Banuelos, M.; Klein, R.; Alexander-Bowman, S.; Rodríguez-Navarro, A. A potassium transporter of the yeast Schwanniomyces occidentalis homologous to the Kup system of Escherichia coli has a high concentrative capacity. *EMBO J.* **1995**, *14*, 3021–3027. [CrossRef]

105. Banuelos, M.A.; Garciadeblas, B.; Cubero, B.; Rodriguez-Navarro, A. Inventory and functional characterization of the HAK potassium transporters of rice. *Plant Physiol.* **2002**, *130*, 784–795. [CrossRef]

106. Fu, H.-H.; Luan, S. AtKUP1: A dual-affinity K^+ transporter from Arabidopsis. *Plant Cell* **1998**, *10*, 63–73.

107. Quintero, F.J.; Blatt, M.R. A new family of K^+ transporters from Arabidopsis that are conserved across phyla. *Febs Lett.* **1997**, *415*, 206–211. [CrossRef]

108. Rigas, S.; Debrosses, G.; Haralampidis, K.; Vicente-Agullo, F.; Feldmann, K.A.; Grabov, A.; Dolan, L.; Hatzopoulos, P. TRH1 encodes a potassium transporter required for tip growth in Arabidopsis root hairs. *Plant Cell* **2001**, *13*, 139–151. [CrossRef]

109. Durell, S.R.; Guy, H.R. Structural models of the KtrB, TrkH, and Trk1, 2 symporters based on the structure of the KcsA K^+ channel. *Biophys. J.* **1999**, *77*, 789–807. [CrossRef]

110. Laurie, S.; Feeney, K.A.; Maathuis, F.J.M.; Heard, P.J.; Brown, S.J.; Leigh, R.A. A role for HKT1 in sodium uptake by wheat roots. *Plant J.* **2002**, *32*, 139–149. [CrossRef]

111. Rodríguez-Navarro, A.; Rubio, F. High.-affinity potassium and sodium transport systems in plants. *J. Exp. Bot.* **2006**, *57*, 1149–1160. [CrossRef]

112. Locascio, A.; Andrés-Colás, N.; Mulet, J.M.; Yenush, L. Saccharomyces cerevisiae as a Tool to Investigate Plant. Potassium and Sodium Transporters. *Int. J. Mol. Sci.* **2019**, *20*, 2133. [CrossRef]

113. Tada, Y. The HKT Transporter Gene from Arabidopsis, AtHKT1; 1, Is Dominantly Expressed in Shoot Vascular Tissue and Root Tips and Is Mild Salt Stress-Responsive. *Plants* **2019**, *8*, 204. [CrossRef]

114. Garciadeblás, B.; Senn, M.E.; Bañuelos, M.A.; Rodríguez-Navarro, A. Sodium transport and HKT transporters: The rice model. *Plant J.* **2003**, *34*, 788–801. [CrossRef]

115. Horie, T.; Yoshida, K.; Nakayama, H.; Yamada, K.; Oiki, S.; Shinmyo, A. Two types of HKT transporters with different properties of Na^+ and K^+ transport in Oryza sativa. *Plant J.* **2001**, *27*, 129–138. [CrossRef]

116. Ren, Z.-H.; Gao, J.-P.; Li, L.-G.; Cai, X.-L.; Huang, W.; Chao, D.-Y.; Zhu, M.-Z.; Wang, Z.-Y.; Luan, S.; Lin, H.-X. A rice quantitative trait locus for salt tolerance encodes a sodium transporter. *Nat. Genet.* **2005**, *37*, 1141. [CrossRef]

117. Golldack, D.; Su, H.; Quigley, F.; Kamasani, U.R.; Muñoz-Garay, C.; Balderas, E.; Popova, O.V.; Bennett, J.; Bohnert, H.J.; Pantoja, O. Characterization of a HKT-type transporter in rice as a general alkali cation transporter. *Plant J.* **2002**, *31*, 529–542. [CrossRef]

118. Haruta, M.; Gray, W.M.; Sussman, M.R. Regulation of the plasma membrane proton pump (H^+-ATPase) by phosphorylation. *Curr. Opin. Plant Biol.* **2015**, *28*, 68–75. [CrossRef]

119. Chanroj, S.; Padmanaban, S.; Czerny, D.D.; Jauh, G.-Y.; Sze, H. K^+ transporter AtCHX17 with its hydrophilic C tail localizes to membranes of the secretory/endocytic system: Role in reproduction and seed set. *Mol. Plant* **2013**, *6*, 1226–1246. [CrossRef]

120. Blumwald, E.; Poole, R.J. Na^+/H^+ antiport in isolated tonoplast vesicles from storage tissue of Beta vulgaris. *Plant Physiol.* **1985**, *78*, 163–167. [CrossRef]

121. Barkla, B.J.; Blumwald, E. Identification of a 170-kDa protein associated with the vacuolar Na^+/H^+ antiport of Beta vulgaris. *Proc. Natl. Acad. Sci. USA* **1991**, *88*, 11177–11181. [CrossRef]

122. Garbarino, J.; DuPont, F.M. NaCl induces a Na^+/H^+ antiport in tonoplast vesicles from barley roots. *Plant Physiol.* **1988**, *86*, 231–236. [CrossRef]

123. Gaxiola, R.A.; Rao, R.; Sherman, A.; Grisafi, P.; Alper, S.L.; Fink, G.R. The Arabidopsis thaliana proton transporters, AtNhx1 and Avp1, can function in cation detoxification in yeast. *Proc. Natl. Acad. Sci. USA* **1999**, *96*, 1480–1485. [CrossRef] [PubMed]

124. Zhang, P.; Senge, M.; Dai, Y. Effects of salinity stress at different growth stages on tomato growth, yield, and water-use efficiency. *Commun. Soil Sci. Plant Anal.* **2017**, *48*, 624–634. [CrossRef]

125. Chakraborty, K.; Bose, J.; Shabala, L.; Shabala, S. Difference in root K^+ retention ability and reduced sensitivity of K^+-permeable channels to reactive oxygen species confer differential salt tolerance in three Brassica species. *J. Exp. Bot.* **2016**, *67*, 4611–4625. [CrossRef] [PubMed]

126. Nguyen, N.T.; Vu, H.T.; Nguyen, T.T.; Nguyen, L.-A.T.; Nguyen, M.-C.D.; Hoang, K.L.; Nguyen, K.T.; Quach, T.N. Co-expression of Arabidopsis AtAVP1 and AtNHX1 to Improve Salt Tolerance in Soybean. *Crop Sci.* **2019**, *59*, 1133–1143. [CrossRef]

127. Fukuda, A.; Nakamura, A.; Tagiri, A.; Tanaka, H.; Miyao, A.; Hirochika, H.; Tanaka, Y. Function, intracellular localization and the importance in salt tolerance of a vacuolar Na^+/H^+ antiporter from rice. *Plant Cell Physiol.* **2004**, *45*, 146–159. [CrossRef] [PubMed]

128. Padmanaban, S.; Czerny, D.D.; Levin, K.A.; Leydon, A.R.; Su, R.T.; Maugel, T.K.; Zou, Y.; Chanroj, S.; Cheung, A.Y.; Johnson, M.A. Transporters involved in pH and K^+ homeostasis affect pollen wall formation, male fertility, and embryo development. *J. Exp. Bot.* **2017**, *68*, 3165–3178. [CrossRef]

129. Maresova, L.; Sychrova, H. Arabidopsis thaliana CHX17 gene complements the kha1 deletion phenotypes in Saccharomyces cerevisiae. *Yeast* **2006**, *23*, 1167–1171. [CrossRef]

130. Song, C.-P.; Guo, Y.; Qiu, Q.; Lambert, G.; Galbraith, D.W.; Jagendorf, A.; Zhu, J.-K. A probable Na^+ $(K^+)/H^+$ exchanger on the chloroplast envelope functions in pH homeostasis and chloroplast development in Arabidopsis thaliana. *Proc. Natl. Acad. Sci. USA* **2004**, *101*, 10211–10216. [CrossRef]

131. Halfter, U.; Ishitani, M.; Zhu, J.-K. The Arabidopsis SOS_2 protein kinase physically interacts with and is activated by the calcium-binding protein SOS_3. *Proc. Natl. Acad. Sci. USA* **2000**, *97*, 3735–3740. [CrossRef]

132. Ren, X.L.; Qi, G.N.; Feng, H.Q.; Zhao, S.; Zhao, S.S.; Wang, Y.; Wu, W.H. Calcineurin B-like protein CBL 10 directly interacts with AKT 1 and modulates K^+ homeostasis in Arabidopsis. *Plant J.* **2013**, *74*, 258–266. [CrossRef]

133. Senadheera, P.; Singh, R.; Maathuis, F.J. Differentially expressed membrane transporters in rice roots may contribute to cultivar dependent salt tolerance. *J. Exp. Bot.* **2009**, *60*, 2553–2563. [CrossRef] [PubMed]

134. Uozumi, N.; Kim, E.J.; Rubio, F.; Yamaguchi, T.; Muto, S.; Tsuboi, A.; Bakker, E.P.; Nakamura, T.; Schroeder, J.I. The Arabidopsis HKT1 gene homolog mediates inward Na^+ currents in Xenopus laevis oocytes and Na^+ uptake in Saccharomyces cerevisiae. *Plant Physiol.* **2000**, *122*, 1249–1260. [CrossRef] [PubMed]

135. Rus, A.; Yokoi, S.; Sharkhuu, A.; Reddy, M.; Lee, B.-h.; Matsumoto, T.K.; Koiwa, H.; Zhu, J.-K.; Bressan, R.A.; Hasegawa, P.M. AtHKT1 is a salt tolerance determinant that controls Na^+ entry into plant roots. *Proc. Natl. Acad. Sci. USA* **2001**, *98*, 14150–14155. [CrossRef] [PubMed]

136. Oomen, R.J.; Benito, B.; Sentenac, H.; Rodríguez-Navarro, A.; Talón, M.; Véry, A.A.; Domingo, C. HKT2; 2/1, a K^+-permeable transporter identified in a salt-tolerant rice cultivar through surveys of natural genetic polymorphism. *Plant J.* **2012**, *71*, 750–762. [CrossRef] [PubMed]

137. Suzuki, K.; Costa, A.; Nakayama, H.; Katsuhara, M.; Shinmyo, A.; Horie, T. OsHKT2; 2/1-mediated Na^+ influx over K+ uptake in roots potentially increases toxic Na^+ accumulation in a salt-tolerant landrace of rice Nona Bokra upon salinity stress. *J. Plant Res.* **2016**, *129*, 67–77. [CrossRef] [PubMed]

138. Rodríguez-Rosales, M.P.; Gálvez, F.J.; Huertas, R.; Aranda, M.N.; Baghour, M.; Cagnac, O.; Venema, K. Plant NHX cation/proton antiporters. *Plant Signal. Behav.* **2009**, *4*, 265–276.

139. Zhang, H.-X.; Hodson, J.N.; Williams, J.P.; Blumwald, E. Engineering salt-tolerant Brassica plants: Characterization of yield and seed oil quality in transgenic plants with increased vacuolar sodium accumulation. *Proc. Natl. Acad. Sci. USA* **2001**, *98*, 12832–12836. [CrossRef]

140. Yin, X.-Y.; Yang, A.-F.; Zhang, K.-W.; Zhang, J.-R. Production and analysis of transgenic maize with improved salt tolerance by the introduction of AtNHX1 gene. *Acta Bot. Sin.-Engl. Ed.* **2004**, *46*, 854–861.

141. Ohta, M.; Hayashi, Y.; Nakashima, A.; Hamada, A.; Tanaka, A.; Nakamura, T.; Hayakawa, T. Introduction of a Na^+/H^+ antiporter gene from Atriplex gmelini confers salt tolerance to rice. *Febs Lett.* **2002**, *532*, 279–282. [CrossRef]

142. LÜ, S.Y.; JING, Y.X.; SHEN, S.H.; ZHAO, H.Y.; MA, L.Q.; ZHOU, X.J.; REN, Q.; LI, Y.F. Antiporter gene from Hordum brevisubulatum (Trin.) link and its overexpression in transgenic tobaccos. *J. Integr. Plant Biol.* **2005**, *47*, 343–349.

143. Zhang, H.-X.; Blumwald, E. Transgenic salt-tolerant tomato plants accumulate salt in foliage but not in fruit. *Nat. Biotechnol.* **2001**, *19*, 765. [CrossRef] [PubMed]

144. Xue, Z.-Y.; Zhi, D.-Y.; Xue, G.-P.; Zhang, H.; Zhao, Y.-X.; Xia, G.-M. Enhanced salt tolerance of transgenic wheat (*Tritivum aestivum* L.) expressing a vacuolar Na^+/H^+ antiporter gene with improved grain yields in saline soils in the field and a reduced level of leaf Na^+. *Plant Sci.* **2004**, *167*, 849–859. [CrossRef]

145. Shi, H.; Zhu, J.-K. SOS_4, a pyridoxal kinase gene, is required for root hair development in Arabidopsis. *Plant Physiol.* **2002**, *129*, 585–593. [CrossRef] [PubMed]

146. Yokoi, S.; Bressan, R.A.; Hasegawa, P.M. Salt stress tolerance of plants. *Jircas Work. Rep.* **2002**, *23*, 25–33.

147. Silva, P.; Gerós, H. Regulation by salt of vacuolar H^+-ATPase and H^+-pyrophosphatase activities and Na^+/H^+ exchange. *Plant Signal. Behav.* **2009**, *4*, 718–726. [CrossRef] [PubMed]

148. Hall, D.; Evans, A.; Newbury, H.; Pritchard, J. Functional analysis of CHX21: A putative sodium transporter in Arabidopsis. *J. Exp. Bot.* **2006**, *57*, 1201–1210. [CrossRef]

149. Shi, H.; Ishitani, M.; Kim, C.; Zhu, J.-K. The Arabidopsis thaliana salt tolerance gene SOS_1 encodes a putative Na^+/H^+ antiporter. *Proc. Natl. Acad. Sci. USA* **2000**, *97*, 6896–6901. [CrossRef]

150. Wegner, L.H.; De Boer, A.H. Properties of two outward-rectifying channels in root xylem parenchyma cells suggest a role in K^+ homeostasis and long-distance signaling. *Plant Physiol.* **1997**, *115*, 1707–1719. [CrossRef]

151. Lacan, D.; Durand, M. Na^+-K^+ exchange at the xylem/symplast boundary (its significance in the salt sensitivity of soybean). *Plant Physiol.* **1996**, *110*, 705–711. [CrossRef]

152. Berthomieu, P.; Conéjéro, G.; Nublat, A.; Brackenbury, W.J.; Lambert, C.; Savio, C.; Uozumi, N.; Oiki, S.; Yamada, K.; Cellier, F. Functional analysis of AtHKT1 in Arabidopsis shows that Na^+ recirculation by the phloem is crucial for salt tolerance. *EMBO J.* **2003**, *22*, 2004–2014. [CrossRef]

153. Davenport, R.J.; Muñoz-mayor, A.; Jha, D.; Essah, P.A.; Rus, A.; Tester, M. The Na^+ transporter AtHKT1; 1 controls retrieval of Na^+ from the xylem in Arabidopsis. *Plant Cell Environ.* **2007**, *30*, 497–507. [CrossRef]

154. Mäser, P.; Eckelman, B.; Vaidyanathan, R.; Horie, T.; Fairbairn, D.J.; Kubo, M.; Yamagami, M.; Yamaguchi, K.; Nishimura, M.; Uozumi, N. Altered shoot/root Na^+ distribution and bifurcating salt sensitivity in Arabidopsis by genetic disruption of the Na^+ transporter AtHKT1. *Febs Lett.* **2002**, *531*, 157–161.

155. Horie, T.; Hauser, F.; Schroeder, J.I. HKT transporter-mediated salinity resistance mechanisms in Arabidopsis and monocot crop plants. *Trends Plant Sci.* **2009**, *14*, 660–668. [CrossRef] [PubMed]

156. Zhang, W.-D.; Wang, P.; Bao, Z.; Ma, Q.; Duan, L.-J.; Bao, A.-K.; Zhang, J.-L.; Wang, S.-M. SOS_1, HKT1; 5, and NHX1 synergistically modulate Na^+ homeostasis in the halophytic grass Puccinellia tenuiflora. *Front. Plant Sci.* **2017**, *8*, 576.

157. Liu, Q.; Liu, R.; Ma, Y.; Song, J. Physiological and molecular evidence for Na^+ and Cl^- exclusion in the roots of two Suaeda salsa populations. *Aquat. Bot.* **2018**, *146*, 1–7. [CrossRef]

158. Xu, B.; Waters, S.; Byrt, C.S.; Plett, D.; Tyerman, S.D.; Tester, M.; Munns, R.; Hrmova, M.; Gilliham, M. Structural variations in wheat HKT1; 5 underpin differences in Na^+ transport capacity. *Cell. Mol. Life Sci.* **2018**, *75*, 1133–1144. [CrossRef] [PubMed]

159. Roy, S.; Chakraborty, U. Role of sodium ion transporters and osmotic adjustments in stress alleviation of Cynodon dactylon under NaCl treatment: A parallel investigation with rice. *Protoplasma* **2018**, *255*, 175–191. [CrossRef]

160. Zhu, M.; Zhou, M.; Shabala, L.; Shabala, S. Physiological and molecular mechanisms mediating xylem Na^+ loading in barley in the context of salinity stress tolerance. *Plant Cell Environ.* **2017**, *40*, 1009–1020. [CrossRef]

161. Batistic, O.; Kudla, J. Integration and channeling of calcium signaling through the CBL calcium sensor/CIPK protein kinase network. *Planta* **2004**, *219*, 915–924. [CrossRef]

162. Kudla, J.; Xu, Q.; Harter, K.; Gruissem, W.; Luan, S. Genes for calcineurin B-like proteins in Arabidopsis are differentially regulated by stress signals. *Proc. Natl. Acad. Sci. USA* **1999**, *96*, 4718–4723. [CrossRef]

163. Kolukisaoglu, Ü.; Weinl, S.; Blazevic, D.; Batistic, O.; Kudla, J. Calcium sensors and their interacting protein kinases: Genomics of the Arabidopsis and rice CBL-CIPK signaling networks. *Plant Physiol.* **2004**, *134*, 43–58. [CrossRef] [PubMed]

164. Zhang, H.; Yin, W.; Xia, X. Calcineurin B-Like family in Populus: Comparative genome analysis and expression pattern under cold, drought and salt stress treatment. *Plant Growth Regul.* **2008**, *56*, 129–140. [CrossRef]

165. Chen, X.-F.; Gu, Z.-M.; Feng, L.; Zhang, H.-S. Molecular analysis of rice CIPKs involved in both biotic and abiotic stress responses. *Rice Sci.* **2011**, *18*, 1–9. [CrossRef]

166. Weinl, S.; Kudla, J. The CBL–CIPK Ca^{2+}-decoding signaling network: Function and perspectives. *New Phytol.* **2009**, *184*, 517–528. [CrossRef] [PubMed]

167. Jung, H.-J.; Kayum, M.A.; Thamilarasan, S.K.; Nath, U.K.; Park, J.-I.; Chung, M.-Y.; Hur, Y.; Nou, I.-S. Molecular characterisation and expression profiling of calcineurin B-like (CBL) genes in Chinese cabbage under abiotic stresses. *Funct. Plant Biol.* **2017**, *44*, 739–750. [CrossRef]

168. Crozet, P.; Margalha, L.; Confraria, A.; Rodrigues, A.; Martinho, C.; Adamo, M.; Elias, C.A.; Baena-González, E. Mechanisms of regulation of SNF1/AMPK/SnRK1 protein kinases. *Front. Plant Sci.* **2014**, *5*, 190. [CrossRef]

169. Albrecht, V.; Ritz, O.; Linder, S.; Harter, K.; Kudla, J. The NAF domain defines a novel protein–protein interaction module conserved in Ca^{2+}-regulated kinases. *Embo J.* **2001**, *20*, 1051–1063. [CrossRef]

170. Guo, Y.; Halfter, U.; Ishitani, M.; Zhu, J.-K. Molecular characterization of functional domains in the protein kinase SOS_2 that is required for plant salt tolerance. *Plant Cell* **2001**, *13*, 1383–1400. [CrossRef]

171. Gong, D.; Guo, Y.; Jagendorf, A.T.; Zhu, J.-K. Biochemical characterization of the Arabidopsis protein kinase SOS_2 that functions in salt tolerance. *Plant Physiol.* **2002**, *130*, 256–264. [CrossRef]

172. Bhattacharyya, M.; Stratton, M.M.; Going, C.C.; McSpadden, E.D.; Huang, Y.; Susa, A.C.; Elleman, A.; Cao, Y.M.; Pappireddi, N.; Burkhardt, P. Molecular mechanism of activation-triggered subunit exchange in Ca^{2+}/calmodulin-dependent protein kinase II. *Elife* **2016**, *5*, e13405. [CrossRef]

173. Ohta, M.; Guo, Y.; Halfter, U.; Zhu, J.-K. A novel domain in the protein kinase SOS_2 mediates interaction with the protein phosphatase 2C ABI2. *Proc. Natl. Acad. Sci. USA* **2003**, *100*, 11771–11776. [CrossRef] [PubMed]

174. Sánchez-Barrena, M.J.; Fujii, H.; Angulo, I.; Martínez-Ripoll, M.; Zhu, J.-K.; Albert, A. The structure of the C-terminal domain of the protein kinase $AtSOS_2$ bound to the calcium sensor $AtSOS_3$. *Mol. Cell* **2007**, *26*, 427–435.

175. Akaboshi, M.; Hashimoto, H.; Ishida, H.; Saijo, S.; Koizumi, N.; Sato, M.; Shimizu, T. The crystal structure of plant-specific calcium-binding protein AtCBL2 in complex with the regulatory domain of AtCIPK14. *J. Mol. Biol.* **2008**, *377*, 246–257. [CrossRef] [PubMed]

176. Sánchez-Barrena, M.; Martínez-Ripoll, M.; Albert, A. Structural biology of a major signaling network that regulates plant abiotic stress: The CBL-CIPK mediated pathway. *Int. J. Mol. Sci.* **2013**, *14*, 5734–5749. [CrossRef] [PubMed]

177. Nagae, M.; Nozawa, A.; Koizumi, N.; Sano, H.; Hashimoto, H.; Sato, M.; Shimizu, T. The crystal structure of the novel calcium-binding protein AtCBL2 from Arabidopsis thaliana. *J. Biol. Chem.* **2003**, *278*, 42240–42246. [CrossRef]

178. Sánchez-Barrena, M.J.; Martínez-Ripoll, M.; Zhu, J.-K.; Albert, A. The structure of the Arabidopsis thaliana SOS_3: Molecular mechanism of sensing calcium for salt stress response. *J. Mol. Biol.* **2005**, *345*, 1253–1264. [CrossRef]

179. Mao, J.; Manik, S.; Shi, S.; Chao, J.; Jin, Y.; Wang, Q.; Liu, H. Mechanisms and physiological roles of the CBL-CIPK networking system in Arabidopsis thaliana. *Genes* **2016**, *7*, 62. [CrossRef]

180. Qiu, Q.-S.; Guo, Y.; Dietrich, M.A.; Schumaker, K.S.; Zhu, J.-K. Regulation of SOS_1, a plasma membrane Na^+/H^+ exchanger in Arabidopsis thaliana, by SOS_2 and SOS_3. *Proc. Natl. Acad. Sci. USA* **2002**, *99*, 8436–8441. [CrossRef]

181. Zhu, J.-K.; Liu, J.; Xiong, L. Genetic analysis of salt tolerance in Arabidopsis: Evidence for a critical role of potassium nutrition. *Plant Cell* **1998**, *10*, 1181–1191. [CrossRef]

182. Li, M.O.; Sanjabi, S.; Flavell, R.A. Transforming growth factor-β controls development, homeostasis, and tolerance of T cells by regulatory T cell-dependent and-independent mechanisms. *Immunity* **2006**, *25*, 455–471. [CrossRef]

183. Tang, R.-J.; Zhao, F.-G.; Garcia, V.J.; Kleist, T.J.; Yang, L.; Zhang, H.-X.; Luan, S. Tonoplast CBL–CIPK calcium signaling network regulates magnesium homeostasis in Arabidopsis. *Proc. Natl. Acad. Sci. USA* **2015**, *112*, 3134–3139. [CrossRef] [PubMed]

184. Ho, C.-H.; Lin, S.-H.; Hu, H.-C.; Tsay, Y.-F. CHL1 functions as a nitrate sensor in plants. *Cell* **2009**, *138*, 1184–1194. [CrossRef] [PubMed]

185. Fuglsang, A.T.; Guo, Y.; Cuin, T.A.; Qiu, Q.; Song, C.; Kristiansen, K.A.; Bych, K.; Schulz, A.; Shabala, S.; Schumaker, K.S. Arabidopsis protein kinase PKS5 inhibits the plasma membrane H+-ATPase by preventing interaction with 14-3-3 protein. *Plant Cell* **2007**, *19*, 1617–1634. [CrossRef] [PubMed]

186. Manik, S.; Shi, S.; Mao, J.; Dong, L.; Su, Y.; Wang, Q.; Liu, H. The calcium sensor CBL-CIPK is involved in plant's response to abiotic stresses. *Int. J. Genom.* **2015**, *2015*, 1–10. [CrossRef] [PubMed]

187. Thoday-Kennedy, E.L.; Jacobs, A.K.; Roy, S.J. The role of the CBL–CIPK calcium signalling network in regulating ion transport in response to abiotic stress. *Plant Growth Regul.* **2015**, *76*, 3–12. [CrossRef]

188. Cho, J.H.; Lee, J.H.; Park, Y.K.; Choi, M.N.; Kim, K.-N. Calcineurin B-like protein CBL10 directly interacts with TOC34 (Translocon of the Outer membrane of the Chloroplasts) and decreases its GTPase activity in Arabidopsis. *Front. Plant Sci.* **2016**, *7*, 1911. [CrossRef]

189. de la Torre, F.; Gutiérrez-Beltrán, E.; Pareja-Jaime, Y.; Chakravarthy, S.; Martin, G.B.; del Pozo, O. The tomato calcium sensor Cbl10 and its interacting protein kinase Cipk6 define a signaling pathway in plant immunity. *Plant Cell* **2013**, *25*, 2748–2764. [CrossRef]

190. Cheong, Y.H.; Pandey, G.K.; Grant, J.J.; Batistic, O.; Li, L.; Kim, B.G.; Lee, S.C.; Kudla, J.; Luan, S. Two calcineurin B-like calcium sensors, interacting with protein kinase CIPK23, regulate leaf transpiration and root potassium uptake in Arabidopsis. *Plant J.* **2007**, *52*, 223–239. [CrossRef]

191. Li, L.; Kim, B.G.; Cheong, Y.H.; Pandey, G.K.; Luan, S. A Ca^{2+} signaling pathway regulates a K(+) channel for low-K response in Arabidopsis. *Proc. Natl. Acad. Sci. USA* **2006**, *103*, 12625–12630. [CrossRef]

192. D'Angelo, C.; Weinl, S.; Batistic, O.; Pandey, G.K.; Cheong, Y.H.; Schültke, S.; Albrecht, V.; Ehlert, B.; Schulz, B.; Harter, K.; et al. Alternative complex formation of the Ca^{2+}-regulated protein kinase CIPK1 controls abscisic acid-dependent and independent stress responses in Arabidopsis. *Plant J.* **2006**, *48*, 857–872.

193. Hashimoto, K.; Eckert, C.; Anschutz, U.; Scholz, M.; Held, K.; Waadt, R.; Reyer, A.; Hippler, M.; Becker, D.; Kudla, J. Phosphorylation of calcineurin B-like (CBL) calcium sensor proteins by their CBL-interacting protein kinases (CIPKs) is required for full activity of CBL-CIPK complexes toward their target proteins. *J. Biol. Chem.* **2012**, *287*, 7956–7968. [CrossRef] [PubMed]

194. Kudla, J.; Batistič, O.; Hashimoto, K. Calcium signals: The lead currency of plant information processing. *Plant Cell* **2010**, *22*, 541–563. [CrossRef] [PubMed]

195. Quintero, F.J.; Martinez-Atienza, J.; Villalta, I.; Jiang, X.; Kim, W.-Y.; Ali, Z.; Fujii, H.; Mendoza, I.; Yun, D.-J.; Zhu, J.-K. Activation of the plasma membrane Na/H antiporter Salt-Overly-Sensitive 1 (SOS_1) by phosphorylation of an auto-inhibitory C-terminal domain. *Proc. Natl. Acad. Sci. USA* **2011**, *108*, 2611–2616. [CrossRef] [PubMed]

196. Luo, Q.; Wei, Q.; Wang, R.; Zhang, Y.; Zhang, F.; He, Y.; Zhou, S.; Feng, J.; Yang, G.; He, G. BdCIPK31, a calcineurin B-like protein-interacting protein kinase, regulates plant response to drought and salt stress. *Front. Plant Sci.* **2017**, *8*, 1184. [CrossRef]

197. Deng, X.; Zhou, S.; Hu, W.; Feng, J.; Zhang, F.; Chen, L.; Huang, C.; Luo, Q.; He, Y.; Yang, G. Ectopic expression of wheat TaCIPK14, encoding a calcineurin B-like protein-interacting protein kinase, confers salinity and cold tolerance in tobacco. *Physiol. Plant.* **2013**, *149*, 367–377.

198. Jin, X.; Sun, T.; Wang, X.; Su, P.; Ma, J.; He, G.; Yang, G. Wheat CBL-interacting protein kinase 25 negatively regulates salt tolerance in transgenic wheat. *Sci. Rep.* **2016**, *6*, 28884. [CrossRef]

6

Sodium Azide Priming Enhances Waterlogging Stress Tolerance in Okra (*Abelmoschus esculentus* L.)

Emuejevoke D. Vwioko [1], Mohamed A. El-Esawi [2,3,*], Marcus E. Imoni [1],
Abdullah A. Al-Ghamdi [4], Hayssam M. Ali [4], Mostafa M. El-Sheekh [2], Emad A. Abdeldaym [5]
and Monerah A. Al-Dosary [4]

[1]   Department of Plant Biotechnology, Faculty of Life Sciences, University of Benin, P.O. Box 1154, Benin City,
      Nigeria; emuejevoke.vwioko@yahoo.com (E.D.V.); marcus.imoni@yahoo.com (M.E.I.)
[2]   Botany Department, Faculty of Science, Tanta University, Tanta 31527, Egypt;
      mostafaelsheikh@science.tanta.edu.eg
[3]   Sainsbury Laboratory, University of Cambridge, Cambridge CB2 1LR, UK
[4]   Botany and Microbiology Department, College of Science, King Saud University, P.O. Box 2455,
      Riyadh 11451, Saudi Arabia; abdaalghamdi@ksu.edu.sa (A.A.A.-G.); hayhassan@ksu.edu.sa (H.M.A.);
      almonerah@ksu.edu.sa (M.A.A.-D.)
[5]   Vegetable Crops Department, Faculty of Agriculture, Cairo University, Giza P.O. Box 12613, Egypt;
      emad.abdeldaym@agr.cu.edu.eg
*   Correspondence: mohamed.elesawi@science.tanta.edu.eg

Abstract: Waterlogging stress adversely affects crop growth and yield worldwide. Effect of sodium azide priming on waterlogging stress tolerance of okra plants was investigated. The study was conducted as a field experiment using two weeks old plants grown from 0%, 0.02%, and 0.05% sodium azide (NaN$_3$)-treated seeds. The waterlogging conditions applied were categorized into control, one week, and two weeks. Different growth and reproductive parameters were investigated. Activity and expression of antioxidant enzymes, root anatomy, and soil chemical analysis were also studied. Results showed that sodium azide priming inhibited germination. The germination percentages recorded were 92.50, 85.00, and 65.00 for 0%, 0.02%, and 0.05% NaN$_3$-treated seeds, respectively, nine days after planting. Waterlogging conditions depressed plant height ten weeks after planting. Under waterlogging conditions, NaN$_3$ promoted plant height and number of leaves formed. NaN$_3$ also supported the survival of plants and formation of adventitious roots under waterlogging conditions. Waterlogging conditions negatively affected the redox potential, organic C, N, and P concentrations in the soil but enhanced Soil pH, Fe, Mn, Zn, and SO$_4$. Under waterlogging conditions, NaN$_3$ increased the average number of flower buds, flowers, and fruits produced in comparison to control. Moreover, NaN$_3$ highly stimulated the development of aerenchyma which in turn enhanced the survival of okra plants under waterlogging conditions. NaN$_3$ priming also enhanced the activities and gene expression level of antioxidant enzymes (ascorbate peroxidase, APX; catalase, CAT) under waterlogging conditions. In conclusion, this study demonstrated that NaN$_3$ priming could improve waterlogging stress tolerance in okra.

Keywords: sodium azide; okra; waterlogging stress; antioxidants; gene expression

1. Introduction

Okra (*Abelmoschus esculentus* L.) is one of the economically important vegetable crops grown in tropical and sub-tropical regions of the world [1]. Okra originated in Ethiopia and was then reproduced in the Mediterranean area, North Africa, and India [1]. Environmental stresses negatively affect the growth, yield, and biological activities of plants worldwide [2–6]. In particular, waterlogging conditions influence the growth and yield of okra plants through causing hypoxic or anoxic conditions, which in turn affect various physiological processes in roots, including carbohydrate metabolism, gas exchanges, and water relations [7–9]. The oxygen-deficient soil environments may lead to changes in the composition and decomposition activities of microbes. Waterlogging conditions also affect soil factors such as EC, pH, soil structure, hydraulic conductance, porosity, and organics [10,11]. Plants could adapt to waterlogging conditions via activating their self-defense mechanisms and developing adventitious roots and hypertrophied stem bases with lenticels and aerenchyma cells [7,12]. Such aerenchyma cells could enhance organ porosity and root aeration [13,14]. These morphological features help plants to manage the low oxygen tension within the tissues, prevent anoxia, and maintain root functions and plant survival.

Applications of chemicals to plants, either as foliar or seed treatments, may induce their physiological mechanisms, leading to plant growth stimulation and stress tolerance [7,15,16]. For instance, seed pretreatment with salicylic acid enhances plant growth, antioxidant activities, and tolerance to harsh environmental factors such as heavy metal, herbicides, low temperature and salt stress [17,18]. Ethylene is also described as a signaling molecule in plants and has been projected as capable of inducing survival traits and tolerance under waterlogging conditions via up-regulating the activity of antioxidant enzymes and genes linked to aerenchyma formation, leaf senescence, adventitious roots, and epinasty [7,19–21]. However, ethylene application as a proactive measure for ameliorating envisaged waterlogging condition on a wide scale may not be appreciated. Hence, seed priming techniques may be easier to enhance growth and yield. Sodium azide (NaN_3) has been successfully used for creating genetic variability and enhancing agronomic traits of crop plants. It affects crops based on the concentration applied. Gnanamurthy et al. [22] and Shagufta et al. [23] reported that NaN_3 priming delayed and inhibited the germination of maize and fenugreek, respectively. However, Vwioko and Onobun [24] reported that NaN_3 enhanced the germination percentage and height of okra plants. Al-Qurainy [25] and Zuzana et al. [26] also stated that NaN_3 stimulated the plant height of *Eruca sativa* and *Diospyros lotus*, respectively. On the other hand, Adamu and Aliyu [27] and Gnanamurthy et al. [22] revealed that NaN_3 priming inhibited plant height. NaN_3 priming also regulates various physiological and molecular mechanisms in plants and modulates the activities of catalase, peroxidase, and cytochrome oxidase [28]. Molecular changes induced by NaN_3 treatments produce mutations by base substitution, leading to changes in amino acid sequences. NaN_3 is reckoned to be an efficient reagent that induces a broad and high variation of morphological and yield parameters in cultivated species. However, it is not popularly used to initiate plants tolerance to environmental factors. Environmental stresses such as salinity and water stress [29,30] increase production of free radical in plants, and resistance to the unfavorable conditions often involves stimulation of the antioxidant response. Haq et al. [31], El Kaaby et al. [32], and Kuasha et al. [33] carried out in vitro studies on the ability of NaN_3 to confer salt tolerance in plants. Haq et al. [31] stated that one of the three cultivars of sugarcane studied regenerated plantlets that were salt tolerant, while El Kaaby et al. [32] and Kuasha et al. [33] stated that NaN_3 depressed the responses of the explants of tomato and sugarcane to salinity stress. Salim et al. [34] also studied the effect of NaN_3 on various plant traits, including disease resistance, yield, antioxidant activities, pigmentations, and salinity and drought stress tolerance. However, the role of NaN_3 in regulating waterlogging stress responses has not been studied yet. Therefore, the main aim of the present study was to assess the ability of NaN_3 to induce waterlogging stress tolerance in okra plants.

2. Materials and Methods

2.1. Plant Material and Application of Sodium Azide Treatments

Seeds of okra variety Clemson spineless produced by Technism (Longué-Jumelles, France) were obtained and used in this study. Okra seeds were soaked in sodium azide treatments, i.e., 0%, 0.02%, and 0.05% (*w/v*), at room temperature (27 °C) for 5 h with a continuous gentle stirring. After 5 h, the seeds were removed and washed 5 times with deionized water to remove all traces of NaN_3. NaN_3 treatments were classed as mild (0.02%) and severe (0.05%).

2.2. Soil Preparation for Potted Field Experiment

Top soil (0–15 cm deep) was collected from the Demonstration Farm, Faculty of Agriculture, University of Benin, Nigeria. The soil type is categorized as ultisol. The composite soil sample was air-dried for three weeks and sieved to remove gravel and other particles. Each experimental pot was filled with 5 kg of soil. Thirty-six (36) pots were prepared to make twelve pots for each NaN_3 treatment. The undersides of the experimental pots were not perforated so that they could retain water.

2.3. Sowing of Seeds in Nursery Beds, Transplanting into Experimental Pots, and Acclimatization

Twelve soil nursery beds (measuring 2 feet by 2 feet) were prepared for the sowing seeds. The beds were allocated to the treated seeds, i.e., 0%, 0.02%, and 0.05% NaN_3. The seeds were sown at a depth of 2–3 cm. Germination records were collected every day for two weeks. After two weeks in the nursery beds, four plants were transferred into each experimental pot and taken to the open field. The plants were allowed to acclimatize for another two weeks in the field before flooding condition was introduced.

2.4. Application of Flooding or Waterlogging Conditions

When the plants were four weeks old, flooding of experimental pots with tap water was carried out. Three conditions of flooding or waterlogging were set up; no flooding (NF), one-week flooding (1 WF), and two weeks flooding (2 WF). Flooding of the pots was done up to 2 cm mark above the soil level. The water level was maintained in each pot by topping daily after inspection during the period.

2.5. Growth Parameters Measured

The field data collected were germination percentage, stem girth, plant height, number of leaves formed, survival percentage of plants, number of adventitious roots formed, number of flower buds formed, number of flowers, and number of fruits produced.

2.6. Soil Chemical Analyses

Soil chemical factors like pH, electrolyte conductivity (EC), redox potential (Eh), nitrogen, phosphorus, sulphate, organic carbon, iron, manganese, zinc, and total soluble phenolics were determined using standard methods. The soil analysis was carried out for the soil samples collected after plant harvested. pH, EC, and Eh were estimated in a soil-water slurry (ratio 1:3) [35]. Total nitrogen was estimated following Kjeldahl method [36]. Total soluble phenolic analysis was done based on the modified citrate extraction protocol followed by Folin–Ciocalteau colorimetric methodology [37]. The methodologies of Appiah and Ahenkorah [38] and Ben Mussa et al. [39] were used to determine sulphate content. Phosphorus measurement was conducted following the methodology of Bray and Kurtz [40]. Walkley–Black chromic acid wet oxidation methodology [41] was used to estimate the organic carbon. Iron content was determined following the hydroxylamine and 1,10- phenanthroline protocol [42]. Manganese was determined following the permanganate oxidation procedures [42]. The determination of zinc was carried using atomic absorption spectrophotometer (Shimadzu Europa GmbH, Duisburg, Germany).

2.7. Soil Microflora Counts

Presence of bacteria and fungi in the soil samples was investigated after plant harvest. Serial dilution processes were used in the analysis of soil microflora. Ten grams of the samples were dispensed into sterile beakers and mixed thoroughly with 90 mL sterile distilled water. Each sample was serially diluted from the stock sample and then transferred to the first tube 9 mL of sterilized water to give 10^{-1} dilution, from which further dilution up to 10^{-4} was made. The pour plate method was utilized for inoculation on a sterilized nutrient agar (NA) or potato dextrose agar (PDA), impregnated with antifungal or antibacterial agents for the growth of bacterial or fungal isolates, respectively. Nutrient agar plates were kept for 24–48 hrs at 37 °C for bacterial growth. Potato dextrose agar was incubated at room temperature (30 ± 2 °C) for 3–5 days. Total viable colonies were then counted for the microbial isolates and represented in terms of colony forming units (cfu/g). Viable counts obtained were recorded with reference to the serial dilution used [43,44].

2.8. Root Anatomy

Harvested plant roots were washed and used to make microscopic slides to examine internal tissues. Root sections were immersed in paraffin wax and left to solidify. Sections were cut and dewaxed by clamping in the microtome. Aniline blue stain was applied to the sections to show a clear contrast of air spaces (aerenchyma) formed. Excess stains were removed by ethanol before oven-drying. Following oven-drying, slides were viewed and then photographed using the microscope IRMECO model IM-660 T1 (IRMECO GmbH & Co. KG, Geesthacht, Germany) with a camera connected to PC. Observations were done under X10 objective lens.

2.9. Antioxidant Enzyme Assays

Activities of catalase (CAT) and ascorbate peroxidase (APX) were determined in the leafy tissue of the NF, 1 WF, and 2 WF plants treated with 0%, 0.02%, and 0.05% NaN_3 collected at the tenth week after planting following the method of Zhang and Kirkham [45]. In brief, 0.25g of leafy tissue was homogenized in 3 mL of solution, composed of PBS (50 mM), EDTA (0.2 mM), and 1% PVP, and centrifuged. Supernatants were assayed to detect the absorbance at 290 nm (for APX) and 240 nm (for CAT).

2.10. RNA Isolation, cDNA Synthesis, and Quantitative RT-PCR

Quantitative real-time PCR (qRT-PCR) assay was conducted to evaluate the expression level of antioxidant enzyme-encoding genes (APX, CAT) in the leafy tissue of the NF, 1 WF, and 2 WF plants treated with 0%, 0.02%, and 0.05% NaN_3 collected at the tenth week after planting. Total RNA samples were isolated from the tissue following Qiagen RNeasy Plant Mini kit. DNA removal and cDNA synthesis were performed using Qiagen RNase-Free DNase Set and Qiagen Reverse Transcription kit, respectively. Quantitative RT-PCR was performed following Qiagen QuantiTect SYBR Green PCR kit protocol. PCR conditions, housekeeping gene, and gene-specific primers were used as reported by Vwioko et al. [7]. The primer pair 5'-TGCCCTTCTATTGTGGTTCC-3' and 5'-GATGAGCACACTTTGGAGGA-3' was used for CAT amplification, whereas the primer pair 5'-ACCAATTGGCTGGTGTTGTT-3' and 5'-TCACAAACACGTCCCTCAAA-3' was used for APX amplification. The primer pair 5'-TTCCTTGATGATGCTTGCTC-3' and 5'-TTGACAGCTCTTGGGTGAAG-3' was used for the housekeeping gene (UBQ1) amplification.

2.11. Statistical Analysis

Mean and standard deviation were measured for the data obtained for the different traits measured. Two-way analysis of variance was conducted using NaN_3 treatments and flooding conditions as factors. Tukey's test was conducted to determine the significance of values. Statistical analyses were performed using SPSS ver. 19 (SPSS Inc., Chicago, IL, USA).

3. Results

3.1. Germination of NaN3-Treated Seeds

The germination was first recorded for okra seeds given control (0%) treatments 2 days after planting (2 DAP). Germination was recorded for 0.02 and 0.05% NaN_3-treated seeds three days after planting (3 DAP). Eight days after planting (8 DAP), the highest and least percentage of germination were recorded for 0% and 0.05% NaN₃-treated seeds, respectively (Figure 1). Twenty-four hours delay in germination was recorded for the NaN₃-treated seeds.

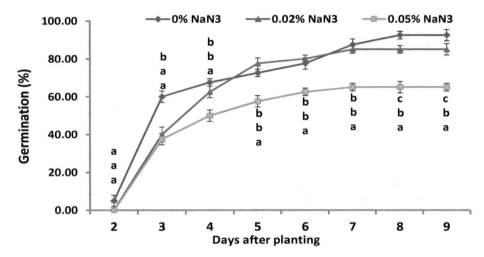

Figure 1. Percentage of germination of NaN₃-treated okra seeds sown in nursery. Values = mean ± SD, $n = 4$. Mean values with similar letters at the same day after planting (DAP) are not significantly different at $p \leq 0.05$.

3.2. Plant Height

Values obtained for plant height showed that non-waterlogged plants produced the highest values irrespective of the NaN₃ treatment given to the seeds ten weeks after planting (10 WAP). For example, mean values obtained for plant height were 31.5, 29.5, and 31.1 cm for 0%, 0.02%, and 0.05%, respectively, under non-waterlogging condition, 10 WAP (Table 1). Under one-week waterlogging condition, the values recorded for 0%, 0.02%, and 0.05% were 15.2, 21.8, and 19.4 cm, respectively, 10 WAP. Similarly, under two weeks waterlogging conditions, the values recorded for 0%, 0.02%, and 0.05% were 16.3, 22.4, and 19.9 cm, respectively, 10 WAP, indicating growth stimulations for plants grown from 0.02% and 0.05% NaN₃-treated seeds.

Table 1. Height (cm) of okra plants grown from NaN₃-treated seeds subjected to different waterlogging conditions four weeks after planting (WAP).

NaN₃ Treatment	Waterlogging Conditions	2 WAP	4 WAP	6 WAP	8 WAP	10 WAP
0%	Non-waterlogging	8.6 [b] ± 0.45	14.7 [e] ± 0.78	18.8 [a] ± 0.28	23.7 [a] ± 1.19	31.5 [a] ± 1.28
	One-week waterlogging	7.7 [c] ± 0.47	17.2 [ab] ± 0.68	17.9 [abc] ± 0.43	18.3 [c] ± 0.25	15.2 [c] ± 0.62
	Two weeks waterlogging	8.3 [b] ± 0.30	15.8 [cd] ± 0.24	16.8 [d] ± 0.94	18.9 [bc] ± 1.05	19.3 [b] ± 0.29
0.02%	Non-waterlogging	9.5 [a] ± 0.62	12.0 [f] ± 0.30	18.0 [abc] ± 0.60	24.5 [a] ± 1.31	29.5 [ab] ± 0.68
	One-week waterlogging	8.3 [bc] ± 0.45	17.9 [ab] ± 0.66	18.4 [ab] ± 0.42	18.8 [bc] ± 0.47	21.8 [d] ± 0.35
	Two weeks waterlogging	7.9 [bc] ± 0.09	17.7 [ab] ± 0.91	18.5 [ab] ± 1.23	19.7 [b] ± 1.30	22.4 [bc] ± 0.45
0.05%	Non-waterlogging	8.1 [bc] ± 0.78	11.9 [f] ± 0.83	17.6 [bcd] ± 0.49	23.5 [a] ± 0.88	31.1 [a] ± 0.98
	One-week waterlogging	7.9 [bc] ± 0.21	17.2 [ab] ± 0.48	17.7 [bcd] ± 0.63	18.0 [c] ± 0.72	19.4 [c] ± 1.60
	Two weeks waterlogging	7.8 [bc] ± 0.22	16.1 [bc] ± 0.28	17.1 [cd] ± 0.20	18.1 [c] ± 0.71	19.9 [c] ± 0.34

Values = mean ± S.D., $n = 4$, WAP = weeks after planting. Mean values with similar letters as superscript in one column are not significantly different at $p \leq 0.05$.

3.3. Stem Girth

The highest stem girth values were obtained for okra plants grown under non-waterlogging conditions (Table 2). Ten weeks after planting, the values recorded for the stem girth of okra plants grown under two-week waterlogging conditions were statistically significant compared to those recorded for plants grown under and non-waterlogging conditions (Table 2).

Table 2. Stem girth (cm) of okra plants grown from NaN_3-treated seeds subjected to different waterlogging conditions four weeks after planting (WAP).

NaN_3 Treatment	Waterlogging Conditions	2 WAP	4 WAP	6 WAP	8 WAP	10 WAP
0%	Non-waterlogging	0.81 [a] ± 0.02	0.95 [c] ± 0.05	1.10 [b] ± 0.08	1.25 [a] ± 0.05	1.35 [a] ± 0.05
	One-week waterlogging	0.80 [a] ± 0.01	1.02 [b] ± 0.09	1.07 [b] ± 0.09	1.15 [b] ± 0.05	1.27 [b] ± 0.05
	Two weeks waterlogging	0.85 [a] ± 0.05	1.17 [a] ± 0.05	1.27 [a] ± 0.05	1.27 [a] ± 0.05	1.27 [b] ± 0.05
0.02%	Non-waterlogging	0.76 [b] ± 0.05	1.02 [b] ± 0.09	1.10 [c] ± 0.08	1.25 [a] ± 0.05	1.37 [a] ± 0.05
	One-week waterlogging	0.89 [a] ± 0.09	1.12 [a] ± 0.05	1.17 [b] ± 0.05	1.25 [a] ± 0.05	1.37 [a] ± 0.05
	Two weeks waterlogging	0.80 [b] ± 0.08	1.15 [a] ± 0.05	1.27 [a] ± 0.05	1.27 [a] ± 0.05	1.30 [b] ± 0.00
0.05%	Non-waterlogging	0.75 [b] ± 0.05	1.05 [b] ± 0.05	1.17 [b] ± 0.05	1.30 [a] ± 0.08	1.45 [a] ± 0.05
	One-week waterlogging	0.82 [a] ± 0.07	1.07 [b] ± 0.05	1.20 [b] ± 0.08	1.20 [b] ± 0.08	1.32 [b] ± 0.09
	Two weeks waterlogging	0.85 [a] ± 0.05	1.17 [a] ± 0.09	1.27 [a] ± 0.05	1.27 [a] ± 0.05	1.32 [b] ± 0.05

Values = mean ± S.D., $n = 4$, WAP = weeks after planting. Mean values with similar letters as superscript in one column are not significantly different at $p \leq 0.05$.

3.4. Number of Leaves Formed, Number of Adventitious Roots Produced, and Percentage of Survival of Plants

The total number of leaves formed per plant recorded indicated that the plants grown under non-waterlogging condition produced the highest number of leaves 10 WAP. The combination of waterlogging conditions and NaN_3 treatments gave higher values for number of leaves formed than when the waterlogging condition is applied only (Table 3). For example, total number of leaves under non-waterlogging conditions were 16, 16.5, and 16.5 for 0%, 0.02%, and 0.05%, respectively. Whereas in one-week waterlogging conditions, values were 13, 14, and 15 for plants grown from 0%, 0.02%, and 0.05% NaN_3-treated seeds.

Table 3. Number of leaves, average number of adventitious roots produced, and survival percentage of okra plants grown from NaN_3-treated seeds under waterlogging conditions 10 WAP.

NaN_3 Treatment	Waterlogging Conditions	No. Leaves per Plant	No. Adventitious Roots per Plant	Survival Percentage
0%	Non-waterlogging	16.0 [a] ± 2.30	0 [c]	100.0 [a] ± 0.00
	One-week waterlogging	13.0 [b] ± 1.10	10.7 [b] ± 7.18	33.3 [b] ± 27.22
	Two weeks waterlogging	12.0 [b] ± 0.00	13.0 [a] ± 8.67	25.0 [b] ± 16.67
0.02%	Non- waterlogging	16.5 [a] ± 1.00	0 [c]	100.0 [a] ± 0.00
	One-week waterlogging	14.0 [b] ± 1.60	15.5 [b] ± 1.29	50.0 [b] ± 19.25
	Two weeks waterlogging	13.0 [b] ± 1.15	21.0 [a] ± 0.81	33.3 [c] ± 0.00
0.05%	Non- waterlogging	16.5 [a] ± 1.00	0 [c]	100.0 [a] ± 0.00
	One-week waterlogging	15.0 [b] ± 1.15	18.0 [b] ± 1.63	50.0 [b] ± 19.25
	Two weeks waterlogging	13.0 [c] ± 1.15	22.2 [a] ± 1.25	50.0 [b] ± 19.25

Values = mean ± S.D., $n = 4$, WAP = weeks after planting. Mean values with similar letters as superscript in one column are not significant different at $p \leq 0.05$.

Plants did not form adventitious roots under non-waterlogging conditions. However, the production of adventitious roots was observed in plants subjected to waterlogging condition. Plants subjected to two weeks of waterlogging condition initiated higher numbers of adventitious roots (Table 3). Furthermore, plants grown from 0.05% NaN_3-treated seeds produced the highest number of adventitious roots recorded. The combination of NaN_3 concentration and waterlogging condition supported the greater production of adventitious roots in okra.

Ten weeks after planting, the number of plants that survived the waterlogging conditions is shown in Table 3. Higher percentage of survival was recorded with the combination of sodium azide and waterlogging condition. For example, under two weeks waterlogging condition, the percentage of survival of okra plants were 25, 33.3, and 50 for 0%, 0.02%, and 0.05% NaN_3-treated seeds, respectively. Similarly, for one-week waterlogging condition, percentage of survival of okra plants were 33.3, 50, and 50 for 0%, 0.02%, and 0.05% NaN_3-treated seeds, respectively.

3.5. Number of Flower Buds, Flowers, and Fruits Produced

The number of flower buds, flowers, and fruits are shown in Table 4. The waterlogging condition caused a decrease in all the reproductive parameters considered. For example, the average number of flower buds recorded for plants grown from control seeds (0.00% NaN_3 treatment) were 5.5, 2.75, and 1.75 for NF, 1 WF, and 2 WF conditions, respectively. Similarly, average number of flowers recorded for the same plants were 5, 2, and 1, respectively. Moreover, the average number of fruits recorded for the same plants were 4.5, 1.25, and 0.5, respectively. The average number of flower buds, flowers, and fruits recorded for plants grown from 0.05% NaN_3-treated seeds and subjected to waterlogging conditions were higher than those recorded for non-treated plants.

Table 4. Average number of flower buds, flowers, and fruits formed per plant of okra grown from NaN3 treated seeds subjected to waterlogging conditions ten weeks after planting.

NaN$_3$ Treatment	Waterlogging Conditions	Number of Flower Buds	Number of Flowers	Number of Fruits
0%	Non-waterlogging	5.5 [a] ± 0.57	5.0 [a] ± 0.81	4.5 [a] ± 0.57
	One-week waterlogging	2.7 [b] ± 1.25	2.0 [b] ± 1.41	1.2 [b] ± 0.95
	Two weeks waterlogging	1.7 [b] ± 1.25	1.0 [b] ± 0.81	0.5 [b] ± 0.57
0.02%	Non-waterlogging	5.0 [a] ± 0.81	5.0 [a] ± 0.81	3.5 [a] ± 1.29
	One-week waterlogging	2.5 [b] ± 0.57	1.7 [b] ± 0.50	1.5 [b] ± 1.00
	Two weeks waterlogging	1.7 [b] ± 0.5	1.2 [b] ± 0.95	1.2 [b] ± 0.95
0.05%	Non-waterlogging	5.0 [a] ± 0.81	4.5 [a] ± 1.29	3.7 [a] ± 1.89
	One-week waterlogging	3.2 [b] ± 0.95	2.5 [b] ± 0.57	2.2 [b] ± 0.91
	Two weeks waterlogging	2.2 [b] ± 0.95	1.5 [b] ± 0.57	1.5 [b] ± 0.57

Values = mean ± S.D., $n = 4$. Values with similar letters as superscript are not significantly different.

3.6. Soil Microflora Counts

The average values obtained for bacteria and fungi counts are shown in Table S1. The bacterial counts were higher than fungal counts in all soil samples analyzed. The bacterial count values were higher in soils collected from waterlogging condition, while the fungal count values were higher in soils collected from non-waterlogging condition. Soils collected from two-week waterlogging conditions gave the least fungal counts.

3.7. Soil Chemical Analysis

There were clear differences in many of the soil chemical parameters analyzed between soil samples collected from non-waterlogging and waterlogging experimental pots (Table S2). The differences in values obtained shows a regular pattern. For example, pH values for NF were 6.0–6.1 while higher values were recorded for 1 WF and 2 WF. Redox potential (Eh) values were consistently higher for NF than 1 WF and 2 WF. Soil Eh ranged from 23.60–24.10 for NF and 7.2–7.4 for 1 WF and 2 WF. The highest values of sulphate ion (SO_4) concentrations and electrolyte conductivity (EC) readings were observed in 1 WF soil samples. Mean values for non-treated soil EC were 228, 413, and 125 µS/cm for NF, 1 WF, and 2 WF, respectively. Similarly, mean values for SO_4 concentration in non-treated soil were 0.52, 1.13, and 0.80 mg/Kg for NF, 1 WF, and 2 WF, respectively. Organic carbon, total nitrogen and available phosphorus contents in soil followed the same reduction pattern under one- and two-week waterlogging conditions. Approximately, 10-fold reductions in organic carbon and total nitrogen contents were observed under waterlogging conditions. The records for soil metallic factors like Fe, Zn, and Mn showed the same pattern where the values were higher in soil samples collected from one- and two-week waterlogging conditions. Mean values obtained for Fe were 116.3, 242.1, and 243.3 mg/kg for NF, 1 WF, and 2 WF, respectively, for soil samples collected from pots where 0% NaN_3 plants were grown. The mean values recorded for Zn in soil samples collected from pots containing 0% NaN_3 plants were 14.2, 22.7, and 35.4 mg/kg for NF, 1 WF, and 2 WF, respectively. The mean values of Mn in the same soil samples were 1.34, 9.68, and 12.9 mg/kg for NF, 1 WF, and 2 WF, respectively. The mean values of total phenol content show low variation.

3.8. Anatomy of Okra Roots

There were structural differences in the anatomy of okra root sections obtained from non-waterlogged and waterlogged plants (Figures 2–4). The presence of air channels (lacunae) was conspicuously absent in non-waterlogged root sections (Figure 2). The development of aerenchyma in the cortex and stele were very conspicuous in root sections of plants subjected to waterlogging conditions (Figures 3 and 4). Furthermore, the aerenchyma cells observed in root sections of waterlogged plants were larger in plants grown from 0.05% NaN$_3$-treated seeds than those from 0.02% NaN$_3$-treated seeds (Figures 3 and 4). This suggests an explanation for the higher percentage of survival recorded for plants grown from 0.05% NaN$_3$-treated seeds. The walls of the aerenchyma cells are thick to prevent their collapse.

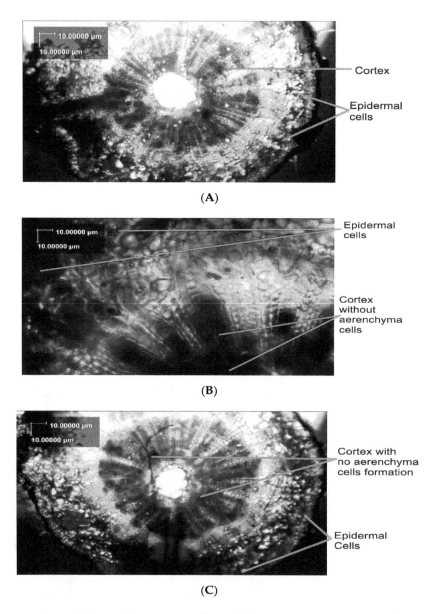

Figure 2. Root sections of okra plants grown from different concentrations of NaN$_3$-treated seeds show no aerenchyma cells formed under non-waterlogging conditions. (**A**) 0% NaN$_3$, (**B**) 0.02% NaN$_3$, (**C**) 0.05% NaN$_3$.

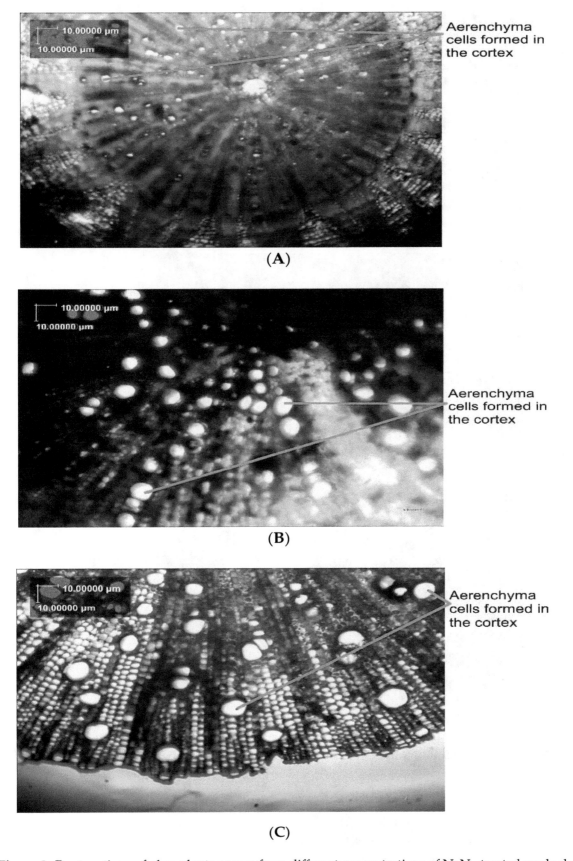

Figure 3. Root sections of okra plants grown from different concentrations of NaN$_3$-treated seeds show aerenchyma cells formed under one-week waterlogging conditions. (**A**) 0% NaN$_3$, (**B**) 0.02% NaN$_3$, (**C**) 0.05% NaN$_3$.

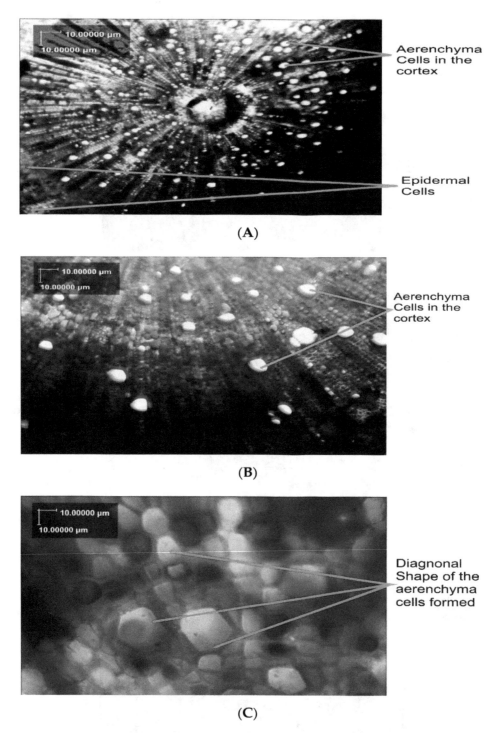

Figure 4. Root sections of okra plants grown from different concentrations of NaN_3-treated seeds show aerenchyma cells formed under two-week waterlogging conditions. (**A**) 0% NaN_3, (**B**) 0.02% NaN_3, (**C**) 0.05% NaN_3.

3.9. Antioxidant Enzymes Activity and Gene Expression Analyses

The effects of the waterlogging condition and NaN_3 treatments on the activities and expression levels of antioxidant enzymes (APX, CAT) in the leaf tissues were investigated. The activity and expression level of APX enzyme were significantly enhanced in plants exposed to waterlogging and sodium azide treatments with respect to non-treated (control) plants (Figure 5). Additionally, under waterlogging conditions, the activity and expression level of CAT enzyme were slightly enhanced in plants treated with sodium azide, as compared to non-treated plants (Figure 5).

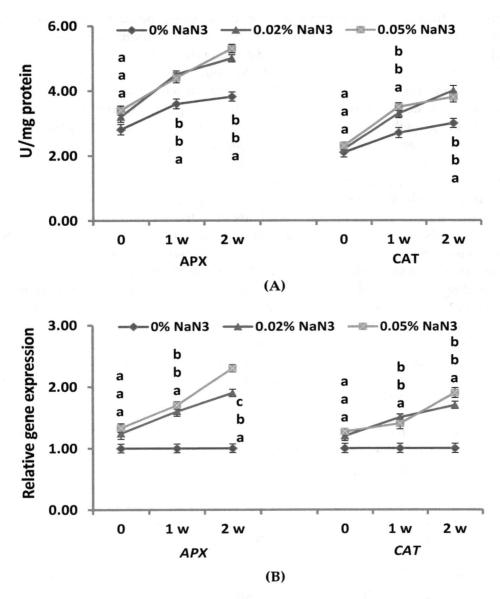

Figure 5. Activity (**A**) and gene expression levels (**B**) of APX and CAT enzymes in okra plants grown from NaN3-treated seeds under waterlogging conditions ten weeks after planting (WAP). Values = mean ± SD, n = 4. Mean values with similar letters at the same WAP are not significantly different at $p \leq 0.05$.

4. Discussion

Waterlogging stress has adverse impacts on crop development and productivity. Waterlogging-induced oxygen depletion results in changes in plant morphology and metabolism. Waterlogging conditions also cause inhibition of photosynthesis, leaf chlorophyll degradation, and early leaf senescence [46]. Negative impacts of flooding might be due to the reduced level of gas diffusion in water, which does not allow terrestrial plants to survive for a long period. Plants develop specific traits to improve gas exchange and cope with waterlogging conditions. These traits include formation of adventitious roots and aerenchyma cells, as well as elongation of stem root juncture above the water surface. These efficiently ameliorate the stress-induced hypoxic or anoxic conditions. The presence of aerenchyma cells facilitates exchange of gases between aerial and submerged plant parts [47]. Kawai et al. [48] proposed that the development of aerenchyma in tissues and organs decreases the number of cells requiring oxygen for respiration. However, the development of adaptive traits to waterlogging stress is species-dependent [7,49,50]. Enhanced formation of aerenchyma was observed upon treating rice plants with exogenous ethylene [14].

In the present study, NaN$_3$ treatments enhanced waterlogging stress tolerance and aerenchyma formation in okra. The results also showed that NaN$_3$ treatments affected okra germination. NaN$_3$-caused seed germination inhibition has also been reported in different plant species [22,23,51,52]. However, NaN$_3$ stimulated the germination of okra plants [24]. This germination inhibition was dependent on the concentration of NaN$_3$ used as seed treatment. Three days after planting (3 DAP), germination has been recorded in all NaN$_3$ treatments applied. Under waterlogging conditions, NaN$_3$ promoted okra growth 10 WAP, indicating that plants grown from 0.02% NaN$_3$-seed treatments exhibited better performance than those grown from 0.05% NaN$_3$-seed treatments. These findings were in a harmony with that reported by Al-Qurainy [25] and Zuzana et al. [26] who stated that NaN$_3$ could stimulate the plant growth and height of *Eruca sativa* and *Diospyros lotus*, respectively. Moreover, the difference in the number of leaves formed under waterlogging and non-waterlogging conditions was significant. Plants grown from NaN$_3$-treated seeds formed more leaves than those from non-treated seeds. Additionally, plants that were grown from 0.02% and 0.05% NaN$_3$-treated seeds produced a greater number of adventitious roots under waterlogging conditions. The emergence of adventitious roots is preceded by epidermal cell death at the nodes of submerged rice plants [47]. The activities leading to epidermal cell death for the emergence of adventitious roots occurred more in plants grown from NaN$_3$-treated seeds. Waterlogging conditions negatively affected the reproductive parameters recorded for okra plants in the current study. These findings are in harmony with that reported by Vwioko et al. [7]. Plants grown from 0.05% NaN$_3$-treated seeds formed a higher number of buds than plants produced from 0% NaN$_3$-treated seeds subjected to two-week waterlogging conditions. Plants grown from 0.05% NaN$_3$-treated seeds also produced more fruits than the control plants under two-week waterlogging conditions.

Waterlogging conditions cause depletion of soil oxygen due to microbial respiration. The reduction of soil oxygen urges anaerobic microorganisms to shift to alternative electron acceptors for their metabolic requirements [53]. Bacteria and fungi ratio in soil community are altered whenever there are soil inundations. Soil bacteria and fungi have a critical role in decomposition and nutrient cycling [54]. In the current investigation, microbial count results exhibited an increase in the bacteria populations and reduction in the fungi populations. The decrease in fungi populations has been previously reported [53,55–57]. Therefore, under waterlogging conditions, fungi presence is less prevalent than bacteria. Fungi require aerobic conditions to thrive but are inhibited by the scarcity of oxygen in the flooded soil environments. Fungi germinate from spores under flooding slowly, resulting in a decreased colonization. Unger et al. [53] suggested that some microbial groups may thrive well under flooded conditions. Gram-positive bacteria showed higher levels compared to Gram-negative bacteria under waterlogging conditions. Mentzer et al. [57] reported that flooding exhibited greater effect than nutrient loading on the microbial community and profoundly altered the composition and functional components.

Water copiously influences several physicochemical processes in soil, particularly under flooded conditions. This begins with the cutoff of oxygen supply to soil environments under waterlogging stress. The lack of oxygen promotes anaerobic metabolism by microbes through utilizing a decomposable organic matter. A reduction in soil redox potential and an increase in pH are recorded [58]. The soil Eh data recorded in a soil-water suspension rightly predicts the level of transformations present in the waterlogged soil [59]. Other important chemical changes in flooded soils indicate the prevalence of reduced forms of nitrogen, oxygen, iron, manganese, or sulphur in soil [53]. There are changes in phase or solubility because of redox reactions. For example, nitrate-nitrogen is transformed into gaseous forms (N$_2$, NO$_2$, N$_2$O) and lost, resulting in nitrogen depletion of soil [60]. In the present study, the soil chemical analysis showed that waterlogging conditions increased pH towards neutral, reduced soil Eh, organic carbon, total nitrogen and available phosphorus. These soil factors indicate higher reduction-oxidation reactions in soils under waterlogging conditions. These patterns of chemical environments and transformations are suspected to favor the tolerant bacteria for their higher counts recorded in waterlogged soil samples. The chemical environments attained under waterlogging soil conditions met the metabolic needs of tolerant bacteria. The decomposition of complex organic compounds is slow under anoxic conditions and in some cases leads to detection of higher amounts of phenolics [53] in waterlogged than in non-waterlogged soils. The present study does not reveal changes in the total phenolics of soil samples, suggesting that either the soil is devoid of complex organics for microbes to degrade under waterlogging conditions, or the microbes utilized readily available forms of carbons that are root exudates. Carbon enters the soil profile via the decomposition of plant residue on the surface or via root exudates in the upper soil horizon [53].

In the current study, root anatomy showed some peculiar features with waterlogged plants. Plants did not develop air-chambers in the cortex and stele regions under non-waterlogging conditions. However, plants subjected to waterlogging conditions formed aerenchyma cells. Further examination of the micrographs showed that plants grown from NaN_3-treated seeds produce more aerenchyma cells than those grown from untreated seeds. It was evident that 0.05% NaN_3-treated seeds produce plants with the highest aerenchyma development and increased with increasing the duration of waterlogging conditions. The formation of aerenchyma in the root as an adaptive trait contributed to the survival of okra plants exposed to waterlogging conditions. Furthermore, under waterlogging conditions, the activities and expression levels of APX and CAT enzymes were enhanced in plants treated with NaN_3 compared to non-treated plants in the present study. The survival of plants in stressed environments might be attributed to the induction of expression levels of antioxidant compounds. Salim et al. [34] reported that NaN_3-treated seeds produce mutant plants that showed higher antioxidation capacities than the normal plants. Moreover, Jeng et al. [61] revealed that these mutants induced increased antioxidant capacities through the generation of scavenging metabolics (DPPH, LPI ability, FRAP, and ABTS radical scavenging activities) than the wild type. In addition, the antioxidant enhancements could be linked to the accumulation of phenolics, anthocyanin, and proanthocyanidins at higher levels in the seed coats. These results are in harmony with that reported by Elfeky et al. [62] who stated that *Helianthus annus* plants grown from NaN_3-treated seeds initiated and induced higher antioxidant capacities than those grown from untreated seeds via increasing carotenoids, peroxidase activity, and protein content. In conclusion, sodium azide priming could enhance waterlogging stress tolerance in okra plants through enhancing the growth and reproductive parameters, inducing the formation of adventitious roots and aerenchyma cells, and increasing the activities and gene expression level of antioxidant enzymes.

Author Contributions: M.A.E.-E. and E.D.V. designed and performed the experiments, analyzed the data, and wrote and revised the manuscript. M.E.I., A.A.A.-G., H.M.A., E.A.A., and M.A.A.-D. helped with analysis and revision of the manuscript. M.M.E.-S. revised the manuscript. All the authors approved the final version of the manuscript.

References

1.   Gemede, H.F.; Ratta, N.; Haki, G.D.; Woldegiorgis, A.Z.; Bey, F. Nutritional Quality and Health Benefits of Okra (*Abelmoschus esculentus*): A Review. *Int. J. Nut. Food Sci.* **2015**, *4*, 208–215. [CrossRef]

2.   El-Esawi, M.A.; Al-Ghamdi, A.A.; Ali, H.M.; Alayafi, A.A.; Witczak, J.; Ahmad, M. Analysis of Genetic Variation and Enhancement of Salt Tolerance in French Pea. *Int. J. Mol. Sci.* **2018**, *19*, 2433. [CrossRef] [PubMed]

3.   El-Esawi, M.A.; Al-Ghamdi, A.A.; Ali, H.M.; Ahmad, M. Overexpression of *AtWRKY30* Transcription Factor Enhances Heat and Drought Stress Tolerance in Wheat (*Triticum aestivum* L.). *Genes* **2019**, *10*, 163. [CrossRef] [PubMed]

4.   El-Esawi, M.A.; Elkelish, A.; Elansary, H.O.; Ali, H.M.; Elshikh, M.; Witczak, J.; Ahmad, M. Genetic transformation and hairy root induction enhance the antioxidant potential of *Lactuca serriola* L. *Oxid. Med. Cell. Longev.* **2017**, *2017*. [CrossRef]

5.   El-Esawi, M.A.; Alayafi, A.A. Overexpression of Rice *Rab7* Gene Improves Drought and Heat Tolerance and Increases Grain Yield in Rice (*Oryza sativa* L.). *Genes* **2019**, *10*, 56. [CrossRef] [PubMed]

6.   El-Esawi, M.A.; Alayafi, A.A. Overexpression of *StDREB2* Transcription Factor Enhances Drought Stress Tolerance in Cotton (*Gossypium barbadense* L.). *Genes* **2019**, *10*, 142. [CrossRef]

7.   Vwioko, E.; Adinkwu, O.; El-Esawi, M.A. Comparative physiological, biochemical and genetic responses to prolonged waterlogging stress in okra and maize given exogenous ethylene priming. *Front. Physiol.* **2017**, *8*, 632. [CrossRef]

8.   Heschbach, C.; Mult, S.; Kreuzwieser, J.; Kopriva, S. Influence of anoxia on whole plant sulphur nutrition of flooding tolerant poplar (*Populus tremula* × *P. alba*). *Plant Cell Environ.* **2005**, *28*, 167–175. [CrossRef]

9.   Herrera, A.; Tezara, W.; Marin, O.; Rengifo, E. Stomatal and non-stomatal limitations of photosynthesis in trees of a tropical seasonally flooded forest. *Physiol. Plant.* **2008**, *134*, 41–48. [CrossRef]

10. Syversten, J.P.; Zablotowicz, R.M.; Smith, M.L. Soil-temperature and flooding effects on two species of citrus. 1. Plant growth and hydraulic conductivity. *Plant Soil* **1983**, *72*, 3–12.

11. Setter, T.L.; Waters, I.; Sharma, S.K.; Singh, K.N.; Kulshreshtha, N.; Yaduvanshi, N.P.S.; Ram, P.C.; Singh, B.N.; Rane, J.; McDonald, G.; et al. Review of wheat improvement for waterlogging tolerance in Australia and India: The importance of anaerobiosis and element toxicities associated with different soils. *Ann. Bot.* **2009**, *103*, 221–235. [CrossRef] [PubMed]

12. Calvo-Polanco, M.; Senorans, J.; Zwiazek, J.J. Role of adventitious roots in water relations of tamarack (*Larix* laricina) seedlings exposed to flooding. *BMC Plant Biol.* **2012**, *12*, 99–107. [CrossRef] [PubMed]

13. Sauter, M. Root responses to flooding. *Curr. Opin. Plant Biol.* **2013**, *16*, 282–286. [CrossRef] [PubMed]

14. Takahashi, H.; Yamauchi, T.; Colmer, T.; Nakazono, M. Aerenchyma Formation in Plants. In *Low-Oxygen Stress in Plants, Oxygen Sensing and Adaptive Responses to Hypoxia*, 1st ed.; Van Dongen, J.T., Licausi, F, Eds.; Plant Cell Monographs; Springer: New York, NY, USA, 2014; Volume 21, pp. 247–265.

15. Janda, T.; Szalai, G.; Tari, I.; Paldi, E. Hydroponic treatments with salicylic acid decreases the effects of chilling injury in maize (*Zea mays* L.) plants. *Planta* **1999**, *208*, 175–180. [CrossRef]

16. Rajasekaran, L.R.; Blake, T.J. New plant growth regulators protect photosynthesis and enhance growth under drought of jack pine seedlings. *J. Plant Growth Reg.* **1999**, *18*, 171–181. [CrossRef]

17. Gondor, O.K.; Pál, M.; Darkó, É.; Janda, T.; Szalai, G. Salicylic Acid and Sodium Salicylate Alleviate Cadmium Toxicity to Different Extents in Maize (*Zea mays* L.). *PLoS ONE* **2016**, *11*, e0160157. [CrossRef]

18. Vwioko, E.D. Performance of soybean (*Glycine max* L.) in salt-treated soil environment following salicylic acid mitigation. *NISEB J.* **2013**, *13*, 44–49.

19. Jackson, M.B. Ethylene-promoted elongation: An adaptation to submergence stress. *Ann. Bot.* **2008**, *101*, 229–248. [CrossRef]

20. Vidoz, M.L.; Loreti, E.; Mensuali, A.; Alpi, A.; Perata, P. Hormonal interplay during adventitious root formation in flooded tomato plants. *Plant J.* **2011**, *63*, 551–562. [CrossRef]

21. Sasidharan, R.; Voesenek, L.A.C.J. Ethylene-mediated acclimations to flooding stress. *Plant Physiol.* **2015**, *169*, 3–12. [CrossRef]

22. Gnanamurthy, S.; Dhanavel, D.; Girija, M.; Pavadai, P.; Bharathi, T. Effect of chemical mutagenesis on quantitative traits of maize (*Zea mays* (L.). *Int. J. Res. Bot.* **2012**, *2*, 34–36.

23. Shagufta, B.; Aijaz, A.W.; Irshad, A.N. Mutagenic sensitivity of gamma rays, EMS and sodium azide in *Trigonella foenumgraecum* L. *Sci. Res. Rep.* **2013**, *3*, 20–26.

24. Vwioko, D.E.; Onobun, E. Vegetative response of ten accessions of *Abelmoschus esculentus* (L) Moench. treated with sodium azide. *J. Life Sci. Res. Dis.* **2015**, *2*, 13–24.

25. Al-Qurainy, F. Effects of sodium azide on growth and yield traits of *Eruca sativa* (L.). *World Appl. Sci. J.* **2009**, *7*, 220–226.

26. Zuzana, K.; Katarína, R.; Elena, Z.; Maria, L.B.; Ján, B. Sodium azide induced morphological and molecular changes in persimmon (*Diospyros lotus* L.). *Agriculture* **2012**, *58*, 57–64.

27. Adamu, A.K.; Aliyu, H. Morphological effects of sodium azide on tomato (*Lycopersicon esculentum* Mill.). *Sci. World J.* **2007**, *2*, 9–12.

28. Gruszka, D.; Szarejko, L.; Maluszynski, M. Sodium azide as a mutagen. In *Plant Mutation Breeding and Biotechnology*; Shu, Q., Forster, B.P., Nakagawa, H., Eds.; CABI Publishing Company: Wallingford, UK, 2012; pp. 159–166.

29. Kravchik, M.; Bernstein, N. Effects of salinity on the transcriptome of growing maize leaf cells points at differential involvement of the antioxidative response in cell growth restriction. *BMC Genom.* **2013**, *16*, 14–24.

30. Mittler, R. Oxidative stress, antioxidant and stress tolerance: A review. *Trends Plant Sci.* **2002**, *7*, 405–410. [CrossRef]

31. Haq, I.U.; Memon, S.; Gill, N.P.; Rajput, M.T. Regeneration of plantlets under NaCl-stress from NaN_3 treated sugarcane explants. *Afr. J. Biotechnol.* **2011**, *10*, 16152–16156.

32. El Kaaby, E.A.J.; Al-Ajeel, S.A.; Al-Anny, J.A.; Al-Aubaidy, A.A.; Ammar, K. Effect of the chemical mutagen sodium azide on plant regeneration of two tomato cultivars under salinity stress condition in vitro. *J. Life Sci.* **2015**, *9*, 25–31. [CrossRef]

33. Kuasha, M.; Nasiruddin, K.M.; Hassan, L. Effects of sodium azide on callus in sugarcane. *Discovery* **2016**, *52*, 1683–1688.

34. Salim, K.; Fahad, A.-Q.; Firoz, A. Sodium azide: A chemical mutagen for enhancement of agronomic traits of crop plants. *Int. J. Sci. Tech.* **2009**, *4*, 1–21.

35. Ademoroti, C.A. *Standard Methods for Water and Effluent Analysis*, 1st ed.; Foludex Press Ltd.: Ibadan, Nigeria, 1996.

36. Bremner, J.M. Determination of nitrogen in soil by the Kjeldahl method. *J. Agric. Sci.* **1960**, *55*, 11–33. [CrossRef]

37. Blum, U. Benefits of citrate over EDTA for extracting phenolics from soils and plant debris. *J. Chem. Ecol.* **1997**, *23*, 347–362. [CrossRef]

38. Appiah, M.R.; Ahenkorah, Y. Determination of available sulphate in some soils of Ghana considering five extraction methods. *Biol. Fertil. Soils* **1989**, *8*, 80–86. [CrossRef]

39. Ben Mussa, S.A.; Elferjani, H.S.; Haroun, F.A.; Abdelnabi, F.F. Determination of available nitrate, phosphate and sulphate in soil samples. *Int. J. PharmTech Res.* **2009**, *1*, 598–604.

40. Bray, R.H.; Kurtz, L.T. Determination of total organic carbon and available phosphorus in soils. *Soil Sci.* **1945**, *59*, 39–48. [CrossRef]

41. Bremner, J.M.; Jenkinson, D.S. Determination of organic carbon in soil. I. oxidation by dichromate of organic matter in soil and plant materials. *J. Soil Sci.* **1960**, *11*, 394–402. [CrossRef]

42. Islam, M.S.; Halim, M.A.; Safiullah, S.; Islam, M.S.; Islam, M.M. Analysis of organic matter, ion and manganese in soil of arsenic affected Singair Area, Bangladesh. *Res. J. Environ. Toxicol.* **2009**, *3*, 31–35.

43. Harrigan, W.F.; McCance, M.E. *Laboratory Methods in Foods and Dairy Microbiology*, 8th ed.; Academic Press: London, UK, 1990.

44. Holt, J.G.; Sneath, P.H.; Krieg, N.R. *Bergey's Manual of Determinative Bacteriology*, 9th ed.; Lippincott, Williams and Wilkins Publishers: Baltimore, MD, USA, 2002; p. 787.

45. Zhang, J.; Kirkham, M.B. Enzymatic Responses of the Ascorbate-Gluta-thione Cycle to Drought in Sorghum and Sunflower Plants. *Plant Sci.* **1996**, *113*, 139–147. [CrossRef]

46. Zou, X.; Hu, C.; Zeng, L.; Xu, M.; Zhang, X. A comparison of screening methods to identify waterlogging tolerance in the field in *Brassica napus* (L.) during plant ontogeny. *PLoS ONE* **2014**, *9*, e89731. [CrossRef] [PubMed]

47. Steffens, B.; Geske, T.; Sauter, M. Aerenchyma formation in the rice stem and its promotion by H_2O_2. *New Phytol.* **2011**, *190*, 369–378. [CrossRef] [PubMed]

48. Kawai, M.; Samarajeewa, P.K.; Barrero, R.A.; Nishigushi, M.; Uchimiya, H. Cellular dissection of the degradation pattern of cortical cell death during aerenchyma formation of rice roots. *Planta* **1998**, *204*, 277–287. [CrossRef]

49. Fukao, T.; Xu, K.; Ronald, P.C.; Bailey-Serres, J. A variable cluster of ethylene response factor-like genes regulates metabolic and developmental acclimation responses to submergence in rice. *Plant Cell* **2006**, *18*, 2021–2034. [CrossRef] [PubMed]

50. Hattori, Y.; Nagai, K.; Furukawa, S.; Song, X.-J.; Kawano, R.; Sakakibara, H.; Wu, J.; Matsumoto, T.; Yoshimura, A.; Kitano, H.; et al. The ethylene response factors SNORKEL 1 and SNORKEL 2 allow rice to adapt to deep water. *Nature* **2009**, *460*, 1026–1030. [CrossRef]

51. Mensah, J.K.; Obadoni, B. Effects of sodium azide on yield parameters of groundnut (*Arachis hypogaea* L.). *Afr. J. Biotechnol.* **2007**, *6*, 668–671.

52. Nakweti, R.K.; Franche, C.; Ndiku, S.L. Effects of sodium azide (NaN_3) on seeds germination, plantlets growth and in vitro antimalarial activities of *Phyllantus odontadenius* Mull. *Arg. Amer. J. Exp. Agric.* **2015**, *5*, 226–238.

53. Unger, I.M.; Kennedy, A.C.; Muzika, R.-M. Flooding effects on soil microbial communities. *Appl. Soil Ecol.* **2009**, *42*, 1–8. [CrossRef]

54. Suzuki, C.; Kunito, T.; Aono, T.; Liu, C.-T.; Oyaizu, H. Microbial indices of soil fertility. *J. Appl. Microbiol.* **2005**, *98*, 1062–1074. [CrossRef]

55. Bossio, D.A.; Scow, K.M. Impacts of carbon and flooding on soil microbial communities: Phospholipid fatty acid profiles and substrate utilization patterns. *Microb. Ecol.* **1998**, *35*, 265–378. [CrossRef]

56. Drenovsky, R.E.; Vo, D.; Graham, K.J.; Scow, K.M. Soil water content and organic carbon availability are major determinants of soil microbial community composition. *Microb. Ecol.* **2004**, *48*, 424–430. [CrossRef] [PubMed]

57. Mentzer, J.L.; Goodman, R.M.; Balser, T.C. Microbial responses over time to hydrologic and fertilization treatments in a simulated wet prairie. *Plant Soil* **2006**, *284*, 85–100. [CrossRef]

58. Stover, R.H. Flooding of soil for disease control. In *Soil Disinfection*; Mulder, D., Ed.; Elsevier: Amsterdam, The Netherlands, 1979.

59. Labuda, S.Z.; Vetchinnikov, A.A. Soil susceptibility on reduction as an index of soil properties applied in the investigation upon soil devastation. *Ecol. Chem. Eng.* **2011**, *18*, 333–344.

60. Vepraskas, M.J.; Faulkner, S.P. Redox chemistry of hydric soils. In *Wetlands Soils: Genesis, Hydrology, Landscapes and Classification*; Richardson, J.L., Vepraskas, M.J., Eds.; Lewis Publishers: Boca Raton, FL, USA, 2001.

61. Jeng, T.L.; Tseng, T.H.; Lai, C.C.; Wu, M.T.; Sung, J.M. Antioxidative charactrisation of NaN_3- induced common bean mutants. *Food Chemistry.* **2010**, *119*, 1006–1011. [CrossRef]

62. Elfeky, S.; Abo-Hamad, S.; Saad-Allah, K.M. Physiological impact of sodium azide on *Helianthus annus* seedlings. *Int. J. Agron. Agric. Res.* **2014**, *4*, 102–109.

7

Substrate Application of 5-Aminolevulinic Acid Enhanced Low-Temperature and Weak-Light Stress Tolerance in Cucumber (*Cucumis sativus* L.)

Ali Anwar [1,2,†], Jun Wang [1,†], Xianchang Yu [1], Chaoxing He [1] and Yansu Li [1,*]

[1]   Institute of Vegetables and Flowers, Chinese Academy of Agricultural Sciences, Beijing 100081, China; anwarsnu@aol.com (A.A.); wangjun01@caas.cn (J.W.); xcyu1962@163.com (X.Y.); hechaoxing@caas.cn (C.H.)

[2]   Graduate School of International Agricultural Technology and Crop Biotechnology Institute/Green Bio Science & Technology, Seoul National University, Pyeongchang 25354, Korea

*   Correspondence: liyansu@caas.cn

†   These authors contributed equally to this work.

Abstract: 5-Aminolevulinic acid (ALA) is a type of nonprotein amino acid that promotes plant stress tolerance. However, the underlying physiological and biochemical mechanisms are not fully understood. We investigated the role of ALA in low-temperature and weak-light stress tolerance in cucumber seedlings. Seedlings grown in different ALA treatments (0, 10, 20, or 30 mg ALA·kg^{-1} added to substrate) were exposed to low temperature (16/8 °C light/dark) and weak light (180 µmol·m^{-2}·s^{-1} photosynthetically active radiation) for two weeks. Treatment with ALA significantly alleviated the inhibition of plant growth, and enhanced leaf area, and fresh and dry weight of the seedlings under low-temperature and weak-light stress. Moreover, ALA increased chlorophyll (Chl) *a*, Chl *b*, and Chl *a+b* contents. Net photosynthesis rate, stomatal conductance, transpiration rate, photochemical quenching, non-photochemical quenching, actual photochemical efficiency of photosystem II, and electron transport rate were significantly increased in ALA-treated seedlings. In addition, ALA increased root activity and antioxidant enzyme (superoxide dismutase, peroxidase, and catalase) activities, and reduced reactive oxygen species (hydrogen peroxide and superoxide radical) and malondialdehyde accumulation in the root and leaf of cucumber seedlings. These findings suggested that ALA incorporation in the substrate alleviated the adverse effects of low-temperature and weak-light stress, and improved Chl contents, photosynthetic capacity, and antioxidant enzyme activities, and thus enhanced cucumber seedling growth.

Keywords: ALA; abiotic stress; chlorophyll; photosynthesis; antioxidant enzyme

1. Introduction

Cucumber (*Cucumis sativus* L.), a member of the *Cucurbitaceae* family, is an important vegetable widely cultivated and consumed around the world [1]. Plants are challenged by numerous environmental stresses (e.g., high and low temperatures, salinity, light, drought, and heavy metal stress) that affect plant growth and productivity [2,3]. Low temperature and low-light stress are the most critical environmental factors that influence cucumber production in a solar greenhouse [2,4]. Plant exposed to low temperature and light stress exhibit a number of physiological and biochemical abnormalities, including reduction in chlorophyll biosynthesis, photosynthetic capacity, carbohydrate and nitrogen metabolism, nutrient uptake and accumulation, and overproduction of reactive oxygen species (ROS) [5]. Accumulation of ROS negatively affects enzyme activities, biosynthesis of carbohydrates, DNA, and proteins, and other biochemical activities, thus leading to oxidative stress [4,5]. In addition, ROS influence the expression of a number of genes involved in diverse processes such as growth, cell cycle, programmed cell death, abiotic stress responses, pathogen attack response, systemic signaling, and development [6]. The antioxidant defense system, which includes superoxide dismutase (SOD), peroxidase (POD), catalase (CAT), glutathione reductase (GR), and ascorbate peroxidase (APX), plays a crucial role in normalizing the production of ROS, thereby protecting plants from abiotic stresses [6,7].

5-Aminolevulinic acid (ALA) is an essential biosynthetic precursor and is considered to be a plant growth regulator [8,9]. The compound is a key precursor in the biosynthesis of porphyrin compounds, such as chlorophyll, heme, and plant hormones [10]. In addition, ALA is involved in photosynthesis regulation under abiotic stress. Exogenous ALA application increases chlorophyll accumulation and chlorophyll fluorescence indices in lettuce and oilseed rape [11,12]. It has recently been reported that ALA regulates the expression level of *fructose-1,6-bisphosphatase* (*FBP*), *triose-3-phosphate isomerase* (*TPI*), and *ribulose-1,5-bisphosphate carboxylase/oxygenase small subunit* (*RBCS*), which activate the Calvin cycle of photosynthesis under drought stress [13]. In a previous study we observed that ALA regulates endogenous hormone and nutrient accumulation in cucumber to induce low-temperature stress tolerance [10]. It is also reported that ALA is involved in the chlorophyll biosynthesis pathway under salinity stress [14], and induces antioxidant enzyme activities and endogenous hormone accumulation under low-temperature stress in cucumber seedlings [10]. Previous studies demonstrated that foliar application of ALA may confer plant tolerance to diverse abiotic stresses, such as chilling, high temperature, salinity, drought, weak light, and heavy metals [14–16]. ALA influences a variety of physiological and biochemical activities of plants in response to abiotic stresses, including chlorophyll biosynthesis, nutrient uptake, and antioxidant enzyme activities [14,16]. Furthermore, ALA induces abiotic stress tolerance through activation of numerous types of transcription factors, signal transduction, and chlorophyll and carbohydrate biosynthesis [9]. These findings suggest that ALA can broadly reduce the detrimental effects of environmental influence.

During winter vegetable cultivation, plants are frequently exposed to low temperature and weak light intensity (predominantly clouds or fog), which can negatively influence production. Therefore, this study was designed to investigate the role of ALA in response to a combination (low temperature and weak light) of stresses on cucumber seedling growth, chlorophyll contents, photosynthetic capacity, antioxidant enzyme activity, and ROS accumulation. The information generated from this study will improve our understanding of responses to both stresses and will be useful for security of winter vegetable production.

2. Materials and Methods

2.1. Plant Material and Experimental Setup

Cucumber (*Cucumis sativus* 'Zhongnong 26') seeds were soaked in water at 55 °C for 2–3 h, and then germinated on moist gauze in the dark at 28 °C. The germinated seeds were transplanted into plug trays containing nursery substrate supplemented with ALA (Sigma, St Louis, MO, USA) and incubated at 28/18 °C (light/dark) under 70%–75% humidity and 300–350 $\mu mol \cdot m^{-2} \cdot s^{-1}$ photosynthetically active radiation for 14 h. The experiment consisted of four treatments based on different concentrations of ALA (applied as kg^{-1} substrate):

CK,   Control (no ALA)
T1,   10 mg ALA
T2,   20 mg ALA
T3,   30 mg ALA

The ALA concentrations were mixed with a constant weight of substrate (kg). The substrates were used to fill a 32-cell seedling tray and a germinated seed was sown in each cell. At the first leaf (fully expanded) stage, the seedlings were transferred to a controlled artificial chamber under 16/8 °C (day/night), photosynthetically active radiation of 180 $\mu mol \cdot m^{-2} \cdot s^{-1}$, and a photoperiod of 12 h for 21 d before sampling. The seedlings were irrigated at two-day intervals with Hoagland's solution to fulfil nutritional demand. Each treatment consisted of four replicates.

2.2. *Measurement of Plant Growth Parameters*

Plant height, stem diameter, and fresh weight were measured with a ruler, vernier caliper, and electronic balance, respectively [10]. Fresh samples were placed in an oven at 105 °C for 30 min, and then dried at 75 °C [1]. Root vitality was determined using the triphenyl tetrazolium chloride method [2]. The seedling vigor index was calculated using the following formula [2].

$$\text{Seedling Vigor Index} = \left(\frac{Hypocotyl\ Diameter}{Plant\ Height} + \frac{Root\ Dry\ Weight}{Shoot\ Dry\ Weight} \right) \times Total\ Dry\ Weight \qquad (1)$$

2.2.1. Chlorophyll, Photosynthesis, and Chlorophyll Fluorescence Measurements

Chlorophyll (Chl) contents were determined using an ethanol extraction method, as previously described [2]. Net photosynthetic rate (P_n), stomatal conductance (G_s), transpiration rate (T_r), and intercellular CO_2 concentration (C_i) on the fourth fully expanded leaf from the shoot tip were measured using a portable photosynthesis system (Li-6400XT, LI-COR, Inc., Lincoln, NE, USA).

The portable photosynthesis system was also used for measurement of chlorophyll fluorescence. The fourth fully expanded leaves from the shoot tip were adapted in the dark for 30 min prior to measurement. The maximum photochemical efficiency of photosystem II (F_v/F_m), maximum antenna conversion efficiency (F_v'/F_m'), photochemical quenching (qP), nonphotochemical quenching (NPQ), the actual photochemical efficiency of photosystem II (ΦPSII), and electron transport rate (ETR) were calculated [14].

2.2.2. Determination of Root Activity

The root activity was determined by TTC (Triphenyltetrazolium chloride) reduction method [2]. Briefly, 0.5 g fresh collected roots were cut into 2 cm length and put in 10 mL 0.5 mM PBS buffer containing 0.4% TTC (w/v) and incubate for one hour at 37 °C. The reaction was stopped by using 2 mL H_2SO_4 (1 mol/L) for 15 min, and then remove all solutions, and then add 10 mL 95% ethanol and incubate for 24 h at room temperature (until root turn white). The absorbance was read at 485 nm using spectrophotometer.

Calculation formula:

$$\text{Tetrazole reduction strength (µg/gFW.h)} = (OD + 0.0035)/4*h*W*0.0022 \ (h = 4, W = 0.4\sim0.5) \qquad (2)$$

2.3. *Measurement of $O_2{}^{\cdot-}$, H_2O_2, and Malondialdehyde Contents*

Superoxide radical ($O_2{}^{\cdot-}$) and hydrogen peroxide (H_2O_2) contents were determined using assay kits (COMINBIO) with a UV-1800 spectrophotometer (Shimadzu, Kyoto, Japan) in accordance with the manufacturer's instructions. Malondialdehyde (MDA) content was measured using the thiobarbituric acid method [2].

2.4. *Activities of Antioxidant Enzymes*

Fresh leaves (~0.5 g) were quickly ground with a pestle in an ice-cold mortar with 4 mL of 50 mM phosphate buffer solution (pH 7.8). The homogenate was centrifuged at 10,500 rpm for 20 min. The supernatant was used to determine the activities of antioxidant enzymes. Superoxide dismutase (SOD) activity was measured, with some modifications, based on 50% inhibition of the photochemical reduction of nitro blue tetrazolium at 560 nm. Peroxidase (POD) activity was measured as the increase in absorbance at 470 nm using the method and catalase (CAT) activity was measured as the decline in absorbance at 240 nm [2].

2.5. Statistical Analysis

Each treatment consisted of four independent biological replicates and the entire experiment was repeated three times. The data were statistically analyzed using analysis of variance (ANOVA), and individual treatments were compared using the least significant difference test (LSD; $p = 0.05$) as implemented in Statistix 8.1 software.

3. Results

3.1. Exogenous ALA Application Promoted Cucumber Seedlings Growth

Application of ALA to the substrate significantly increased plant height, stem diameter, leaf area, fresh and dry weight, and seedling vigor index in cucumber seedlings, which were significantly reduced under low-temperature and weak-light stress (Figure 1). Compared with the control treatment (CK), the plant height, stem diameter, leaf area, fresh and dry weight, and seedling vigor index of cucumber seedlings increased by 14.4%, 13.4%, 63.1%, 54.3%, 54.5%, and 53.8%, respectively, in the T2 treatment. Growth parameters in the T1 and T3 treatments were not statistically different, but were significantly higher than those of the CK (Figure 1). The results suggested that ALA application alleviated the detrimental effects of the combined stress of low temperature and weak light, and thus enhanced cucumber seedling growth.

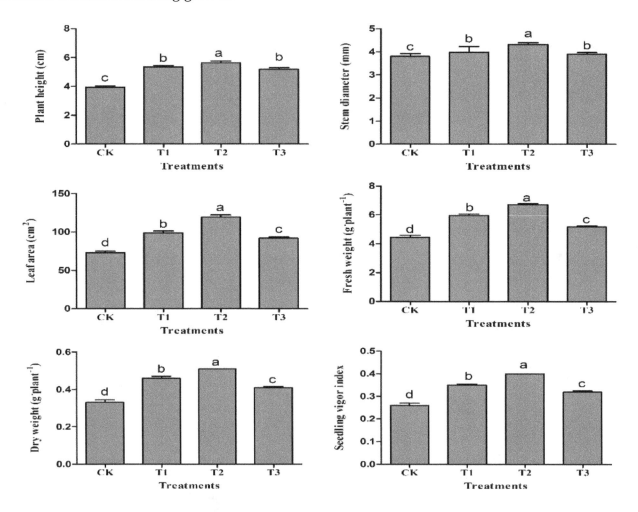

Figure 1. Effect of exogenous ALA application on growth of cucumber seedlings under low-temperature and weak-light stress. Data are the means of four replicates; error bars indicate the standard deviation. Treatments with the same lower-case letter are not significantly different (least significant difference test, $p = 0.05$).

3.2. Exogenous ALA Application Enhanced Root Activity of Cucumber Seedlings

Root activity represents overall vigor, including root metabolic processes, enzyme activities, and water and nutrient absorption and uptake processes, thus it is considered to be an important index for plant response to environmental variables. The present results suggested that root vitality of cucumber seedlings was negatively affected by combined low-temperature and weak-light stress (Figure 2). The ALA-treated seedlings showed significantly enhanced tolerance to low-temperature and weak-light stress, and resulted in a significant increment in root vitality of the cucumber seedlings. Maximum root activity was observed in the T2 treatment and the lowest vitality was recorded in the CK.

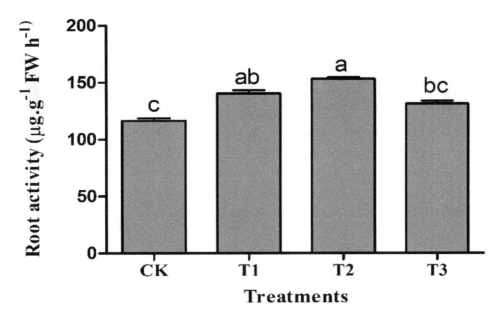

Figure 2. Effect of exogenous ALA application on root activity under low-temperature and weak-light stress. Data are the means of four replicates; error bars indicate the standard deviation. Treatments with the same lower-case letter are not significantly different (least significant difference test, $p = 0.05$).

3.3. Exogenous ALA Application Increased Chlorophyll Content of Cucumber Seedlings

Chlorophylls (Chl) are extremely sensitive to abiotic stress and quickly degrade under an extreme stress intensity, which ultimately reduces photosynthetic capacity. The present results showed that low-temperature and weak-light stress induced a significant decrease in Chl a, Chl b, and Chl $a+b$ contents, whereas the Chl a/b ratio was unchanged among the ALA treatments (Figure 3). Compared with the CK, the contents of Chl a, Chl b, and Chl $a+b$ were increased by 22.14%, 28.26%, and 23.59% respectively, in the T2 treatment, by 9.40%, 13.04%, and 10.25% in the T1 and 6.04%, 8.70%, and 6.70% in the T3 treatment (Figure 3). The differences in contents between the T1 and T2 treatments were non-significant, but were higher significantly than those of the CK and lower than those of the T2 treatment. The results showed that exogenous ALA increased Chl contents to reduce the harmful effect of low-temperature and weak-light stress.

The photosynthetic capacity was significantly enhanced by exogenous ALA application and low-temperature and weak-light stress. Significant increases in P_n, G_s, and T_r by 16.50%, 128.57%, and 148.54%, respectively, were observed compared with ALA-treated seedlings (T2; Figure 4). Similarly, the T1 and T3 treatments resulted in a significant increment in photosynthetic parameters compared with those of the CK. The C_i was slightly increased in ALA-treated seedlings, but the difference with the CK was non-significant. These findings indicated that ALA regulated chlorophyll contents and resulted in improved photosynthesis under combined low-temperature and weak-light stress.

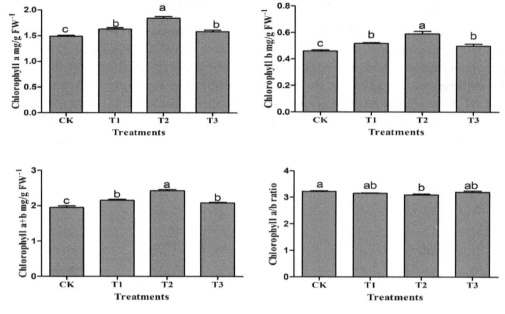

Figure 3. Effect of exogenous ALA application on chlorophyll contents of cucumber seedlings under low-temperature and weak-light stress. Data are the means of four replicates; error bars indicate the standard deviation. Treatments with the same lower-case letter are not significantly different (least significant difference test, $p = 0.05$).

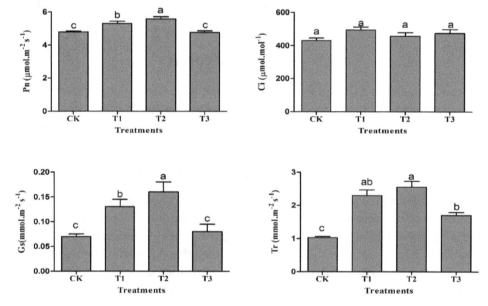

Figure 4. Effect of exogenous ALA on photosynthesis of cucumber seedlings under low-temperature and weak-light stress. Data are the means of four replicates; error bars indicate the standard deviation. Treatments with the same lower-case letter are not significantly different (least significant difference test, $p = 0.05$). P_n, net photosynthetic rate; G_s, stomatal conductance; C_i, intercellular CO_2 concentration; T_r, transpiration rate.

3.4. Effects of exogenous ALA Application on Chlorophyll Fluorescence

Chlorophyll fluorescence analysis is an important and commonly used technique to investigate the plant photosynthetic capacity and response to stress. The present results indicated that F_v/F_m and F_v'/F_m' were non-significantly different among all treatments (Table 1). In general, ALA-treated seedlings showed a significant in qP, ΦPSII, and ETR compared with the CK. However, the opposite trend was observed for NPQ. These findings suggested that ALA played a significant role in abiotic stress tolerance and protected the photosynthetic machinery.

Table 1. Effects of exogenous ALA application in the substrate on chlorophyll fluorescence parameters of cucumber seedlings under low-temperature and weak-light stress.

Treatment	F_v/F_m	F_v'/F_m'	qP	NPQ	ΦPSII	ETR
CK	0.60 ± 0.02a	0.48 ± 0.09a	0.28 ± 0.07d	0.54 ± 0.06a	0.20 ± 0.02c	20.62 ± 2.74bc
T1	0.61 ± 0.03a	0.46 ± 0.04a	0.55 ± 0.08b	0.38 ± 0.06c	0.26 ± 0.03b	21.73 ± 2.76ab
T2	0.61 ± 0.02a	0.44 ± 0.02a	0.63 ± 0.09a	0.37 ± .040c	0.30 ± 0.02a	23.82 ± 1.92a
T3	0.60 ± 0.01a	0.46 ± 0.03a	0.43 ± 0.05c	0.47 ± 0.04b	0.21 ± 0.02c	19.22 ± 1.48c

Data are the means of four replicates ± standard deviation. Treatments with the same lower-case letter within a column are not significantly different (least significant difference test, $p = 0.05$). F_v/F_m, maximum photochemical efficiency of photosystem II; F_v'/F_m', maximum antenna conversion efficiency; qP, photochemical quenching; NPQ, non-photochemical quenching; ΦPSII, actual photochemical efficiency of photosystem II; ETR, electron transport rate.

3.5. Exogenous ALA Application Promoted Antioxidant Enzyme Activities

Overproduction of ROS and accumulation of MDA result in damage to chlorophylls, protein biosynthesis, and DNA, which ultimately results in oxidative stress. Plants have evolved a defense system (antioxidant enzymes), which control ROS overproduction under abiotic stress. In the present study, activities of antioxidant enzyme (SOD, POD, and CAT) were significantly increased in response to ALA treatment compared with those of the CK (Figure 5). The T2 exogenous ALA treatment significantly increased SOD, POD, and CAT activities by 83.91%, 20.27%, and 27.96%, in leaves and 74.58%, 63.97%, and 56.53% in roots, respectively, compared with activities in the CK. The POD activity was significantly higher in T3 leaves and roots compared with those observed in the CK (Figure 5). These findings suggested that exogenous ALA may regulate the plant defense system to reduce the adverse effects of combined low-temperature and weak-light stress.

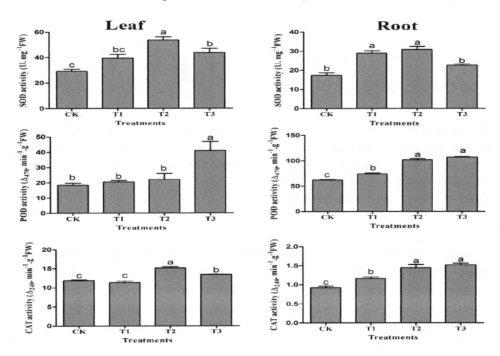

Figure 5. Effect of exogenous ALA application on antioxidant enzyme activities under low-temperature and weak-light stress in cucumber seedlings. Data are the means of four replicates; error bars indicate the standard deviation. Treatments with the same lower-case letter are not significantly different (least significant difference test, $p = 0.05$). SOD, superoxide dismutase; POD, peroxidase; CAT, catalase.

3.6. Exogenous ALA Application Reduced $O_2^{·-}$, H_2O_2, and MDA Accumulation

Plant exposure to abiotic stress leads to overproduction of ROS and accumulation of MDA, which are highly reactive and toxic, and affect a variety of physiological and biochemical activities. The ROS and MDA contents were significantly higher in roots and leaves of the CK (Figure 6). The $O_2^{·-}$ content in the leaves and roots of CK seedlings was 1.27 and 2.39 μmol g^{-1} FW, respectively, and 0.41 and 1.24 μmol g^{-1} FW in the T2 treatment. The H_2O_2 content in CK leaves and roots were 7.86 and 2.66 μmol g^{-1} FW, respectively, compared with 5.54 and 2.36 μmol g^{-1} FW, respectively, in the T2 treatment. The MDA content was significantly higher in the CK and decreased significantly in the T2 treatment (Figure 6). The ROS and MDA contents were significantly lower in the T1 and T2 treatments compared with those of the CK, but the differences were non-significant for T2. These findings revealed that ALA application plays an important role in stabilizing ROS accumulation and biosynthesis under combined low-temperature and weak-light stress.

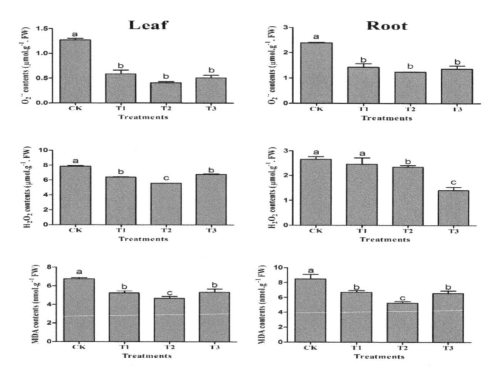

Figure 6. Effect of exogenous ALA application on reactive oxygen species and malondialdehyde contents under low-temperature and weak-light stress in cucumber seedlings. Data are the means of four replicates; error bars indicate the standard deviation. Treatments with the same lower-case letter are not significantly different (least significant difference test, $p = 0.05$). $O_2^{·-}$, superoxide radical; H_2O_2, hydrogen peroxide; MDA, malondialdehyde.

4. Discussion

Low-temperature and weak-light stress damage a variety of plant physiological and biochemical metabolic processes, and hence reduce yield [5]. ALA is a vital precursor of the tetrapyrrole biosynthesis pathway, and is considered to be a plant growth regulator and to regulate plant defense mechanisms to mitigate the harmful effects of abiotic stress [9,10]. As reported previously, low temperature has adverse impacts on cucumber seedlings, and results in a significant reduction in Chl accumulation and photosynthetic capacity [12,13]. Exogenous ALA application is involved in regulation of endogenous hormones, chlorophyll and nutrient accumulation, and the plant defense system, and significantly reduces the harmful effects of low temperature and improves cucumber seedling growth [10]. In the present study, the combined stress of low temperature and weak light imposed significant negative effects on cucumber seedling growth (Figure 1). These results were similar to those of previous studies, which reported that ALA stimulates the plant defense system to alleviate the harmful effects of salinity and low-temperature stress and promoted cucumber seedling growth [10,14].

Chlorophyll is highly sensitive to abiotic stress and is quickly degraded, and is also considered to be an indicator of chloroplast development and photosynthesis proficiency. Abiotic stresses increase degradation of Chl [1], which ultimately affects photosynthetic capacity [17–19]. Previous studies have reported that ALA improves Chl accumulation, photosynthetic capacity, and nutrient uptake to reduce the harmful impacts of salinity stress in *Brassica napus* L. [8,20]. In the present study Chl contents (Chl *a*, Chl *b*, and Chl *a+b*) were significantly increased in ALA-treated seedlings (Figure 3). These findings suggested that ALA increased Chl protection under combined low-temperature and weak-light stress. A recent study reported that exogenous ALA application regulates the ALA metabolic pathway and the transcript level of downstream genes (*HEMA1*, *HEMH*, *CHLH*, *POR*, and *CAO*), and ALA accumulation under salinity stress [14]. The activities of glutamyl-tRNA reductase and glutamate-1-semiadelhyde 2,1-aminomutase, which catalyze ALA biosynthesis [20,21], are improved in ALA-treated plants under abiotic stress [22–24]. Transcriptome analysis suggested that ALA regulates thousands of genes that are involved in Chl biosynthesis (*ChlD*, *ChlH*, and *Chl1-1*), photosynthesis, cell cycle, transaction factors, and defense-related genes [11,25,26]. The results are supported by previous findings that ALA activates chlorophyll biosynthesis and accumulation in bluegrass in response to osmotic stress [25]. These findings help to elucidate the specific role of ALA in the Chl biosynthesis pathway and stimulation of Chl biosynthesis-related gene expression and enzyme activities, thus enhancing Chl accumulation under low-temperature and weak-light stress.

In the present study, photosynthesis capacity and chlorophyll fluorescence are significantly affected by the combined stress of low temperature and weak light (Figures 3 and 4), and were enhanced significantly in ALA-treated seedlings. These findings indicate that ALA reduced the toxic effects of low-temperature and weak-light stress. ALA regulates photosynthesis-related parameters and transcript levels of *RBCS*, *TPI*, *FBP*, *fructose-1,6-bisphosphate aldolase*, and *transketolase* under drought stress in rapeseed plants [13]. In tomato plants, plasma membrane intrinsic proteins (PIPs) genes, such as *PIP1* and *PIP2*, are regulated by exogenous ALA treatment. ALA not only enhances salinity stress tolerance, but also stimulates chlorophyll accumulation, chlorophyll fluorescence, and photosynthetic capacity [13]. In our previous study, we reported that exogenous ALA increases endogenous hormone accumulation, especially of 24-epibrassinolide, which regulates plant defense mechanisms, photosynthesis-related enzymes (such as Rubisco) and increasing the expression level of *rca*, *rbcS*, and *rbcl* involved in photosynthesis [27–29]. Transcriptome analysis suggested that photosystem II oxygen-evolving enhancer protein, photosystem I subunit, light-harvesting chlorophyll protein complex I and II, ferredoxin, P_n, T_r, ΦPSII, ETR, and qP, are upregulated under ALA treatment [25]. ALA is a crucial precursor in the biosynthesis of all porphyrin compounds, such as chlorophyll, heme, and phytohormones [8,12,13]. These findings indicated that exogenous ALA application stimulated the biosynthesis pathway, that enhanced chlorophyll (Figure 3), photosynthetic capacity (Figure 4), and chlorophyll fluorescence (Table 1), and reduced the detrimental effects of low-temperature and weak-light stress.

Plants increase ROS and MDA accumulation under exposure to abiotic stress, which is highly toxic and causes damaging impacts on chlorophyll, lipid, protein, and carbohydrate biosynthesis [6]. To alleviate these harmful effects, plants have evolved a defense system to scavenge these toxic and reactive species through antioxidation of enzymatic and nonenzymatic systems, which leads to damage and may cause cell death [6,30]. In the present study, low temperature and weak light significantly reduced antioxidant enzyme activities and increased $O_2^{\bullet-}$, H_2O_2 and MDA accumulation, whereas the opposite trend was observed under ALA treatment (Figure 6). Root vitality is an indicator of the overall physiological and biochemical vigor of roots [2], which are extremely sensitive to abiotic stress. Root activity decreased rapidly, and ultimately affected water and nutrient uptake, thus causing negative effects on chlorophyll, photosynthesis, enzyme activities, and growth under abiotic stress [10]. In present study, ALA diminished the detrimental influence of low-temperature and weak-light stress and increase root activity (Figure 2). Previous studies have reported that ALA plays an important role in upregulation of plant defense mechanisms under abiotic stresses [10,14,25]. Antioxidant enzymes (SOD, POD, and CAT) are involved directly in scavenging $O_2^{\bullet-}$ and H_2O_2, and catalyzing their conversion to H_2O and O_2. The current results showed that exogenous ALA enhanced activities of the antioxidant enzymes SOD, POD, and CAT in leaves and roots of cucumber under low temperature and weak light (Figure 5). In cucumber seedlings, significantly enhanced activities of SOD, POD, CAT, APX (Ascorbate peroxidase), and GR (Glutathione reductase), and reduced ROS and MDA accumulation, are observed under ALA treatment combined with low-temperature stress [10]. Previous studies have reported that ALA activates the plant defense system and defense-related genes, such as genes encoding SOD, POD, CAT, and APX, in rice and strawberry under osmotic and photodynamic stresses and reduce overproduction of ROS and MDA [31–33]. ALA is a precursor of heme biosynthesis, and CAT, POD, and APX contain a heme prosthetic group [14], which might be the reason that antioxidant enzyme activities were stimulated in ALA-treated seedlings (Figure 5). A number of defense-related genes, such as those encoding ascorbate/glutathione, CAT, and POD, are upregulated in ALA-treated bluegrass seedlings under osmotic stress. These findings are in line with those of previous studies, in which exogenous ALA upregulated antioxidant enzyme activities and reduced ROS and MDA accumulation in cucumber seedlings under low-temperature stress [8,10,12,33]. Thus, it can be concluded that exogenous ALA application increased tolerance to low-temperature and weak-light stress, and stabilized ROS and MDA accumulation, thus enhancing cucumber seedling growth (Figure 1).

5. Conclusions

The present results have demonstrated that exogenous ALA application to cucumber alleviates growth inhibition by stimulating the plant defense system and stabilizing ROS accumulation, thus enhancing tolerance to low-temperature and weak-light stress. ALA is involved in chlorophyll biosynthesis and accumulation to enhance photosynthetic capacity, and may be involved in carbohydrate and amino acid biosynthesis, which contributes to improved plant growth under low-temperature and weak-light stress. This study provides novel evidence of the potent roles of ALA and provides insight into the ALA regulatory mechanism in conjunction with low-temperature and weak-light stress. ALA was applied through the substrate and induced a distinct response to combined low-temperature and weak-light stress. The results will be helpful for off-seasonal and protected vegetable production in a greenhouse.

Author Contributions: X.Y., A.A. and Y.L. conceived and designed the experiments. A.A. and J.W. performed the experiments. A.A. analyzed the data and wrote the manuscript. X.Y., C.H. and Y.L. contributed in reagents/materials/analysis tools. Y.L. and C.H. review the manuscript. All authors have read and agreed to the published version of the manuscript.

References

1. Huang, S.; Li, R.; Zhang, Z.; Li, L.; Gu, X.; Fan, W.; Lucas, W.J.; Wang, X.; Xie, B.; Ni, P.; et al. The genome of the cucumber, *Cucumis sativus* L. *Nat. Genet.* **2009**, *41*, 1275–1281. [CrossRef]
2. Anwar, A.; Bai, L.; Miao, L.; Liu, Y.; Li, S.; Yu, X.; Li, Y. 24-Epibrassinolide Ameliorates Endogenous Hormone Levels to Enhance Low-Temperature Stress Tolerance in Cucumber Seedlings. *Int. Mol. Sci.* **2018**, *19*, 2497. [CrossRef] [PubMed]
3. Anwar, A.; Liu, Y.; Dong, R.; Bai, L.; Yu, X.; Li, Y. The physiological and molecular mechanism of brassinosteroid in response to stress: A review. *Biol. Res.* **2018**, *51*, 46. [CrossRef] [PubMed]
4. Xia, X.J.; Wang, Y.J.; Zhou, Y.H.; Tao, Y.; Mao, W.H.; Shi, K.; Asami, T.; Chen, Z.; Yu, J.Q. Reactive oxygen species are involved in brassinosteroid-induced stress tolerance in cucumber. *Plant Physiol.* **2009**, *150*, 801–814. [CrossRef] [PubMed]
5. Shu, S.; Tang, Y.; Yuan, Y.; Sun, J.; Zhong, M.; Guo, S. The role of 24-epibrassinolide in the regulation of photosynthetic characteristics and nitrogen metabolism of tomato seedlings under a combined low temperature and weak light stress. *Plant Physiol. Bioch.* **2016**, *107*, 344–353. [CrossRef] [PubMed]

6.	Gill, S.S.; Tuteja, N. Reactive oxygen species and antioxidant machinery in abiotic stress tolerance in crop plants. *Plant Physiol. Bioch.* **2010**, *48*, 909–930. [CrossRef]

7.	Xi, Z.; Wang, Z.; Fang, Y.; Hu, Z.; Hu, Y.; Deng, M.; Zhang, Z. Effects of 24-epibrassinolide on antioxidation defense and osmoregulation systems of young grapevines (*V. vinifera* L.) under chilling stress. *Plant Growth Regul.* **2013**, *71*, 57–65. [CrossRef]

8.	Naeem, M.S.; Jin, Z.L.; Wan, G.L.; Liu, D.; Liu, H.B.; Yoneyama, K.; Zhou, W.J. 5-Aminolevulinic acid improves photosynthetic gas exchange capacity and ion uptake under salinity stress in oilseed rape (*Brassica napus* L.). *Plant Soil* **2010**, *332*, 405–415. [CrossRef]

9.	Wu, Y.; Liao, W.; Dawuda, M.M.; Hu, L.; Yu, J. 5-Aminolevulinic acid (ALA) biosynthetic and metabolic pathways and its role in higher plants: a review. *Plant Growth Regul.* **2019**, *87*, 357–374. [CrossRef]

10.	Anwar, A.; Yan, Y.; Liu, Y.; Li, Y.; Yu, X. 5-Aminolevulinic Acid Improves Nutrient Uptake and Endogenous Hormone Accumulation, Enhancing Low-Temperature Stress Tolerance in Cucumbers. *Int. Mol. Sci.* **2018**, *19*, 3379. [CrossRef]

11.	Aksakal, O.; Algur, O.; Aksakal, F.; Aysin, F. Exogenous 5-aminolevulinic acid alleviates the detrimental effects of UV-B stress on lettuce (*Lactuca sativa* L) seedlings. *Acta Physiol. Plant.* **2017**, *39*. [CrossRef]

12.	Liu, D.; Wu, L.; Naeem, M.S.; Liu, H.; Deng, X.; Xu, L.; Zhang, F.; Zhou, W. 5-Aminolevulinic acid enhances photosynthetic gas exchange, chlorophyll fluorescence and antioxidant system in oilseed rape under drought stress. *Acta Physiol. Plant.* **2013**, *35*, 2747–2759. [CrossRef]

13.	Liu, D.; Hu, L.Y.; Ali, B.; Yang, A.G.; Wan, G.L.; Xu, L.; Zhou, W.J. Influence of 5-aminolevulinic acid on photosynthetically related parameters and gene expression in *Brassica napus* L. under drought stress. *Soil Sci. Plant Nutr.* **2016**, *62*, 254–262. [CrossRef]

14.	Wu, Y.; Jin, X.; Liao, W.; Hu, L.; Dawuda, M.M.; Zhao, X.; Tang, Z.; Gong, T.; Yu, J. 5-Aminolevulinic Acid (ALA) Alleviated Salinity Stress in Cucumber Seedlings by Enhancing Chlorophyll Synthesis Pathway. *Front. Plant Sci.* **2018**, *9*, 635. [CrossRef] [PubMed]

15.	Wang, L.J.; Jiang, W.B.; Huang, B.J. Promotion of 5-aminolevulinic acid on photosynthesis of melon (*Cucumis melo*) seedlings under low light and chilling stress conditions. *Physiol. Plant.* **2004**, *121*, 258–264. [CrossRef]

16.	An, Y.; Feng, X.; Liu, L.; Xiong, L.; Wang, L. ALA-Induced Flavonols Accumulation in Guard Cells Is Involved in Scavenging H_2O_2 and Inhibiting Stomatal Closure in Arabidopsis Cotyledons. *Front. Plant Sci.* **2016**, *7*, 1713. [CrossRef]

17.	Pandey, S.; Fartyal, D.; Agarwal, A.; Shukla, T.; James, D.; Kaul, T.; Negi, Y.K.; Arora, S.; Reddy, M.K. Abiotic Stress Tolerance in Plants: Myriad Roles of Ascorbate Peroxidase. *Front. Plant Sci.* **2017**, *8*, 581. [CrossRef]

18.	Jin, S.H.; Li, X.Q.; Wang, G.G.; Zhu, X.T. Brassinosteroids alleviate high-temperature injury in *Ficus concinna* seedlings via maintaining higher antioxidant defence and glyoxalase systems. *AoB PLANTS* **2015**, *7*. [CrossRef]

19.	Ogweno, J.O.; Song, X.S.; Shi, K.; Hu, W.H.; Mao, W.H.; Zhou, Y.H.; Yu, J.Q.; Nogués, S. Brassinosteroids Alleviate Heat-Induced Inhibition of Photosynthesis by Increasing Carboxylation Efficiency and Enhancing Antioxidant Systems in *Lycopersicon esculentum*. *J. Plant Growth Regul.* **2008**, *27*, 49–57. [CrossRef]

20.	Tanaka, Y.; Tanaka, A.; Tsuji, H. Effects of 5-Aminolevulinic Acid on the Accumulation of Chlorophyll b and Apoproteins of the Light-Harvesting Chlorophyll a/b-Protein Complex of Photosystem II. *Plant Cell Physiol.* **1993**, *34*, 465–472.

21.	Korkmaz, A.; Korkmaz, Y.; Demirkıran, A.R. Enhancing chilling stress tolerance of pepper seedlings by exogenous application of 5-aminolevulinic acid. *Environ. Exp. Bot.* **2010**, *67*, 495–501. [CrossRef]

22.	Kwon, S.W.; Sohn, E.J.; Kim, D.W.; Jeong, H.J.; Kim, M.J.; Ahn, E.H.; Kim, Y.N.; Dutta, S.; Kim, D.-S.; Park, J. Anti-inflammatory effect of transduced PEP-1-heme oxygenase-1 in Raw 264.7 cells and a mouse edema model. *Biochem. Bioph. Res. Co.* **2011**, *411*, 354–359. [CrossRef] [PubMed]

23.	Nunkaew, T.; Kantachote, D.; Kanzaki, H.; Nitoda, T.; Ritchie, R. Effects of 5-aminolevulinic acid containing supernatants from selected Rhodopseudomonas palustris strains on rice growth under NaCl stress, with mediating effects on chlorophyll, photosynthetic electron transport and antioxidative enzymes. *Electron. J. Biotechn.* **2014**, *17*, 1. [CrossRef]

24.	Tsuchiya, T.; Akimoto, S.; Mizoguchi, T.; Watabe, K.; Kindo, H.; Tomo, T.; Tamiaki, H.; Mimuro, M. Artificially produced [7-formyl]-chlorophyll d functions as an antenna pigment in the photosystem II isolated from the chlorophyllide a oxygenase-expressing *Acaryochloris marina*. *BBA-Bioenergetics* **2012**, *1817*, 1285–1291. [CrossRef]

25. Niu, K.; Ma, H. The positive effects of exogenous 5-aminolevulinic acid on the chlorophyll biosynthesis, photosystem and calvin cycle of Kentucky bluegrass seedlings in response to osmotic stress. *Environ. Exp. Bot.* **2018**, *155*, 260–271. [CrossRef]

26. Zhao, Y.Y.; Yan, F.; Hu, L.P.; Zhou, X.T.; Zou, Z.R.; Cui, L.R. Effects of exogenous 5-aminolevulinic acid on photosynthesis, stomatal conductance, transpiration rate, and PIP gene expression of tomato seedlings subject to salinity stress. *Genet. Mol. Res.* **2015**, *14*, 6401–6412. [CrossRef]

27. Wei, L.J.; Deng, X.G.; Zhu, T.; Zheng, T.; Li, P.X.; Wu, J.Q.; Zhang, D.W.; Lin, H.H. Ethylene is Involved in Brassinosteroids Induced Alternative Respiratory Pathway in Cucumber (*Cucumis sativus* L.) Seedlings Response to Abiotic Stress. *Front. Plant Sci.* **2015**, *6*, 982. [CrossRef]

28. Choudhary, S.P.; Yu, J.Q.; Yamaguchi-Shinozaki, K.; Shinozaki, K.; Tran, L.S. Benefits of brassinosteroid crosstalk. *Trends Plant Sci.* **2012**, *17*, 594. [CrossRef]

29. Xia, X.-J.; Huang, L.-F.; Zhou, Y.-H.; Mao, W.-H.; Shi, K.; Wu, J.-X.; Asami, T.; Chen, Z.; Yu, J.-Q. Brassinosteroids promote photosynthesis and growth by enhancing activation of Rubisco and expression of photosynthetic genes in *Cucumis sativus*. *Planta* **2009**, *230*, 1185. [CrossRef]

30. Zhu, T.; Deng, X.; Zhou, X.; Zhu, L.; Zou, L.; Li, P.; Zhang, D.; Lin, H. Ethylene and hydrogen peroxide are involved in brassinosteroid-induced salt tolerance in tomato. *Sci. Rep.* **2016**, *6*, 35392. [CrossRef]

31. Phung, T.H.; Jung, S. Differential antioxidant defense and detoxification mechanisms in photodynamically stressed rice plants treated with the deregulators of porphyrin biosynthesis, 5-aminolevulinic acid and oxyfluorfen. *Biochem. Bioph. Res. Co.* **2015**, *459*, 346–351. [CrossRef] [PubMed]

32. Cai, C.; He, S.; An, Y.; Wang, L. Exogenous 5-aminolevulinic acid improves strawberry tolerance to osmotic stress and its possible mechanisms. *Physiol. Plantarum* **2019**. [CrossRef] [PubMed]

33. Anwar, A.; Li, Y.; He, C.; Yu, X. 24-Epibrassinolide promotes NO_3^- and NH_4^+ ion flux rate and NRT1 gene expression in cucumber under suboptimal root zone temperature. *BMC Plant Biol.* **2019**, *19*, 225.

8

Response to the Cold Stress Signaling of the Tea Plant (*Camellia sinensis*) Elicited by Chitosan Oligosaccharide

Yingying Li, Qiuqiu Zhang, Lina Ou, Dezhong Ji, Tao Liu, Rongmeng Lan, Xiangyang Li and Linhong Jin *

State Key Laboratory Breeding Base of Green Pesticide and Agricultural Bioengineering, Key Laboratory of Green Pesticide and Agricultural Bioengineering, Ministry of Education, Guizhou University, Huaxi District, Guiyang 550025, China; gs.yingyingli17@gzu.edu.cn (Y.L.); gs.zhangqq18@gzu.edu.cn (Q.Z.); gs.lnou17@gzu.edu.cn (L.O.); gs.dzji19@gzu.edu.cn (D.J.); gs.taoliu18@gzu.edu.cn (T.L.); gs.rmlan19@gzu.edu.cn (R.L.); xyli1@gzu.edu.cn (X.L.)
* Correspondence: lhjin@gzu.edu.cn

Abstract: Cold stress caused by a low temperature is a significant threat to tea production. The application of chitosan oligosaccharide (COS) can alleviate the effect of low temperature stress on tea plants. However, how COS affects the cold stress signaling in tea plants is still unclear. In this study, we investigated the level of physiological indicators in tea leaves treated with COS, and then the molecular response to the cold stress of tea leaves treated with COS was analyzed by transcriptomics with RNA-Sequencing (RNA-Seq). The results show that the activity of superoxide dismutase (SOD) activity, peroxidase (POD) activity, content of chlorophyll and soluble sugar in tea leaves in COS-treated tea plant were significantly increased and that photosynthesis and carbon metabolism were enriched. Besides, our results suggest that COS may impact to the cold stress signaling via enhancing the photosynthesis and carbon process. Our research provides valuable information for the mechanisms of COS application in tea plants under cold stress.

Keywords: tea plant; cold stress; chitosan oligosaccharide; physiological response; transcriptome

1. Introduction

The tea plant (*Camellia sinensis* (L.) O. Kuntze) is one of the most important commercial beverage crops in the world and an important revenue source in tea-producing countries [1]. The tea production in over 50 countries has reached over 5.95 million tons on 4.1 million hectares around the world [2]. Among them, the cultivar 'Anji Baicha' is a special green-revertible albino mutant widely cultivated in China, especially in Zhejiang, Hubei and Guizhou provinces, which exhibits periodic albinism during the development of young shoots [3,4]. It is rare and represent precious tea germplasm because of it special flavor, and also has high levels of total amino acids and low levels of polyphenols, which differs from conventional tea [3–8]. In addition, it has a higher commercial value than green tea [4].

The tea plant can grow in different agroclimates and adapted to optimal temperature of 18 to 30 °C and pH ranging from 4.5 to 5.5, but the thermophilic nature of tea plants confines their growth to temperate area [9–11]. Furthermore, tea plants that are exposed to a low temperature, such as a sudden frost in fall or early spring, may be at risk of cold stress [12]. Cold environment can adversely affect tea plants on their growth, development, and spatial distribution with decreasing yield and quality, which is one of the factors restricting the healthy development of the tea industry [13–15]. So, it is significant to explore the ways to improve the cold resistance of tea plants. Some studies have reported that the cold resistance of tea plant can be effectively improved by cultivating cold-resistant tea plant varieties (e.g., Fudingdabai, Shuchazao), cold acclimation of tea plant and the application of exogenous substances [16–19].

Chitosan oligosaccharide (COS) prepared from chitosan, is an environmentally friendly plant growth regulator and stress tolerance inducer [20–24]. Chitosan is a linear polysaccharide composed of β-1,4-glucosamines. The hydrolysis of the glycosidic chitosan chains yields oligosaccharides, including the water-soluble oligochitosan [21,22]. Chitosan and COS have a rich history of being researched for applications in agriculture, primarily for plant defense and yield increase [23,24]. As a natural biocontroller and elicitor of defense responses, COS can boost the innate ability of plants to defend themselves by stimulating secondary metabolite synthesis, and increasing the chlorophyll content and photosynthetic ability [20,21], enrich the soluble sugar in plant [25], and enhancing the activities of antioxidant enzymes [25–27]. COS stimulated the signaling pathways involved in disease resistance in rice [28], and its role in tobacco mosaic virus (TMV) resistance in *Arabidopsis* has been investigated [29]. And studies have shown that COS enhances carbon metabolism, nitrogen metabolism, photosynthesis, and defense against abiotic stress in plants [30]. As reported, COS was able to mitigate the effects of abiotic stresses in plant, including salt, cold and drought [25–27,31,32]. The mechanism of COS in increasing abiotic stress tolerances was summarized as: enhancing the activities of antioxidant enzymes [25], photosynthesis, and stimulate secondary metabolite synthesis [31]. For example, COS has been applied to wheat seedlings for improved chilling tolerance by enhancing antioxidant activities of superoxide dismutase (SOD) and peroxidase (POD) and increasing content of chlorophyll.

These physiological responses of plants elicited by COS are closely related to the regulation of plant gene expression. Transcriptome sequencing has been widely applied to tea plant, which is has the advantage of highly accurate, highly efficient and sensitive profiling in recent years [33]. RNA sequencing (RNA-Seq) technology for measuring transcriptomes of organisms can analyze genes related to abiotic and biotic stress responses, growth, development and metabolites [34–37], to improve our understanding of the molecular mechanism of the tea plant [13–16,38], and RNA-Seq will also be a valuable tool to reveal the role of exogenous substances in tea plant cold resistance.

Though many investigators provided valuable information to cold stress in tea plant, the action mode of COS eliciting responses to cold stress of tea plant is unclear. Therefore, in this report, we studied the effect of exogenous COS on the molecular mechanism of tea plant under low temperature stress. Herein, the physiological parameters of tea plants with and without COS-treatment were compared. The molecular response to cold resistance within tea plant was analyzed by RNA-Seq technology. This research improves the understanding of the cold resistance mechanism of COS-treated tea plant and provides important guidance for COS application under low temperature stress.

2. Materials and Methods

2.1. Plant Materials and Cold Treatments

Two-year-old albino tea cultivar (*Camellia sinensis* (L.) O. Kuntze cv. 'Anji Baicha') were used in the experiment from AnShun County, Guizhou Province, China. Additionally, the tea plants were transplanted into the plastic pot. Plants were grown in a growth chamber at the experimental of Guizhou University, Guizhou Province, China (16 h day/8 h night at 25 °C/20 °C and relative humidity of 70%). After a month, tea plants were treated with 10 mL of following elicitors by surface spraying with sterile distilled water (control, CK), or with 1.25 mL/L COS solution (COS comes from Hainan Zhengye Zhongnong High-tech Co., Ltd., Haikou, Hainan Province, China). After 24 h, the two groups of tea plants were separately maintained in a chamber at −4 and −8 °C at cold treatment for 24 h, with one group maintained under normal room temperature conditions. Three independent biological repeats were collected for each treatment. Fresh leaves from the stable stage (re-greening stage) of chlorophyll development of Anji Baicha were harvested at 24 h and frozen immediately in liquid nitrogen and stored at −80 °C for further study.

2.2. Physiological Response Assay

Physiological indexes of tea leaves (containing 1st, 2nd, 3rd leaf and old leaves), involving the activities of SOD and POD, and content of chlorophyll and soluble sugar, were determined. Additionally, the assay kits used included the SOD assay kit, the POD assay kit, the chlorophyll assay kit, the soluble sugar assay kit (Solarbio, Cat. No. BC0175, BC0095, BC0995, BC0035, respectively, Beijing, China). All assays were performed according to the manufacturer's instructions.

2.3. cDNA Library Construction and Sequencing

We selected tea leaves from control and COS treatment on −4 °C for RNA-Seq analysis. Total RNA was extracted from tea leaves using TRIzol reagent (Invitrogen, Carlsbad, CA, USA) following the manufacturer's instruction. Poly (A) + mRNA was purified with oligo (dT) beads. The mRNA was randomly cut into short fragments using Fragmentation Buffer, which were used as a template for the short fragment mRNA, first-strand cDNA was synthesized with 6 bp random primers, and then the Buffer, dNTPs and DNA polymerase I were added to synthesize the second-strand cDNA. RNA Integrity was confirmed using 1.5% agarose gel. RNA quality was checked by a NanoDrop^{TM} OneC spectrophotometer (Thermo Fisher Scientific, New York, NY, USA). RNA qualified was measured by Qubit^{TM} RNA BR Assay Kit in Qubit$^{®}$ 2.0 (Life Technologies, Carlsbad, CA, USA). The cDNA library construction and Illumina sequencing of the samples were performed using a 150 bp paired-end Illumina Nova-seq 6000 (Illumina, San Diego, CA, USA) by Seqhealth Technology Co., Ltd. (Wuhan, China).

2.4. RNA-Seq Data Analysis

The raw reads were first filtered to obtain the clean reads by removing the adaptor sequences, unknown sequences "N" and low-quality reads using Trimmomatic (version 0.36). After filtering, the clean reads were mapped to the reference genome of Camellia sinensis using STATR software (version 2.5.3a).

2.5. Identification of Differentially Expressed Genes

The expression levels of each gene were calculated and normalized by the corresponding Reads Per Kilobase of transcript per Million mapped reads (RPKM). The RPKM method can eliminate the influence of gene length and sequencing amount differences on gene expression. FeatureCounts (version 1.5.1) was used to count the read numbers mapped to each gene [39]. Additionally, differentially expressed genes (DEGs) were identified with the edge R package (version 3.12.1) [40]. The resulting p-values were adjusted using Benjamini and Hochberg's method for controlling the false discovery rate (FDR). Genes with p-value < 0.05 and a logarithm two-fold change $|log_2FC| > 1$ were defined as DEGs.

2.6. Gene Ontology and KEGG Pathway Analysis

Gene ontology (GO) analysis and Kyoto encyclopedia of genes and genomes (KEGG) enrichment analysis of DEGs were both implemented by KEGG orthology based annotation system (KOBAS) software (version 2.1.1) with p-value < 0.05 to judge statistically significant enrichment [41].

2.7. Quantitative RT-PCR (qRT-PCR) Analysis

To verify the RNA-Seq analysis, we randomly selected five unigenes and used qRT-PCR to confirm their participation in the high-temperature reaction. RT-qPCR was conducted on ABI ViiA^{TM} 7 Real-Time PCR System (Applied Biosystems, Foster, CA, USA) using GoTaq$^{®}$ qPCR Master Mix (Promega, Madison, WI, USA). The PCR amplifications were consisted of 95 °C for 3 min, followed by 40 cycles of 95 °C for 15 s, 60 °C for 30 s, and then 72 °C for 30 s. Gene expression was normalized using the glyceraldehyde-3-phosphate dehydrogenase (GADPH) as an internal reference gene, and the relative changes of gene expression were calculated using the $2^{-\Delta\Delta Ct}$ method. The list of primers is presented in Table S1.

2.8. Statistical Analysis

Data were expressed as the mean ± standard error, and the data were subjected to one-way analysis of variance (ANOVA) ($p < 0.05$) followed by a significant difference test (LSD) using SPSS statistics v17.0 (SPSS Inc., Chicago, IL, USA).

3. Results

3.1. Physiological Parameter Response to a Low Temperature

To analyze the effects of COS on tea plant growth, we measured the change in activity of SOD, and POD enzymes and content of chlorophyll, soluble sugar in COS-treated tea plant and their respond to low temperature stress, with sterile distilled water served as control. As shown in Figure 1, under a low temperature, the tea plant responds to cold stress with all the physiological parameters changed and COS-enhanced freeze protection. As in the control group, a low temperature caused increases in those physiological parameters. As shown in Figure 1A, the enzyme activity of SOD was significantly increased by 24.04% at −4 °C and 32.68% at −8 °C. Similarly, the enzyme activity of POD was significantly increased by 38.05% at −4 °C and 8.81% at −8 °C. Cold stress significantly reduced the chlorophyll content by 20.18% and 21.96% at −4 and −8 °C, respectively (Figure 1C). Moreover, soluble sugar content was significantly increased by 29.87% at −4 °C and 28.16% at −8 °C, respectively (Figure 1D). The results show that cold stress consistently increased SOD and POD activity, and soluble sugar content, when the temperature was switched from 25 °C to −4 °C or −8 °C, but POD activity was highest at −4 °C.

When exogenous COS was used, it consistently enhanced SOD and POD activities, and the soluble sugar content and chlorophyll content in the tea plant. For example, COS improved SOD activity by 11.75% at 25 °C, 25.93% at −4 °C and 9.21% at −8 °C, respectively, as compared with the control. Similarly, POD activity was enhanced by 19.91%, 19.23% and 30.09% on 25 °C, −4 °C and −8 °C, respectively.

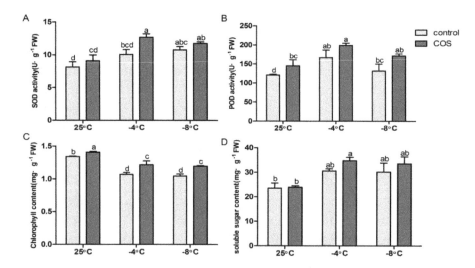

Figure 1. Effect of chitosan oligosaccharide (COS) on physiological parameters of tea leaves. (**A**) Superoxide dismutase (SOD) activity; (**B**) peroxidase (POD); (**C**) chlorophyll content; (**D**) soluble sugar content. The data represent the means ± SD of three replicates samples. Different letters indicate significant differences at $p < 0.05$.

For all the tested parameters, the effects of COS were more pronounced under cold stress. When tea plants were treated with COS combined with cold stress, SOD enhanced by 56.21% and 44.91% at −4 and −8 °C, respectively. Similarly, POD increased 37.26% and 18.04%. The content of soluble sugar also increased by 45.22% and 40.25% at −4 and −8 °C, respectively. Chlorophyll content was decreased by 13.47% and 14.99%, respectively. The results show that COS treatment consistently increased chlorophyll content, but three parameters of SOD, POD and soluble sugar were highest at −4 °C of cold stress combined with COS.

3.2. Transcriptome Sequencing and Assembly

To understand the response of the tea plant to cold stress and the effect of COS on the molecular level, we compared the transcriptomes between COS treatment and the control group at −4 °C by RNA-Seq. Replicate samples of the control group (ConT3_1/2/3) and COS-treatment group (TreT3_1/2/3) were included in this study. We obtained 5.59–6.60 million raw reads in control and 5.79–6.77 million raw reads in the COS-treatment group. After filtering and removing low-quality reads, the clean reads were limited 5.26–6.21 million and 5.45–6.34 million, respectively. Of these clean reads, the GC content was 46.46–47.21% and the Q30 values were over 98.45%. The ratio of total mapped reads between the control and COS-treatment groups was 94.69–94.90% and 94.85–95.20% for *Camellia sinensis* according to the Genome Database. Unique mapped reads were 91.48–92.10% in the control group and 88.02–90.66% in the COS-treatment group (Table 1).

Table 1. Statistical analyses and mapping results of RNA sequencing reads.

Sample	ConT3_1	ConT3_2	ConT3_3	TreT3_1	TreT3_2	TreT3_3
Raw reads	55,965,032	56,476,808	66,044,722	57,864,054	67,743,104	65,453,870
Clean reads	52,619,470	53,061,678	62,155,236	54,555,936	63,422,124	61,416,118
Q30 (%)	98.45	98.45	98.70	98.65	98.55	98.45
GC content (%)	46.60	46.46	46.63	46.82	46.83	47.21
Total reads	44,163,580	43,980,650	52,344,630	45,455,332	52,920,720	51,188,834
Total mapped	41,828,592 (94.71%)	41,644,005 (94.69%)	49,676,907 (94.90%)	43,274,546 (95.20%)	50,292,573 (95.03%)	48,551,418 (94.85%)
Unique mapped	38,522,223 (92.10%)	38,095,551 (91.48%)	45,663,412 (91.92%)	39,232,023 (90.66%)	45,576,183 (90.62%)	42,734,470 (88.02%)

3.3. Differentially Expressed Genes Analysis

In order to verify the correlation of gene expression level between samples, we demonstrated that the biological repeatability between samples was great through spearman correlation analysis based on the RPKM of different samples. Genes with *p*-value < 0.05 and |log$_2$(Foldchange)| > 1 were defined as differentially expressed genes between control and COS. There were identified 4503 differentially expressed genes (DEGs) between the control and COS, including 1605 up-regulated and 2898 down-regulated genes in the leaves of tea plant (Figure 2 and Table S2).

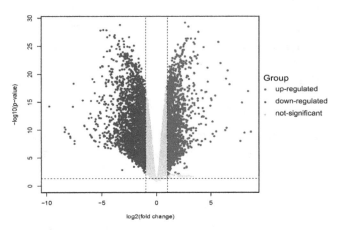

Figure 2. Volcano plot of differentially expressed genes (DEGs) showed up-regulated and down-regulated between control and COS under −4 °C treatment. The red dots represent up-regulated genes, the blue dots represent down-regulated genes, and the gray dots represent no significant difference. The horizontal coordinates indicate the change in multiple expression, the longitudinal coordinates indicate the magnitude of differences.

3.4. Gene Ontology (GO) Annotation

The differentially expressed mRNAs were analyzed by GO enrichment, as shown in Figure 3 and Table S3. The differentially expressed genes were mostly enriched in biological process (Figure 3). In the biological process categorization, functional enrichment mainly focuses on metabolic processes and nutrient synthesis processes, such as "single-organism biosynthetic process" (GO: 0044711), "metabolic process" (GO: 0008152), "carbohydrate metabolic process" (GO: 0005975) and "carbohydrate derivative biosynthetic process" (GO: 1901137). The molecular function category includes the expression of transmembrane transporters and catalytic enzyme-related genes, such as "catalytic activity" (GO: 0003824), "transporter activity" (GO: 0005215), "transmembrane transporter activity" (GO: 0022857), and "ion transmembrane transporter activity" (GO: 0015075). Besides, "serine-type endopeptidase activity" (GO: 0004252) was mostly enriched in the molecular function category.

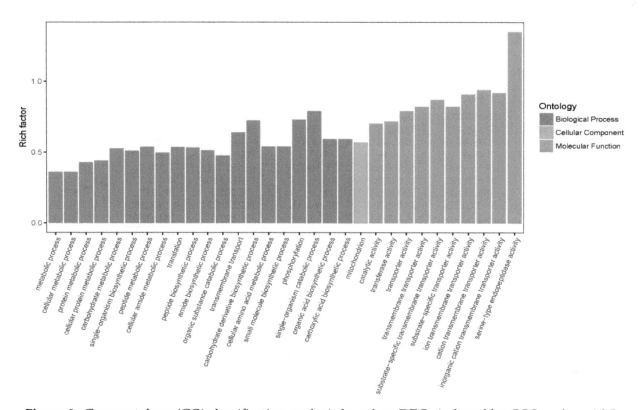

Figure 3. Gene ontology (GO) classification analysis based on DEGs induced by COS under −4 °C treatment. The horizontal coordinates indicate GO terms, the longitudinal coordinates indicate rich factor, rich factor represents the ratio between the number of different genes enriched in the term and the background genes in GO term.

3.5. Kyoto Encyclopedia of Genes and Genomes (KEGG) Pathway Annotation

The KEGG enrichment scatter plot is a graphical representation of the statistical analyses that visualizes the pathway enrichment (Figure 4). The degree of KEGG enrichment was measured in terms of richness factor, p-value, and the number of genes in the pathway. The important enriched pathways with high generation, low p-value and large numbers of genes are shown in the Figure 4 and Table S4. As shown in Figure 4, these enriched pathways, including "photosynthesis" (ko00195), "carbon fixation in photosynthetic organisms" (ko00710), "photosynthesis–antenna proteins" (ko00196), "ribosome" (ko03010), "carbon metabolism" (ko01200).

Compared with the control group, 71 genes were significantly induced to up-regulated by COS treatment, including PSII, PSI, cytochrome b6/f complex, photosynthethic electron transport and F-type ATPase (Table S5). In the carbon metabolism pathway, a total of 77 genes were differentially expressed, including 52 up-regulated and 25 down-regulated (Table S6). A total of 43 genes were assigned to the plant hormone signal transduction pathway, including 16 genes that were up-regulated in auxin, abscisic acid, ethylene, salicylic acid (Table S7). These results suggest that the addition of COS at a low temperature have a complex effect on biological process and metabolism of the tea plant.

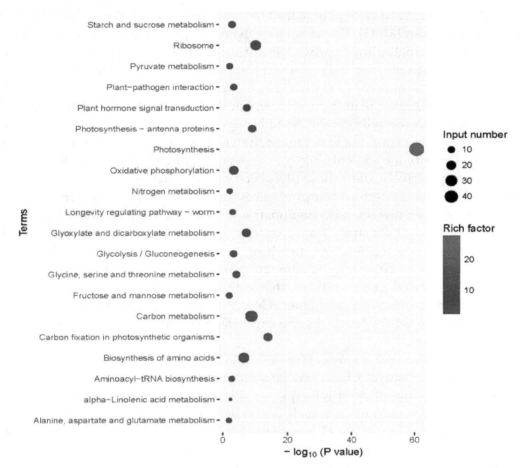

Figure 4. Kyoto Encyclopedia of Genes and Genomes (KEGG) enrichment analysis based on DEGs induced by COS under −4 °C treatment. The significance of enrichment is shown on the horizontal coordinates (represented by $-\log_{10}$ (p-value), the greater the value, the more significant the enrichment), and the KEGG pathway is shown on the longitudinal coordinates. The size of the dots indicates the number of different genes contained in the KEGG pathway, and the color of the dots indicates the degree of rich factor enrichment.

3.6. qRT-PCR Validation of Differentially Expressed Transcripts from RNA-Seq

Five transcripts were randomly selected for qRT-PCR analysis, which used to confirm validity and accuracy the RNA-Seq data. The results show that the trend of qRT-PCR is consistent with the results of RNA-Seq in Figure S1.

4. Discussion

Cold stress affects photosynthetic activities and metabolic functions in plants, which further affected growth, development, and metabolism. It has a negative effect on the yield and quality of tea. Anji Baicha is a temperature-sensitive albino tea cultivar. When the environment temperature is below 20 °C in early spring, the white shoots phenomenon will appear. After about two weeks, the plant gradually turns as green, as does those of common tea cultivars [4–6]. The change of leaf color was mainly due to chloroplast development in the albescent stage, the etioplast–chloroplast transition was blocked, and the accumulation of chlorophyll was inhibited under low temperature [4–8,37]. In this study, we chose Anji Baicha in the stable stage of chlorophyll development as a research object, the results revealed that COS could enhance antioxidant activity, increase accumulation of sugar content and chlorophyll content in tea plant. It is confirmed that COS could play an important role in improving stress tolerance of Anji Baicha.

Cold stress can cause excessive production of reactive oxygen species (ROS), disrupt the normal physiological and metabolic balance of plants, lead to the increase of membrane lipid peroxidation and damage to vital biomolecules [42,43]. Plants have evolved complex mechanisms to combat against the damage induced by ROS, including improve the antioxidant enzymes [44,45]. In this study, under cold stress, the tea plant natively reacted to protect themselves by increasing the activity of SOD and POD enzyme, and the application of COS provided external assistance plant. Chlorophyll content in COS-treated tea plant was higher than in control, which indicated that COS application mitigated the cold-induced decline in chlorophyll content. Soluble sugar can maintain the osmotic balance, and the soluble sugar in COS treated tea plant was higher than that without COS treatment, suggesting that COS can stabilize cell membrane and enhance cold resistance of plant. Those results indicated that the utilization of COS can positively affect these physiological parameters in tea plants, and beneficially regulate the natural defense system and improve growth and developmental processes of tea plants under cold stress. Moreover, this was also demonstrated in wheat seedlings where the application of COS could enhance the activities of antioxidant enzymes and the content of chlorophyll and alleviate the damage of abiotic stress in wheat [25–27,46]. In wheat, COS could enhance the activities of antioxidant enzymes and the content of chlorophyll, alleviate plant the damage of abiotic stress [25–27,46]. These differentially expressed genes indicate that the application of COS has complex effects on metabolism and signaling pathways of tea plants at low temperature. From RNA sequencing, we found that COS significantly altered the level of gene expression involved in photosynthesis and carbon metabolism under cold stress.

The up-regulated differentially expressed genes could be important for the pathology and biological processes of response to cold stress. Chlorophyll content is an important parameter frequently used to indicate chloroplast development, and which is sensitive to abiotic stresses [47]. COS can increase chlorophyll content under cold stress, which is consistent with the observations from RNA-Seq. Compared with the control group, COS treatment may increase the photosynthesis of plants by significantly up-regulating photosystem I (PSI), photosystem II (PSII)-related genes (Table S5). In the PSII core complex, PsbR is an important link, which can stabilize the assembly of the oxygen-evolving complex protein PsbP [48]. In the present study, PsbR was up-regulated, which was consistent with the action of chitosan heptamer response in wheat seedling [49]. Besides, Chlorophyll a/b-binding protein can participate in light uptake, transfer energy to the reaction centers of the photosystem I and photosystem II, and regulate the excitation energy distribution to maintain the structure of the thylakoid membrane [50], and all of 23 chlorophyll a/b-binding protein genes were also up-regulated, which can imply the recovery of photosynthesis activities by COS treatment under cold stress [51]. These results indicate that COS may enhance photosynthesis via the upregulation of related proteins to improve the cold resistance of tea plant.

In the carbon metabolism pathway, genes encoding ribulose bisphosphate carboxylase small subunit (rbcS), phosphoglycerate kinase, glyceraldehyde-3-phosphate dehydrogenase, triosephosphate isomerase were up-regulated significantly (Table S6). RbcS is one of the subunits of Ribulose-1,5-bisphosphate carboxylase/oxygenase (RuBisCo), and the activity of rbcS decreased to inhibit photosynthesis under cold stress [52]. This result was consistent with previous research demonstrating the application of COS to regulate the photosynthetic mechanism and carbon metabolism and thereby the plant growth [53].

During plant development, the response of plants to endogenous and environmental signals is mediated by several hormones, which are involved in almost every aspect of plant growth. For example, plants respond very quickly to auxin, including cell growth and the activation of multiple auxin-responsive genes [53]. Indole-3-acetic acid (GH3) and the ethylene receptor (ETR) were up-regulated genes in the plant hormone signal transduction pathway (Table S7). GH3 is an important response gene of auxin-responsive protein (IAA), which can encode a class of IAA-amido synthetases responsible for balancing endogenous free IAA content and plays an important role in IAA-regulated plant growth and development [54,55]. The ETR responds to ethylene and abscisic acid (ABA) signaling. ETR is the most important ethylene receptor protein in plants, and the lack of ETR will hinder the transduction of ethylene signal cascade reaction, resulting in the insensitivity to ethylene in plant [56–58].

The application COS can improve antioxidant enzyme activities, and the content of chlorophyll and soluble sugar. Besides, compared with the control group, the addition of COS significantly changed the photosynthesis pathway and carbon metabolism of tea plants under low temperature stress, which may contribute to COS' ability to improve the cold tolerance of tea plants. These results may represent that COS participates in the specific regulatory mechanism related to cold adaptation in the cold resistance of Anji Baicha. As for the comparison of cold resistance between Anji Baicha and other tea plants (e.g., Xiaoxueya, Fudingdabai), we are further carrying out relevant experimental verification.

5. Conclusions

In summary, low temperature will impact the key physiological and developmental processes that determine the yield of tea. This study indicates that the utilization of COS can positively affect these physiological parameters in tea plants by improving antioxidant enzyme activities, and the content of chlorophyll and soluble sugar. Hence, COS can beneficially regulate the natural defense system and improve the growth and developmental processes of tea plants under cold stress. With transcriptome sequencing and differentially expressed genes analysis, we identified 1605 up-regulated and 2898 down-regulated genes in COS compared to the control, and photosynthesis and the carbon metabolism pathway of enrichment may play a role in the COS-improved cold resistance of a tea plant. The results may provide the foundation for further research on the regulation mechanism of COS on plant cold tolerance.

Supplementary Materials
Table S1: Primer sequences used for qRT-PCR. Table S2: The list of different expression genes. Table S3: GO enrichment list of different expression genes. Table S4: KEGG pathway enrichment list of different expression genes. Table S5: Differentially expressed genes in photosynthesis related pathway. Table S6: Differentially expressed genes in carbon metabolism pathway. Table S7: Differentially expressed genes in plant hormone signal transduction pathway. Figure S1. Verification of relative expression levels of DEGs in transcriptome date by qRT-PCR between control and COS.

Author Contributions: Y.L. conducted the experiments; Y.L., L.O. and D.J. designed and performed the experiments; Y.L., Q.Z., T.L. and R.L. analyzed the data; X.L. and L.J. conceived and supervised the project. All authors have read and agreed to the published version of the manuscript.

Acknowledgments: We are grateful to Xia Zhou, Guizhou University, for the fruitful discussions and helpful comments on earlier draft.

References

1. Singh, H.R.; Hazarika, P. Biotechnological Approaches for Tea Improvement. In *Biotechnological Progress and Beverage Consumption*; Academic Press: Cambridge, MA, USA, 2020; pp. 111–148.
2. FAOSTAT-Food and Agriculture Organization of the United Nations Statistics Division. Available online: http://faostat3.fao.org/home/E (accessed on 31 December 2018).
3. Ma, C.L.; Chen, L.; Wang, X.; Jin, J.Q.; Ma, J.Q.; Yao, M.Z. Differential expression analysis of different albescent stages of 'Anji Baicha' (*Camellia sinensis* (L.) O. Kuntze) using cDNA microarray. *Sci. Hortic.* **2012**, *148*, 246–254. [CrossRef]
4. Du, Y.Y.; Liang, Y.R.; Wang, H.; Wang, K.R.; Lu, J.L.; Zhang, G.H.; Lin, W.P.; Li, M.; Fang, Q.Y. A study on the chemical composition of albino tea cultivars. *J. Hort. Sci. Biotechnol.* **2006**, *81*, 809–812. [CrossRef]

5.   Cheng, H.; Li, S.F.; Chen, M.; Yu, F.L.; Yan, J.; Liu, Y.M.; Chen, L.A. Physiological and biochemical essence of the extraordinary characters of Anji Baicha. *J. Tea Sci.* **1999**, *19*, 87–92.

6.   Du, Y.Y.; Chen, H.; Zhong, W.L.; Wu, L.Y.; Ye, J.H.; Lin, C.; Zheng, X.Q.; Lu, J.L.; Liang, Y.R. Effect of temperature on accumulation of chlorophylls and leaf ultrastructure of low temperature induced albino tea plant. *Afr. J. Biotechnol.* **2008**, *7*, 1881–1885. [CrossRef]

7.   Feng, L.; Gao, M.J.; Hou, R.Y.; Hu, X.Y.; Zhang, L.; Wan, X.C.; Wei, S. Determination of quality constituents in the young leaves of albino tea cultivars. *Food Chem.* **2014**, *155*, 98–104. [CrossRef]

8.   Wei, K.; Wang, L.Y.; Zhou, J.; He, W.; Zeng, J.M.; Jiang, Y.W.; Cheng, H. Comparison of catechins and purine alkaloids in albino and normal green tea cultivars (*Camellia sinensis* L.) by HPLC. *Food Chem.* **2012**, *130*, 720–724. [CrossRef]

9.   Shen, J.; Wang, Y.; Chen, C.; Ding, Z.; Hu, J.; Zheng, C.; Li, Y. Metabolite profiling of tea (*Camellia sinensis* L.) leaves in winter. *Sci. Hortic.* **2015**, *192*, 1–9. [CrossRef]

10. Wang, L.; Cao, H.; Qian, W.; Yao, L.; Hao, X.; Li, N.; Yang, Y.; Wang, X. Identification of a novel bZIP transcription factor in *Camellia sinensis* as a negative regulator of freezing tolerance in transgenic *Arabidopsis*. *Ann. Bot.* **2017**, *119*, 1195–1209. [CrossRef] [PubMed]

11. Zhang, Q.W.; Li, T.Y.; Wang, Q.S.; LeCompte, J.; Harkess, R.L.; Bi, G.H. Screening tea cultivars for novel climates: Plant growth and leaf quality of *Camellia sinensis* cultivars grown in Mississippi, United States. *Front. Plant Sci.* **2020**, *11*, 280. [CrossRef]

12. Li, X.; Ahammed, G.; Li, Z.; Zhang, L.; Wei, J.; Yan, P.; Zhang, L.; Han, W. Freezing stress deteriorates tea quality of new flush by inducing photosynthetic inhibition and oxidative stress in mature leaves. *Sci. Hortic.* **2018**, *230*, 155–160. [CrossRef]

13. Wang, X.C.; Zhao, Q.Y.; Ma, C.L.; Zhang, Z.H.; Cao, H.L.; Kong, Y.M.; Yue, C.; Hao, X.Y.; Chen, L.; Ma, J.Q.; et al. Global transcriptome profiles of *Camellia sinensis* during cold acclimation. *BMC Genom.* **2013**, *14*, 415. [CrossRef] [PubMed]

14. Yin, Y.; Ma, Q.; Zhu, Z.; Cui, Q.; Chen, C.; Chen, X.; Fang, W.; Li, X. Functional analysis of CsCBF3 transcription factor in tea plant (*Camellia sinensis*) under cold stress. *Plant Growth Regul.* **2016**, *80*, 335–343. [CrossRef]

15. Zhang, Y.; Zhu, X.; Chen, X.; Song, C.; Zou, Z.; Wang, Z.; Wang, M.; Fang, W.; Li, X. Identification and characterization of cold-responsive microRNAs in tea plant (*Camellia sinensis*) and their targets using high-throughput sequencing and degradome analysis. *BMC Plant Biol.* **2014**, *14*, 271. [CrossRef] [PubMed]

16. Yang, Y.J.; Zheng, L.Y.; Wang, X.C. Effect of cold acclimation and ABA on cold hardiness contents of proline in tea plants. *J. Tea Sci.* **2004**, *24*, 177–182.

17. Li, Y.Y.; Wang, X.W.; Ban, Q.Y.; Zhu, X.X.; Jiang, C.J.; Wei, C.L.; Bennetzen, J.L. Comparative transcriptomic analysis reveals gene expression associated with cold adaptation in the tea plant *Camellia sinensis*. *BMC Genom.* **2019**, *20*, 624. [CrossRef] [PubMed]

18. Ban, Q.Y.; Wang, X.W.; Pan, C.; Wang, Y.W.; Kong, L.; Jiang, H.G.; Xu, Y.Q.; Wang, W.Z.; Pan, Y.T.; Li, Y.Y.; et al. Comparative analysis of the response and gene regulation in cold resistant and susceptible tea plants. *PLoS ONE* **2017**, *12*, e0188514. [CrossRef] [PubMed]

19. Li, J.H.; Yang, Y.Q.; Sun, K.; Chen, Y.; Chen, X.; Li, X.H. Exogenous Melatonin Enhances Cold, Salt and Drought Stress Tolerance by Improving Antioxidant Defense in Tea Plant (Camellia sinensis (L.) O. Kuntze). *Molecules* **2019**, *24*, 1826. [CrossRef]

20. Cabrera, J.; Wégria, G.; Onderwater, R.; González, G.; Nápoles, M.; Falcón-Rodríguez, A.; Costales, D.; Rogers, H.; Diosdado, E.; González, S.; et al. Practical use of oligosaccharins in agriculture. *Acta. Hortic.* **2013**, *1009*, 195–212. [CrossRef]

21. Kim, S.; Rajapakse, N. Enzymatic production and biological activities of chitosan oligosaccharides (COS): A review. *Carbohydr. Polym.* **2005**, *62*, 357–368. [CrossRef]

22. Yin, H.; Du, Y.G.; Zhang, J.Z. Low molecular weight and oligomeric chitosans and their bioactivities. *Curr. Top. Med. Chem.* **2009**, *9*, 1546–1559. [CrossRef]

23. Wang, M.Y.; Chen, Y.C.; Zhang, R.; Wang, W.X.; Zhao, X.M.; Du, Y.G.; Yin, H. Effects of chitosan oligosaccharides on the yield components and production quality of different wheat cultivars (*Triticum aestivum* L.) in Northwest China. *Field Crop. Res.* **2015**, *172*, 11–20. [CrossRef]

24. Chatelain, P.G.; Pintado, M.E.; Vasconcelos, M.W. Evaluation of chitooligosaccharide application on mineral accumulation and plant growth in *Phaseolus vulgaris*. *Plant Sci.* **2014**, *215*, 134–140. [CrossRef] [PubMed]

25. Zou, P.; Tian, X.Y.; Dong, B.; Zhang, C.S. Size effects of chitooligomers with certain degrees of polymerization on the chilling tolerance of wheat seedlings. *Carbohydr. Polym.* **2017**, *160*, 194–202. [CrossRef] [PubMed]

26. Ma, L.J.; Li, Y.Y.; Yu, C.M.; Wang, Y.; Li, X.M.; Li, N.; Chen, Q.; Bu, N. Alleviation of exogenous oligochitosan on wheat seedlings growth under salt stress. *Protoplasma* **2012**, *249*, 393–399. [CrossRef] [PubMed]

27. Zou, P.; Li, K.C.; Liu, S.; Xing, R.; Qin, Y.K.; Yu, H.K.; Zhao, M.M.; Li, P.C. Effect of chitooligosaccharides with different degrees of acetylation on wheat seedlings under salt stress. *Carbohydr. Polym.* **2015**, *126*, 62–69. [CrossRef] [PubMed]

28. Yang, A.M.; Yu, L.; Chen, Z.; Zhang, S.X.; Shi, J.; Zhao, X.Z.; Yang, Y.Y.; Hu, D.Y.; Song, B.A. Label-free quantitative proteomic analysis of chitosan oligosaccharide-treated rice infected with southern rice black-streaked dwarf virus. *Viruses* **2017**, *9*, 115. [CrossRef]

29. Jia, X.C.; Meng, Q.S.; Zeng, H.H.; Wang, W.X.; Yin, H. Chitosan oligosaccharide induces resistance to tobacco mosaic virus in *Arabidopsis* via the salicylic acid-mediated signalling pathway. *Sci. Rep.* **2016**, *6*, 26144–26155. [CrossRef]

30. Ahmed, K.; Khan, M.; Siddiqui, H.; Jahan, A. Chitosan and its oligosaccharides, a promising option for sustainable crop production-a review. *Carbohydr. Polym.* **2020**, *227*, 115331. [CrossRef]

31. Cheplick, S.; Sarkar, D.; Bhowmik, P.C.; Shetty, K. Improved resilience and metabolic response of transplanted blackberry plugs using chitosan oligosaccharide elicitor treatment. *Can. J. Plant Sci.* **2017**, *98*, 717–731. [CrossRef]

32. Zeng, D.F.; Luo, X.R. Physiological effects of chitosan coating on wheat growth and activities of protective enzyme with drought tolerance. *Open J. Soil Sci.* **2012**, *2*, 282–288. [CrossRef]

33. Hu, Z.H.; Tang, B.; Wu, Q.; Zheng, J.; Leng, P.S.; Zhang, K.Z. Transcriptome sequencing analysis reveals a difference in monoterpene biosynthesis between scented *Lilium* 'Siberia' and unscented *Lilium* 'Novano'. *Front. Plant Sci.* **2017**, *8*, 1351. [CrossRef] [PubMed]

34. Wang, W.D.; Xin, H.H.; Wang, M.L.; Ma, Q.P.; Wang, L.; Kaleri, N.A.; Wang, Y.H.; Li, X.H. Transcriptomic analysis reveals the molecular mechanisms of drought-stress-induced decreases in *Camellia sinensis* leaf quality. *Front. Plant Sci.* **2016**, *7*, 385. [CrossRef] [PubMed]

35. Hao, X.Y.; Tang, H.; Wang, B.; Yue, C.; Wang, L.; Zeng, J.M.; Yang, Y.J.; Wang, X.C. Integrative transcriptional and metabolic analyses provide insights into cold spell response mechanisms in young shoots of the tea plant. *Tree Physiol.* **2018**, *38*, 1655–1671. [CrossRef] [PubMed]

36. Paul, A.; Jha, A.; Bhardwaj, S.; Singh, S.; Shankar, R.; Kumar, S. RNA-seq-mediated transcriptome analysis of actively growing and winter dormant shoots identifies non-deciduous habit of evergreen tree tea during winters. *Sci. Rep.* **2014**, *4*, 5932. [CrossRef] [PubMed]

37. Li, C.F.; Xu, Y.X.; Ma, J.Q.; Jin, J.Q.; Huang, D.J.; Yao, M.Z.; Ma, C.L.; Chen, L. Biochemical and transcriptomic analyses reveal different metabolite biosynthesis profiles among three color and developmental stages in 'Anji Baicha' (*Camellia sinensis*). *BMC Plant Biol.* **2016**, *16*, 195. [CrossRef] [PubMed]

38. Wei, C.L.; Yang, H.; Wang, S.B.; Zhao, J.; Liu, C.; Gao, L.P.; Xia, E.H.; Lu, Y.; Tai, Y.L.; She, G.B.; et al. Draft genome sequence of *Camellia sinensis* var. *sinensis* provides insights into the evolution of the tea genome and tea quality. *Proc. Natl. Acad. Sci. USA* **2018**, *115*, 201719622. [CrossRef] [PubMed]

39. Liao, Y.; Smyth, G.; Shi, W. featureCounts: An efficient general-purpose program for assigning sequence reads to genomic features. *Bioinformatics* **2014**, *30*, 923–930. [CrossRef]

40. Robinson, M.; McCarthy, D.; Smyth, G. edgeR: A bioconductor package for differential expression analysis of digital gene expression data. *Bioinformatics* **2010**, *26*, 139–140. [CrossRef]

41. Xie, C.; Mao, X.; Huang, J.; Ding, Y.; Wu, J.; Dong, S.; Kong, L.; Gao, G.; Li, C.; Wei, L. KOBAS 2.0: A web server for annotation and identification of enriched pathways and diseases. *Nucleic Acids Res.* **2011**, *39*, 316–322. [CrossRef]

42. Apel, K.; Hirt, H. Reactive oxygen species: Metabolism, oxidative stress, and signal transduction. *Annu. Rev. Plant Biol.* **2004**, *55*, 373–399. [CrossRef]

43. Gill, S.S.; Tuteja, N. Reactive oxygen species and antioxidant machinery in abiotic stress tolerance in crop plants. *Plant Physiol. Biochem.* **2010**, *48*, 909–930. [CrossRef] [PubMed]

44. Chen, J.N.; Huang, M.; Cao, F.B.; Pardha-Saradhi, P.; Zou, Y.B. Urea application promotes amino acid metabolism and membrane lipid peroxidation in *Azolla*. *PLoS ONE* **2017**, *12*, e0185230. [CrossRef] [PubMed]

45. Zhou, C.Z.; Zhu, C.; Fu, H.F.; Li, X.Z.; Chen, L.; Lin, Y.L.; Lai, Z.X.; Guo, Y.Q. Genome-wide investigation of superoxide dismutase (SOD) gene family and their regulatory miRNAs reveal the involvement in abiotic stress and hormone response in tea plant (*Camellia sinensis*). *PLoS ONE* **2019**, *14*, e0223609. [CrossRef] [PubMed]

46. Zou, P.; Li, K.C.; Liu, S.; He, X.F.; Xing, R.; Zhang, X.Q.; Li, P.C. Effect of sulfated chitooligosaccharides on wheat seedlings (Triticum aestivum L.) under saltstress. *J. Agric. Food Chem.* **2016**, *64*, 2815–2821. [CrossRef] [PubMed]

47. Anwar, A.; Yan, Y.; Liu, Y.; Li, Y.; Yu, X. 5-aminolevulinic acid improves nutrient uptake and endogenous hormone accumulation, enhancing low-temperature stress tolerance in cucumbers. *Int. J. Mol. Sci.* **2018**, *19*, 3379. [CrossRef] [PubMed]

48. Suorsa, M.; Sirpio, S.; Allahverdiyeva, Y.; Paakkarinen, V.; Mamedov, F.; Styring, S.; Aro, E.M. PsbR, a missing link in the assembly of the oxygen-evolving complex of plant photosystem II. *J. Biol. Chem.* **2006**, *281*, 145–150. [CrossRef]

49. Zhang, X.Q.; Li, K.C.; Xing, R.; Liu, S.; Chen, X.L.; Yang, H.Y.; Li, P.C. MiRNA and mRNA expression profiles reveal insight into the chitosan-mediated regulation of plant growth. *J. Agric. Food Chem.* **2018**, *66*, 3810–3822. [CrossRef] [PubMed]

50. Li, X.W.; Zhu, Y.X.; Chen, C.Y.; Geng, Z.J.; Li, X.Y.; Ye, T.T.; Mao, X.N.; Du, F. Cloning and characterization of two chlorophyll A/B binding protein genes and analysis of their gene family in *Camellia sinensis*. *Sci. Rep.* **2020**, *10*, 4602. [CrossRef]

51. Jiang, X.F.; Zhao, H.; Guo, F.; Shi, X.P.; Ye, C.; Y, P.X.; Liu, B.Y.; Ni, D.J. Transcriptomic analysis reveals mechanism of light-sensitive albinism in tea plant *Camellia sinensis* 'Huangjinju'. *BMC Plant Biol.* **2020**, *20*, 216. [CrossRef]

52. Sharma, A.; Kumar, V.; Shahzad, B.; Ramakrishnan, M.; Sidhu, G.P.S.; Bali, A.S.; Handa, N.; Kapoor, D.; Yadav, P.; Khanna, K.; et al. Photosynthetic response of plants under different abiotic stresses: A review. *J. Plant Growth Regul.* **2020**, *39*, 509–531. [CrossRef]

53. Chamnanmanoontham, N.; Pongprayoon, W.; Pichayangkura, R.; Roytrakul, S.; Chadchawan, S. Chitosan enhances rice seedling growth via gene expression network between nucleus and chloroplast. *Plant Growth Regul.* **2015**, *75*, 101–114. [CrossRef]

54. Abel, S.; Nguyen, M.D.; Theologis, A. The PS-IAA4/5-like family of early inducible mRNAs in *Arabidopsis thaliana*. *J. Mol. Biol.* **1995**, *251*, 533–549. [CrossRef] [PubMed]

55. Feng, S.; Yue, R.; Tao, S.; Yang, Y.; Zhang, L.; Xu, M.; Wang, H.; Shen, C. Genome-wide identification, expression analysis of auxin-responsive GH3 family genes in maize (*Zea mays* L.) under abiotic stresses. *J. Integr. Plant Biol.* **2015**, *57*, 783–795. [CrossRef] [PubMed]

56. Solano, R.; Ecker, J.R. Ethylene gas: Perception, signaling and response. *Curr. Opin. Plant Biol.* **1998**, *1*, 393–398. [CrossRef]

57. Chomczynski, P.; Sacchi, N. Single-step method of RNA isolation by acid guanidinium thiocyanate-phenol-chloroform extraction. *Anal. Biochem.* **1987**, *162*, 156–159. [CrossRef]

58. La Camera, S.; Gouzerh, G.; Dhondt, S.; Hoffmann, L.; Fritig, B.; Legrand, M.; Heitz, T. Metabolic reprogramming in plant innate immunity: The contributions of phenylpropanoid and oxylipin pathways. *Immunol. Rev.* **2004**, *198*, 267–284. [CrossRef]

9

Transcriptomic Analysis of Female Panicles Reveals Gene Expression Responses to Drought Stress in Maize (*Zea mays* L.)

Shuangjie Jia [1,†], Hongwei Li [1,†], Yanping Jiang [1], Yulou Tang [1], Guoqiang Zhao [1], Yinglei Zhang [1], Shenjiao Yang [2], Husen Qiu [2], Yongchao Wang [1], Jiameng Guo [1], Qinghua Yang [1,*] and Ruixin Shao [1,*]

[1] The Collaborative Center Innovation of Henan Food Crops, National Key Laboratory of Wheat and Maize Crop Science, Henan Agricultural University, Zhengzhou 450046, China; jiasj2017@126.com (S.J.); L_hongwei@126.com (H.L.); jiangyanping.up@gmail.com (Y.J.); tyl134679@163.com (Y.T.); z1013468268@163.com (G.Z.); yinglei609@163.com (Y.Z.); wangyongchao723@163.com (Y.W.); guojiameng@hotmail.com (J.G.)

[2] Farmland Irrigation Research Institute, CAAS/National Agro-ecological System Observation and Research Station of Shangqiu, Xinxiang 453002, China; shenjiao@gmail.com (S.Y.); qiuhusen2008@163.com (H.Q.)

* Correspondence: shaoruixin@henau.edu.cn (R.S.); yangqinghua@henau.edu.cn (Q.Y.)

† These authors are equal contribution to this study.

Abstract: Female panicles (FPs) play an important role in the formation of yields in maize. From 40 days after sowing to the tasseling stage for summer maize, FPs are developing and sensitive to drought. However, it remains unclear how FPs respond to drought stress during FP development. In this study, FP differentiation was observed at 20 and 30 days after drought (DAD) and agronomic trait changes of maize ears were determined across three treatments, including well-watered (CK), light drought (LD), and moderate drought (MD) treatments at 20, 25, and 30 DAD. RNA-sequencing was then used to identify differentially expressed genes (DEGs) in FPs at 30 DAD. Spikelets and florets were suppressed in LD and MD treatments, suggesting that drought slows FP development and thus decreases yields. Transcriptome analysis indicated that 40, 876, and 887 DEGs were detected in LD/CK, MD/CK, and MD/LD comparisons. KEGG pathway analysis showed that 'biosynthesis of other secondary metabolites' and 'carbohydrate metabolism' were involved in the LD response, whereas 'starch and sucrose metabolism' and 'plant hormone signal transduction' played important roles in the MD response. In addition, a series of molecular cues related to development and growth were screened for their drought stress responses.

Keywords: transcriptome analysis; summer maize; drought; female panicle

1. Introduction

Under the influence of global warming, changes in climatic conditions are creating unusual weather phenomena worldwide, often imposing drought stress on crops [1,2]. From the agricultural perspective, drought often results in decreased crop productivity and growth [3–5], especially for cereal crops. Maize (*Zea mays* L.) is one of the most important cereal crops and has the most extensive planting area globally [6,7]. One of the most important factors limiting maize growth and development around the world is a lack of water [8–11]. Accordingly, improving tolerance of maize to drought stress is essential for achieving high and stable yields in cereal crops.

As a multidimensional stress, water limitation triggers a wide variety of plant responses; these range from responses at the physiological and biochemical levels to the molecular level [12–16]. When external drought stimuli are perceived and captured by sensors on cell membranes, the signals are transmitted through multiple signal transduction pathways. Then, plant can regulate the expression of drought-responsive genes to protect themselves from the harmful effects of external stimuli [17,18]. The expressed products of drought-responsive genes are mainly proteins involved in signaling cascades and transcriptional regulation (such as protein kinase, protein phosphatase, and transcription factors) and functional proteins [19]. With the rapid development of high-throughput sequencing technologies, transcriptome analyses have been conducted to identify stress-mediated differences at the level of gene expression. Previous research has shown that many significantly differentially regulated genes that were associated with drought tolerance are induced in different organs of maize plants [20–26]. For example, 249 and 3000 differentially expressed genes (DEGs) were involved in root tissues after 6 h of light and severe drought stress, respectively [23]. In leaves, a total of 619 DEGs and 126 transcripts had their expression levels altered by drought stress at flowering time [24]. In tassels, 1902 DEGs were found after 5–7 days of drought stress [25]. In young ears, a total of 1825 DEGs were identified on the 5th day of drought stress at the V9 stage [26].

The panicle stage (from jointing to flowering) is the key stage for panicle differentiation and development in maize, the number of rows per ear and the number of grains per row are dependent on spikelet and floret differentiation at this time [27]. Therefore, the growth and development of female panicles (FPs) play an important role in the formation of maize yields. Although great advances in understanding differentiation of FPs and how drought stress affects genes transcription in FPs have been achieved in the past few decades [26,28–31], so far, progress in understanding the general molecular basis of FP development in response to long-term drought stress across the panicle stage has not been reported.

Accordingly, in this study, maize inbred line PH6WC (6WC) was used as drought-sensitive experimental material [32], soil water was controlled by means of drip irrigation for 30 days, and the gene expression dynamics of developing FPs at 30 days after drought (DAD) were investigated using transcriptomic analysis. The DEGs were identified and assigned to functional categories to reveal the various metabolic pathways in FPs that are involved in responses to long-term drought stress at different levels. Furthermore, differences in transcription factors between treatments were also analyzed. Overall, the exploration and function prediction of drought-response genes in maize FPs represent an efficient approach to improving the molecular breeding of drought-resistance maize cultivars.

2. Materials and Methods

2.1. Plant Material and Growth Conditions

Field experiments were conducted at field experiment stations (34°31′ N, 115°35′ E, 50.7 m above sea level) during the maize growing season (June–October, 2018) in Shangqiu (Henan, China). The maize inbred line 6WC was grown in 9 experimental plots (each plot was 2 m wide, 3.3 m long, 1.8 m deep) which were under a movable awning and filled with luvo-aquic soils, with a 20 cm sand filter layer at the bottom. Maize were planted into four rows, with 40 cm between rows in each plot. Two to three seeds were sown at each acupoint, with subsequent thinning to one seedling conducted at the trifoliate stage (V3). The final stand density was 8 plants m^{-2}. During the jointing and tasseling stages, topdressing fertilizer was applied, and weeds, insects, and diseases were controlled throughout the experiment. The top soil (0–40 cm layer) had a pH (water) of 7.3, mean mineral P content of 3.24 g/kg, and inorganic N at sowing of 3.60 g/kg. The average daily maximum and minimum temperatures of the field experiment during the trial were 32.98 °C and 20.71 °C, respectively.

2.2. Drought Stress Treatments

During FP development, soil moisture was controlled by means of drip irrigation at 80 ± 5% of the field water capacity (FWC) (well-watered, CK), 60 ± 5% of FWC (light drought, LD), and 45 ± 5% of FWC (moderate drought, MD). The meter was checked every morning and evening throughout the growth period to guide adjustments of the soil moisture. When the treated soil moisture dropped towards its lower limit, moderate drip irrigation was carried out until the upper limit level was reached, and the irrigation volume was measured by a water meter. After 30 DAD, the drought treatment plots were rehydrated to the CK level. Other field management measures reflected standard field management practices.

2.3. Measurement of Morphology and Microscopic Observation of Female Panicles

Plant height was measured from the ground to the top of the leaves in their natural growth state at 20, 25, and 30 DAD. The length and width of all the green leaves were measured by ruler in order to calculate leaf area, and the leaf area index (LAI) was determined according to this method [33]. Dry matter accumulation in stalks, leaves, tassels and ears of maize plants were measured at 20, 25, and 30 DAD. After 30 min of defoliation at 105 °C, dry weight was determined after being dried at 75 °C until a constant weight was reached. The percentage of drought limitation was calculated as (T2−T1)/T1. Here, T2 was shoot dry matter under the MD or LD treatment, while T1 was shoot dry matter under the control or LD treatment [34]. FPs were dissected with a dissecting needle at 20 and 27 DAD and analyzed under a stereomicroscope (Guanpujia, SMZ-B2, Beijing, China) to observe the differentiation of developing female inflorescences. There were three biological replicates in each group.

2.4. RNA Isolation and Illumina Sequencing

Three plants with consistent growth were selected from each treatment at 30 DAD, FP bracts were sampled, and the upper, middle, and lower parts of the ears were mixed evenly and then frozen at −80 °C prior to RNA-sequencing (RNA-seq) analysis. Total RNA was extracted using the mirVana miRNA Isolation Kit (Ambion, Inc., Austin, TX, USA). RNA integrity was evaluated using the Agilent 2100 Bioanalyzer (Agilent Technologies, Santa Clara, CA, USA). Shanghai OE Biotech (Shanghai, China) conducted RNA-Seq library construction and high-throughput sequencing based on total RNA from the female inflorescence. The libraries were sequenced on the Illumina sequencing platform (HiSeq 2500l, Illumina, San Diego, CA, USA) with 125 bp paired-end reads. The raw RNA-seq data have been uploaded to NCBI SRA (BioProject ID: PRJNA604094).

2.5. Read Mapping and Differential Expression Analysis

Base calling was conducted using the raw image data generated by sequencing to obtain sequence data, and the called raw data (raw reads) were stored in fastq format. Raw data (raw reads) were processed using Trimmomatic [35]. The reads containing poly-N runs and low-quality reads were removed to obtain the clean reads. Then, the clean reads were mapped to the reference (NCBI_B73_v4) genome [36] using HISAT2 (version 2.2.1.0) [37]. The Fragments Per kb Per Million Reads (FPKM) values were calculated using cufflinks (version 2.2.1) [38,39], followed by differential expression analysis using DESeq (version1.18.0) [40]. Genes with |fold change| > 2 and $p < 0.05$ were identified as differentially expressed genes with p presented as raw p-values rather than FDR adjusted p-values.

2.6. GO and KEGG Enrichment Analysis

The significantly expressed Gene Ontology (GO) terms were selected by GO enrichment analysis according to the GO database (http://geneontology.org/). The differences in the frequency of assignment of GO terms in the DEG set were compared with the expressed genes in the CK, LD, and MD samples ($p < 0.05$). Functional groups encompassing DEGs were identified based on GO analysis, and pathway analysis was conducted according to the Kyoto Encyclopedia of Genes and Genomes (KEGG) database (http://www.genome.jp/kegg/), with manual reannotation based on several databases and a literature search.

150

Abiotic Stress and Plant Responses

2.7. Differential Expression Verification by Quantitative Real-Time PCR (qRT-PCR)

Transcriptome sequencing data were validated by qRT-PCR. Total RNA was reverse transcribed using EasyScript One-Step gDNA Removal and cDNA Synthesis SuperMix (TRANS, Beijing, China). The qRT-PCR experimental methods used HiScript Q RT SuperMix for qPCR (Vazyme, Nanjing, China). The primer sequences were designed using Primer 5 and are listed in Supplementary Materials Table S1. The relative quantification $2^{-\Delta\Delta Ct}$ method was used to calculate the expression level of target genes in different treatments.

2.8. Statistical Analysis

Data collation and graphic rendering were conducted with SIGMAPLOT 14.0 (Systat Software Inc., San Jose, CA, USA), Microsoft Excel, and Microsoft PowerPoint 2016 software (Microsoft Corp., Redmond, WA, USA). All data are expressed as the mean ± SD value from three independent experiments unless otherwise stated. Data were analyzed by one-way ANOVA using Duncan's multiple range test at a $p < 0.05$ significance threshold in SPSS (IBM Corp., Armonk, NY, USA).

3. Results

3.1. Female Panicle Development, Phenotypic Change and Yield Components

Plant height, LAI, dry matter accumulation, and the percentage of drought limitation were measured at 20, 25, and 30 DAD. For example, plant height was reduced by 17.63%, 17.01%, and 16.44% under the LD treatment compared with CK, respectively; furthermore, MD significantly decreased plant height by 25.86%, 26.34%, and 29.73% (Figure 1a), respectively. Leaf area index was significantly decreased under MD at 20, 25, and 30 DAD compared with CK plants, but was not significantly changed at 20 and 30 DAD under LD (Figure 1b). For dry matter accumulation, MD significantly decreased the shoot dry matter at 20, 25, and 30 DAD compared with CK plants, but it was not significantly reduced in the LD vs. CK comparison at 20 and 30 DAD (Figure 1c). In addition, the percentage of drought limitation had the highest absolute values at 20, 25, and 30 DAD after MD treatment. However, the percentage of drought limitation was not significantly affected by LD treatment at 20 and 30 DAD compared with CK (Figure 1d).

From 20 to 30 DAD, according to the book *Corn Growth and Development* [41], FPs may be in the process of differentiation. To explore the responses of FPs to drought stress, maize plants were dissected. The length and diameter of FPs were significantly decreased under the MD treatment at 20 and at 30 DAD; silk was seen in CK and LD plants, but there was no floral differentiation MD plants (Figure 2a). Proportion of dry matter in FP was decreased under LD and MD treatments (Figure 2b). To investigate the effect of drought stress on development and number of mature ears, ear size, and dry matter, yield components were determined (Figure 2c, Supplementary Materials Table S2), which showed that LD and MD treatments significantly decreased ear size and increased the bald tip length. For this reason, the numbers of rows and kernels were reduced by 14.00% and 29.00% under the LD treatment and 19.00% and 43.00%, respectively, under the MD treatment. Therefore, drought resulted in great losses in grain yield of 32.00% and 35.00% under the LD and MD treatments, respectively (Figure 2c, Supplementary Materials Table S2).

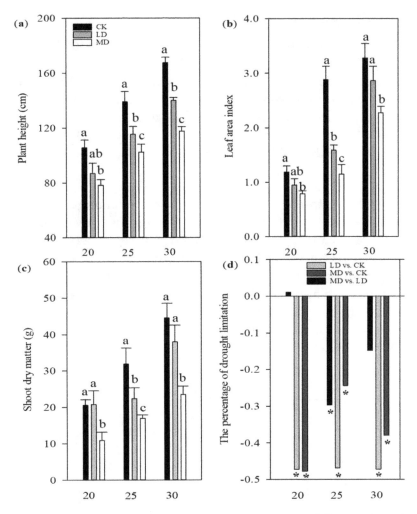

Figure 1. Agronomic traits changes of maize plants in response to different drought stress. Effect of drought stress on (**a**) plant height; (**b**) leaf area index; (**c**) shoot dry matter; and (**d**) the percentage of drought limitation at 20, 25, and 30 days after drought (DAD). Values are the means of the replicates ± sd. Different lowercase letters and * symbols indicate statistical significance of differences at a $p < 0.05$ level. There were three treatments, including well-watered (CK), light drought (LD), and moderate drought (MD), and five biological replicates were sampled for each treatment.

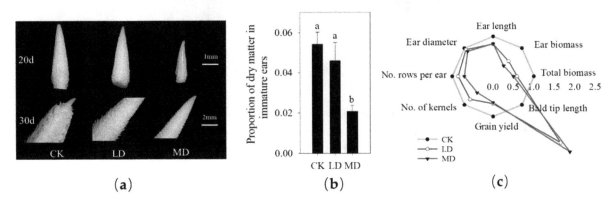

Figure 2. Development of female inflorescence and agronomic traits changes of ears in response to drought stress. (**a**) Contrasting sizes and differentiation of the control and drought-treated plants at 20 and 30 DAD. (**b**) Proportion of dry matter in immature ears at 30 DAD. (**c**) Radar chart showing changes in yield traits for mature maize plants grown under well-watered (CK), light drought (LD), and moderate drought (MD) conditions. (**b**) and (**c**) were calculated with 15 and 10 biological replicates sampled for each treatment.

3.2. Overview of RNA Sequencing and Mapping

A total of 49.42 million raw reads were obtained from PH6WC transcriptome libraries (Supplementary Materials Table S3). More than 96.72% of them (47.80 million clean reads) remained after discarding low-quality reads and reads containing adaptor sequences, which were then used for downstream analyses. The clean reads were mapped to the B73 reference genome (ZmB73_RefGen_v4). Overall, 94.34–94.52% of clean reads from nine samples were mapped onto the reference genome (Supplementary Materials Table S3). On average, approximately 43.93 (91.29%), 43.73 (91.13%), and 43.70 (91.17%) million reads from the CK, LD, and MD treatments, respectively, were uniquely mapped onto the reference genome.

Compared with the CK treatment, only 40 genes ($\log_2$ foldchange > 1 and $p < 0.05$), including nine up-regulated and 31 down-regulated genes, showed significantly differential expression in the LD treatment, and a total of 212 up-regulated and 664 down-regulated genes were identified in the MD treatment. A total of 887 DEGs, including 208 up-regulated and 679 down-regulated genes, were identified in the MD versus LD comparison (Figure 3a). A Venn diagram of the DEGs illustrated that there were 10 common genes that appeared in the LD vs. CK and MD vs. CK comparisons, five genes shared between the LD vs. CK and MD vs. LD comparisons, and 565 genes shared between the MD vs. LD and MD vs. CK comparisons. However, there were no DEGs commonly expressed in all three comparisons (Figure 3b).

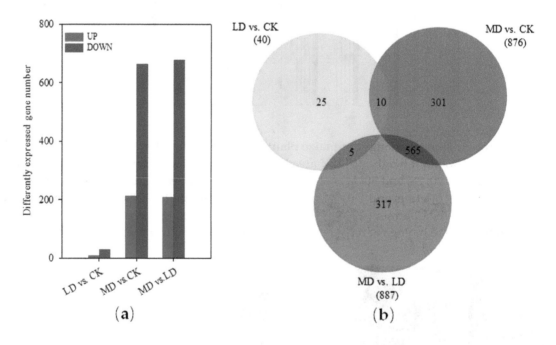

(a)   (b)

Figure 3. Identification and characterization of differentially expressed genes (DEGs) between the drought treatment and control plants. (**a**) The number of DEGs in three comparison groups. (**b**) A Venn diagram comparison summarizing overlaps in differentially expressed genes among the three comparisons.

3.3. Gene Expression Validation by qRT-PCR

To investigate the changes in gene expression at the mRNA level, eight randomly selected genes and three specific genes *cuc2* (LOC103629107), *TE1* (LOC541683), *DLF1* (LOC100037791) were analyzed using quantitative real-time RT-PCR for validation of RNA-seq. The level of expression of the genes amplified is shown in Figure 4.

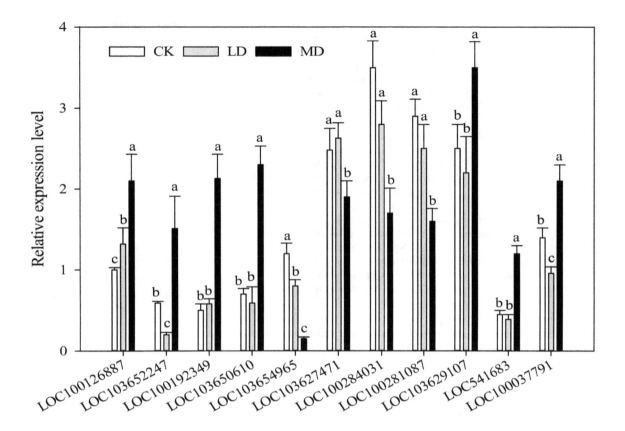

Figure 4. The expression patterns of eleven genes in female panicle tissues under well-watered (CK), light drought (LD), and moderate drought (MD) conditions by qRT-PCR. Values are the mean ± SD of three independent experiments. Maize β-actin expression was used as a control.

Raw data were compared to transcriptomics data (Supplementary Materials Table S4), which closely resembled each other, validating the differential expression of the genes identified as being under drought stress.

3.4. GO Annotation and Enrichment

A total of 26, 629, and 630 DEGs were assigned by GO analysis conducted based on the genes from the LD vs. CK, MD vs. LD, and MD vs. LD comparisons, respectively. The most significantly regulated 20 terms among biological processes from the MD vs. CK and MD vs. LD comparison genes are shown in Figure 5a,b, but there were no significant terms (i.e., terms with gene number > 2 and $p < 0.05$) resulting from the LD vs. CK comparison. The up-regulated terms from the MD vs. CK comparison are involved in "regulation of timing of plant organ formation," "developmental process," and "regulation of cell proliferation." The down-regulated terms "response to water deprivation" and "post-embryonic plant morphogenesis" were also enriched. The up-regulated terms from the MD vs. LD comparison are "regulation of timing of plant organ formation," "regulation of cell proliferation," and "gibberellin biosynthetic process." The down-regulated terms "reductive pentose-phosphate cycle," "phosphate ion homeostasis," and "ethylene-activated signaling pathway" were enriched among genes from the MD vs. LD comparison. The genes associated with GO terms related to development, growth, and responses to stimulus were also significantly different among the three comparisons. These genes that changed in their levels of transcriptional expression were basically the same in the MD vs. CK and MD vs. LD comparison groups, but lower in expression differences in the LD vs. CK comparison group (Figure 5c,d, Supplementary Materials Table S5).

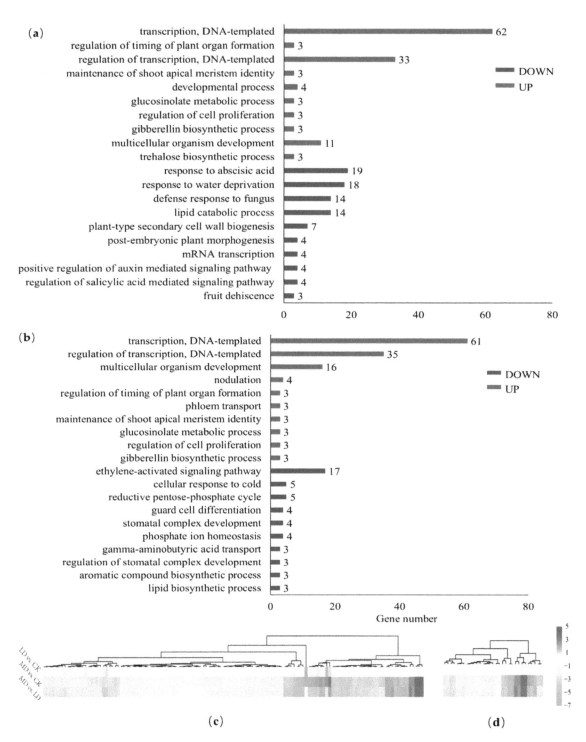

Figure 5. Top 20 biological processes enriched by the up-and down-regulated genes in the (**a**) MD vs. CK and (**b**) MD vs. LD comparisons. Expression pattern of the differentially expressed genes associated with (**c**) development progress and (**d**) growth in the three comparisons. Colors indicate the $\log_2$ fold change values. Red indicates up-regulation, and green indicates down-regulation in that comparison.

3.5. Metabolic Pathways Related to Soil Drought Stress

To further characterize genes affected by drought stress, we performed a KEGG pathway classification analysis to identify functional enrichment of DEGs. Thus, 8, 72, and 74 terms were significantly enriched in the transcriptome profile comparisons of LD vs. CK, MD vs. CK, and MD vs. LD groups (Supplementary Materials Table S6). The significant differences in the top 20 enriched KEGG pathways in the MD vs. CK and MD vs. LD comparisons are shown in Figure 6. In the

MD vs. CK comparison, genes associated with the pathway "Starch and sucrose metabolism" were most enriched followed by those associated with "plant hormone signal transduction" (Figure 6a, Supplementary Materials Table S7). In the MD vs. LD group, "Plant hormone signal transduction," "Starch and sucrose metabolism," and "Glycosphingolipid biosynthesis" were the three most enriched terms (Figure 6a, Supplementary Materials Table S7).

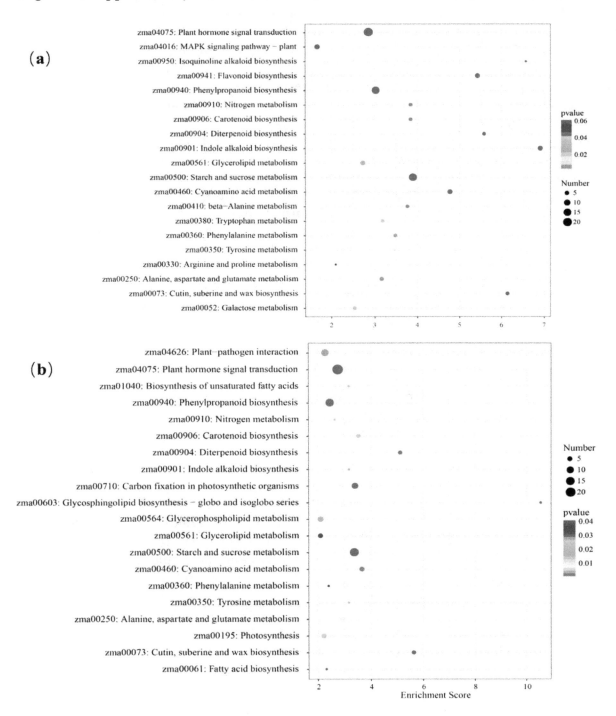

Figure 6. Top 20 enriched KEGG pathways in the (**a**) "MD vs. CK" and (**b**) "MD vs. LD" comparisons. Pathway entries with the corresponding number of genes (among those pathways with more than two genes) are shown, and the corresponding -log$_{10}$ p-value of each entry is sorted in descending order. The number of DEGs in each pathway is positively related to the size of plot, and the p-values shown in red are more significant.

4. Discussion

4.1. Responses of Plant Growth and Female Panicle Differentiation to Soil Drought Stress

The panicle stage is the most important productive stage in corn development, and soil drought stress in this stage can affect the plant growth rate, prolong the growth processes of the panicle stage, hinder the normal differentiation and development of ears, and ultimately lead to decreased crop seed setting rates and yields [42–46]. Further, the drought response depends on the time and intensity of water loss as well as the developmental stage [47,48]. In this study, LD and MD treatments compared with the CK treatment decreased green leaf area and significantly suppressed shoot dry matter accumulation over the prolonged drought treatment, and relative to the LD treatment, the MD treatment affected plant growth much more (Figure 1), which is consistent with previous research by Boonjung et al. [49].

FPs are the precursor to maize ears, and the differentiation and development of FPs mainly occurs from the V9 to VT stages, which include growth cone extension (V9), spikelet differentiation (V11), floret differentiation (V12), and formation of the sexual organs (VT). Developing organs are sensitive to drought, especially during their early phases [50]. When drought occurs between the V9 and VT stages, how does the degree of drought affect the formation of ears? In our study, spikelet and floral differentiation, as observed under stereomicroscope, were significantly inhibited and thus delayed by soil drought (Figure 2). Some studies have shown that the number of kernel rows is determined at the spikelet differentiation stage, and the floret differentiation period is the key period that affects grain number [51–53]. Here, mature ears in the MD and LD treatments were much shorter and thinner than those in the CK treatment; in addition, the bald tip length and number of unfilled grains were both increased under drought treatments. As indicated above, drought affected grain yields as well (Figure 2c).

4.2. Genes Involved in Development and Growth in Response to Soil Drought Stress

Drought treatments affected the expression of genes associated with development and growth of the inflorescence (Figure 4). *Terminal ear 1 (te1)* maize mutants have shortened internodes, abnormal phyllotaxy, leaf pattern defects, and partial feminization of tassels [54]. Similarly, *cup-shaped cotyledon 2 (cuc2)* mutants have been reported to have abnormalities in the regulation of the shoot meristem boundary and formation and subsequent development [55,56]. Here, *cuc2* were up-regulated under moderate drought stress (Figure 4, Supplementary Table S5), combined with developmental change (Figure 2a), implying that the differential expression of the gene under MD treatments may be related to the mature delay of the FPs tissue. *DLF1* was also up-regulated under MD stress (Figure 4, Supplementary Materials Table S5), suggesting that the trans-activator protein encoded by, this gene plays an important role in the signal transduction pathway and the regulation of plant growth at the FP development stage [57].

4.3. Genes Involved in Auxin Signaling in Response to Soil Drought Stress

Auxin is an important phytohormone that is closely related to plant resistance to adverse environmental conditions, and it can induce rapid and transient expression of some genes, including auxin response factor genes (ARF) and primary auxin response genes (Aux/IAA, GH3, SAUR and LBD); the protein products of these genes can specifically bind to ARFs to activate or inhibit downstream gene expression under drought [58–61]. In the current study, auxin signaling genes were involved in the response to drought, as the expression of IAA-conjugating genes (GH3) was up-regulated, and the expression levels of auxin biosynthesis genes were down-regulated after MD stress (Figure 7), leading to the reduction of auxin levels (Supplementary Materials Figure S1a). This implies that drought improves GH3 transcription to help maintain endogenous auxin at an appropriate level in maize [62,63]. However, when the concentration of auxin increases, auxin combines with transport inhibitor response 1 (TIR1), causing Aux/IAA ubiquitination and degradation; then, ARF is released, which further activates the expression of small auxin-up RNA (SAUR) genes [64]. SAUR genes are early auxin-responsive genes involved in plant growth, and the SAUR family regulates a series of cellular, physiological, and developmental processes in response to environmental signals [65–67]. In our study, three SAUR genes were down-regulated under the MD treatment (Figure 7), which may explain MD-induced inhibition of maize growth.

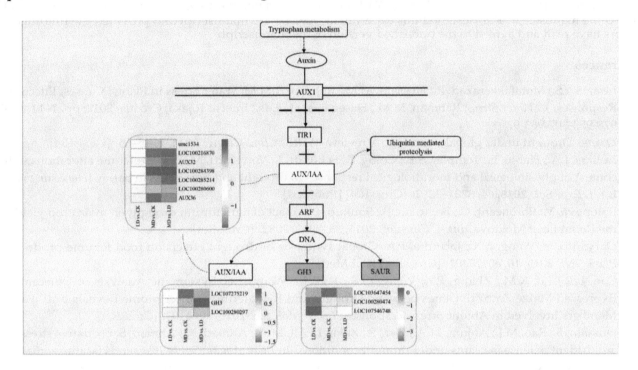

Figure 7. Genes involved in auxin plant hormone signal transduction pathway in the Kyoto Encyclopedia of Genes and Genomes (KEGG). Differentially expressed genes involved in AUX/IAA, GH3, and SAUR were shown by heat-maps, and the number was calculated with log2foldchange in three comparisons. Color of heat-maps represented different fold change. Yellow box means involved significantly differentially expressed genes were mainly up-regulated, and green box means down-regulated.

4.4. Reactive Oxygen Scavenging System and Ion Channel in Response to Soil Drought Stress

Limited water supply enhances the production of reactive oxygen species (ROS) [68,69], and plants are protected by glutathione S-transferase (GST) and other antioxidant enzymes scavenging excessive ROS from the damage caused by ROS [70]. This is because GST comprises a large superfamily of multifunctional protein and participates in ascorbic acid (AsA)/glutathione (GSH) cycling pathways [68]. Here, probable glutathione S-transferase GST12 was significantly down-regulated under the MD treatment compared with the CK treatment. However, GSTU6 (LOC103637303) and GST activity was up-regulated under MD stress (Supplementary Materials Table S8 and Figure S1b), which may scavenge ROS and protect both plant cell membrane structure and protein activity [71,72], implying that GST is involved in responding to drought stress [73,74].

Supplementary Materials
Table S1: The primer sequences for qRT-PCR, Table S2: Ear and grain yield traits at harvest time under CK, LD and MD treatments, Table S3: Number of reads based on RNA-Seq data of CK, LD and MD treatments, Table S4: RNA-Seq expression levels of the eight genes for qRT-PCR, Table S5: Expression pattern of the differently expressed genes about development progress (A) and growth (B) under CK, LD and MD treatments, Table S6: Number of enriched KEGG pathways terms and DEGs in three comparisons (LD vs.CK, MD vs. CK; MD vs. LD), Table S7: Differentially expressed genes (DEGs) in the top 20 enriched KEGG pathways in two comparisons

(MD vs. CK; MD vs. LD), Table S8: Genes significantly enriched in Glutathione metabolism in three comparisons (LD vs.CK, MD vs. CK; MD vs. LD), Figure S1: Effect of drought stress treatments on the content of IAA (a) and the activities of GST (b) in female panicles. CK, well-watered; LD, light drought; MD, moderate drought, five biological replicates were sampled for each treatment.

Author Contributions: S.J. did the experiment and wrote the manuscript. H.L. analyzed the data. Y.J., Y.T., G.Z. and Y.Z. did a part of experiment. S.Y., H.Q., Y.W. and J.G. gave some good suggestions on the manuscript. Q.Y. revised the manuscript. R.S. designed this experiment, revised the manuscript and provided the funding. All authors have read and agreed to the published version of the manuscript.

References

1. Lisar, S.Y.S.; Motafakkerazad, R.; Hossain, M.M.; Rahman, I.M.M. Water Stress in Plants: Causes, Effects and Responses. In *Water Stress*; Rahman, M.M., Hasegawa, H., Eds.; Intech: Rijeka, Croatia, 2012; pp. 1–14. ISBN 978-953-307-963-9.

2. Dai, A. Drought under global warming: A review. *WIREs Clim. Chang.* **2011**, *2*, 45–65. [CrossRef]

3. Rollins, J.A.; Habte, E.; Templer, S.E.; Colby, T.; Schmidt, J.; Von Korff, M. Leaf proteome alterations in the context of physiological and morphological responses to drought and heat stress in barley (*Hordeum vulgare* L.). *J. Exp. Bot.* **2013**, *64*, 3201–3212. [CrossRef] [PubMed]

4. Potopová, V.; Boroneanţ, C.; Boincean, B.; Soukup, J. Impact of agricultural drought on main crop yields in the Republic of Moldova. *Int. J. Climatol.* **2016**, *36*, 2063–2082. [CrossRef]

5. Daryanto, S.; Wang, L.X.; Jacinthe, P.A. Global synthesis of drought effects on food legume production. *PLoS ONE* **2015**, *10*, e0127401. [CrossRef] [PubMed]

6. Cao, L.R.; Lu, X.M.; Zhang, P.Y.; Wang, G.R.; Wei, L.; Wang, T.C. Systematic Analysis of Differentially Expressed Maize ZmbZIP Genes between Drought and Rewatering Transcriptome Reveals bZIP Family Members Involved in Abiotic Stress Responses. *Int. J. Mol. Sci.* **2019**, *20*, 4103. [CrossRef]

7. Hussain, S.; Rao, M.J.; Anjum, M.A.; Ejaz, S.; Zakir, I.; Ali, M.A.; Ahmad, N.; Ahmad, S. Oxidative stress and antioxidant defense in plants under drought conditions. In *Plant Abiotic Stress Tolerance*; Hasanuzzaman, M., Hakeem, K.R., Nahar, K., Alharby, H.F., Eds.; Springer: Cham, Switzerland, 2019; pp. 207–219. ISBN 978-3-030-06117-3.

8. Meeks, M.; Murray, S.; Hague, S.; Hays, D. Measuring maize seedling drought response in search of tolerant germplasm. *Agronomy* **2013**, *3*, 135–147. [CrossRef]

9. Vaughan, M.M.; Block, A.; Christensen, S.A.; Allen, L.H.; Schmelz, E.A. The effects of climate change associated abiotic stresses on maize phytochemical defenses. *Phytochem. Rev.* **2018**, *17*, 37–49. [CrossRef]

10. Calanca, P.P. Effects of abiotic stress in crop production. In *Quantification of Climate Variability, Adaptation and Mitigation for Agricultural Sustainability*; Ahmed, M., Stockle, C.O., Eds.; Springer: Cham, Switzerland, 2017; pp. 165–180. ISBN 978-3-319-32057-1.

11. EL Sabagh, A.; Hossain, A.; Islam, M.S.; Barutcular, S.; Fahad, S.; Ratnasekera, D.; Kumar, N.; Meena, R.S.; Vera, P.; Saneoka, H. Role of osmoprotectants and soil amendments for sustainable soybean (*Glycine max* L.) production under drought condition: A review. *J. Exp. Biol. Agric. Sci.* **2018**, *6*, 32–41. [CrossRef]

12. Yang, M.; Geng, M.Y.; Shen, P.F.; Chen, X.H.; Li, Y.J.; Wen, X.X. Effect of post-silking drought stress on the expression profiles of genes involved in carbon and nitrogen metabolism during leaf senescence in maize (*Zea mays* L.). *Plant Physiol. Biochem.* **2019**, *135*, 304–309. [CrossRef]

13. Chiuta, N.; Mutengwa, C. Response of yellow quality protein maize inbred lines to drought stress at seedling stage. *Agronomy* **2018**, *8*, 287. [CrossRef]

14. Zhan, J.P.; Li, G.S.; Ryu, C.H.; Ma, C.; Zhang, S.S.; Lloyd, A.; Hunter, B.G.; Larkins, B.A.; Drews, G.N.; Wang, X.F.; et al. Opaque-2 regulates a complex gene network associated with cell differentiation and storage functions of maize endosperm. *Plant Cell* **2018**, *30*, 2425–2446. [CrossRef] [PubMed]

15. Jain, D.; Ashraf, N.; Khurana, J.P.; Kameshwari, M.N. The 'Omics' Approach for Crop Improvement Against Drought Stress. In *Genetic Enhancement of Crops for Tolerance to Abiotic Stress: Mechanisms and Approaches*; Rajpal, V.R., Sehgal, D., Kumar, A., Raina, S.N., Eds.; Springer: Cham, Switzerland, 2019; pp. 183–204. ISBN 978-3-319-91955-3.

16. Dastogeer, K.M.G.; Li, H.; Sivasithamparam, K.; Jones, M.G.K.; Wylie, S.J. Fungal endophytes and a virus confer drought tolerance to Nicotiana benthamiana plants through modulating osmolytes, antioxidant enzymes and expression of host drought responsive genes. *Environ. Exp. Bot.* **2018**, *149*, 95–108. [CrossRef]

17. Kaur, G.; Asthir, B. Molecular responses to drought stress in plants. *Biol. Plant* **2017**, *61*, 201–209. [CrossRef]

18. Fang, Y.J.; Xiong, L.Z. General mechanisms of drought response and their application in drought resistance improvement in plants. *Cell. Mol. Life Sci.* **2015**, *72*, 673–689. [CrossRef] [PubMed]

19. Nakashima, K.; Yamaguchi-Shinozaki, K.; Shinozaki, K. The transcriptional regulatory network in the drought response and its crosstalk in abiotic stress responses including drought, cold, and heat. *Front. Plant Sci.* **2014**, *5*, 170. [CrossRef] [PubMed]

20. Feng, F.; Qi, W.W.; Lv, Y.D.; Yan, S.M.; Xu, L.M.; Yang, W.Y.; Yuan, Y.; Chen, Y.H.; Zhao, H.; Song, R.T. OPAQUE11 is a central hub of the regulatory network for maize endosperm development and nutrient metabolism. *Plant Cell* **2018**, *30*, 375–396. [CrossRef]

21. An, Y.X.; Chen, L.; Li, Y.X.; Li, C.H.; Shi, Y.S.; Song, Y.C.; Zhang, D.F.; Li, Y.; Wang, T.Y. Candidate loci for the kernel row number in maize revealed by a combination of transcriptome analysis and regional association mapping. *BMC Plant Biol.* **2019**, *19*, 201. [CrossRef]

22. Wilson, J. *Control of Maize Development by MicroRNA and Auxin Regulated Pathways*; East Carolina University: Greenville, NC, USA, 2018.

23. Opitz, N.; Paschold, A.; Marcon, C.; Malik, W.A.; Lanz, C.; Piepho, H.P.; Hochholdinger, F. Transcriptomic complexity in young maize primary roots in response to low water potentials. *BMC Genomics* **2014**, *15*, 741. [CrossRef]

24. Song, K.; Kim, H.C.; Shin, S.; Kim, K.H.; Moon, J.C.; Kim, J.Y.; Lee, B.M. Transcriptome analysis of flowering time genes under drought stress in maize leaves. *Front. Plant Sci.* **2017**, *8*, 267. [CrossRef]

25. Li, L. *The Major Metabolic Pathways at Maize Developing Young Tassel in Response to Drought Stress and Identification of Drought-Tolerant Candidate SNAC Genes*; Xinjiang Agricultural University: Wulumuqi, Xinjiang, China, 2015.

26. Wang, B.M.; Liu, C.; Zhang, D.F.; He, C.M.; Zhang, J.R.; Li, Z.X. Effects of maize organ-specific drought stress response on yields from transcriptome analysis. *BMC Plant Biol.* **2019**, *19*, 335. [CrossRef]

27. Zhao, M.; Wang, Q.X.; Wang, K.J.; Li, C.H.; Hao, J.P. Maize. In *Crop Cultivation Science: North*; Yu, Z.W., Ed.; China Agriculture Press: Beijing, China, 2003; pp. 69–111. ISBN 9787109179363.

28. Chen, D.Q.; Wang, S.W.; Cao, B.B.; Cao, D.; Leng, G.H.; Li, H.B.; Yin, L.N.; Shan, L.; Deng, X.P. Genotypic variation in growth and physiological response to drought stress and re-watering reveals the critical role of recovery in drought adaptation in maize seedlings. *Front. Plant Sci.* **2016**, *6*, 1241. [CrossRef] [PubMed]

29. Oury, V.; Caldeira, C.F.; Prodhomme, D.; Pichon, J.P.; Gibon, Y.; Turc, O. Is change in ovary carbon status a cause or a consequence of maize ovary abortion in water deficit during flowering? *Plant Physiol.* **2016**, *171*, 997–1008. [CrossRef] [PubMed]

30. Khandagale, S.G.; Dubey, R.B.; Sharma, V.; Khan, R. Response of physiological traits of maize to moisture stress induced at different developmental stages. *Int. J. Chem. Stud.* **2018**, *6*, 2757–2761.

31. Ma, C.Y.; Li, B.; Wang, L.N.; Xu, M.L.; Zhu, L.E.; Jin, H.Y.; Wang, Z.C.; Ye, J.R. Characterization of phytohormone and transcriptome reprogramming profiles during maize early kernel development. *BMC Plant Biol.* **2019**, *19*, 197. [CrossRef]

32. Jia, S.J.; Li, H.W.; Jiang, Y.P.; Zhao, G.Q.; Wang, H.Z.; Yang, S.J.; Yang, Q.H.; Guo, J.M.; Shao, R.X. Effects of drought on photosynthesis and ear development characteristics of maize. *Acta Ecol. Sin.* **2020**, *3*, 1–9.

33. Li, Y.B.; Song, H.; Zhou, L.; Xu, Z.Z.; Zhou, G.S. Tracking chlorophyll fluorescence as an indicator of drought and rewatering across the entire leaf lifespan in a maize field. *Agric. Water Manag.* **2019**, *211*, 190–201. [CrossRef]

34. Xu, Z.Z.; Zhou, G.S.; Shimizu, H. Are plant growth and photosynthesis limited by pre-drought following rewatering in grass? *J. Exp. Bot.* **2009**, *60*, 3737–3749. [CrossRef]

35. Bolger, A.M.; Lohse, M.; Usadel, B. Trimmomatic: A flexible trimmer for Illumina sequence data. *Bioinformatics* **2014**, *30*, 2114–2120. [CrossRef]

36. Jiao, Y.P.; Peluso, P.; Shi, J.H.; Liang, T.; Stitzer, M.C.; Wang, B.; Campbell, M.S.; Stein, J.C.; Wei, X.H.; Chin, C.S.; et al. Improved maize reference genome with single-molecule technologies. *Nature* **2017**, *546*, 524–527. [CrossRef]

37. Kim, D.; Langmead, B.; Salzberg, S.L. HISAT: A fast spliced aligner with low memory requirements. *Nat. Methods* **2015**, *12*, 357–360. [CrossRef]

38. Roberts, A.; Pimentel, H.; Trapnell, C.; Pachter, L. Identification of novel transcripts in annotated genomes using RNA-Seq. *Bioinformatics* **2011**, *27*, 2325–2329. [CrossRef] [PubMed]

39. Roberts, A.; Trapnell, C.; Donaghey, J.; Rinn, J.L.; Pachter, L. Improving RNA-Seq expression estimates by correcting for fragment bias. *Genome Biol.* **2011**, *12*, R22. [CrossRef] [PubMed]

40. Anders, S.; Huber, W. *Differential Expression of RNA-Seq Data at the Gene Level—the DESeq Package*; European Molecular Biology Laboratory: Heidelberg, Germany, 2012.

41. Abendroth, L.J.; Elmore, R.W.; Boyer, M.J.; Marlay, S.K. Vegetative Stages (VE to VT). In *Corn Growth and Development*; Iowa State University: Ames, IA, USA, 2011; pp. 13–27.

42. Mueller, N.D.; Gerber, J.S.; Johnston, M.; Ray, D.K.; Ramankutty, N.; Foley, J.A. Closing yield gaps through nutrient and water management. *Nature* **2012**, *490*, 254–257. [CrossRef]

43. Zhang, X.B.; Lei, L.; Lai, J.S.; Zhao, H.M.; Song, W.B. Effects of drought stress and water recovery on physiological responses and gene expression in maize seedlings. *BMC Plant Biol.* **2018**, *18*, 68. [CrossRef] [PubMed]

44. Hayano-Kanashiro, C.; Calderón-Vázquez, C.; Ibarra-Laclette, E.; Herrera-Estrella, L.; Simpson, J. Analysis of gene expression and physiological responses in three Mexican maize landraces under drought stress and recovery irrigation. *PLoS ONE* **2009**, *4*, e7531. [CrossRef] [PubMed]

45. Huo, Y.J.; Wang, M.P.; Wei, Y.Y.; Xia, Z.L. Overexpression of the maize psbA gene enhances drought tolerance through regulating antioxidant system, photosynthetic capability, and stress defense gene expression in tobacco. *Front. Plant Sci.* **2016**, *6*, 1223. [CrossRef] [PubMed]

46. Avramova, V.; AbdElgawad, H.; Vasileva, I.; Petrova, A.S.; Holek, A.; Mariën, J.; Asard, H.; Beemster, G.T.S. High antioxidant activity facilitates maintenance of cell division in leaves of drought tolerant maize hybrids. *Front. Plant Sci.* **2017**, *8*, 84. [CrossRef]

47. Witt, S.; Galicia, L.; Lisec, J.; Cairns, J.; Tiessen, A.; Araus, J.L.; Palacios-Rojas, N.; Fernie, A.R. Metabolic and phenotypic responses of greenhouse-grown maize hybrids to experimentally controlled drought stress. *Mol. Plant* **2012**, *5*, 401–417. [CrossRef]

48. Bartels, D.; Souer, E. Molecular responses of higher plants to dehydration. In *Plant Responses to Abiotic Stress*; Hirt, H., Shinozaki, K., Eds.; Springer: Berlin/Heidelberg, Germany, 2004; pp. 9–38. ISBN 9783-540-20037-6.

49. Boonjung, H.; Fukai, S. Effects of soil water deficit at different growth stages on rice growth and yield under upland conditions. 2. Phenology, biomass production and yield. *Field Crops Res.* **1996**, *48*, 47–55. [CrossRef]

50. Su, Z.; Ma, X.; Guo, H.H.; Sukiran, N.L.; Guo, B.; Assmann, S.M.; Ma, H. Flower development under drought stress: Morphological and transcriptomic analyses reveal acute responses and long-term acclimation in Arabidopsis. *Plant Cell* **2013**, *25*, 3785–3807. [CrossRef]

51. Gonzalez, V.H.; Lee, E.A.; Lukens, L.L.; Swanton, C.J. The relationship between floret number and plant dry matter accumulation varies with early season stress in maize (*Zea mays* L.). *Field Crops Res.* **2019**, *238*, 129–138. [CrossRef]

52. Nielsen, R.L. *Ear Size Determination in Corn*; Corny News Network Articles; Purdue University: West Lafayette, IN, USA, 2007.

53. Hu, X.J.; Wang, H.W.; Diao, X.Z.; Liu, Z.F.; Li, K.; Wu, Y.J.; Liang, Q.J.; Wang, H.; Huang, C.L. Transcriptome profiling and comparison of maize ear heterosis during the spikelet and floret differentiation stages. *BMC Genomics* **2016**, *17*, 959. [CrossRef]

54. Jeffares, D.C. *Molecular Genetic Analysis of the Maize Terminal Ear1 Gene and in Silico Analysis of Related Genes*; Massey University: Palmerston North, New Zealand, 2001.

55. Vroemen, C.W.; Mordhorst, A.P.; Albrecht, C.; Kwaaitaal, M.A.C.J.; Vries, S.C. The CUP-SHAPED COTYLEDON3 gene is required for boundary and shoot meristem formation in Arabidopsis. *Plant Cell* **2003**, *15*, 1563–1577. [CrossRef]

56. Hibara, K.; Karim, M.R.; Takada, S.; Taoka, K.; Furutani, M.; Aida, M.; Tasaka, M. Arabidopsis CUP-SHAPED COTYLEDON3 regulates postembryonic shoot meristem and organ boundary formation. *Plant Cell* **2006**, *18*, 2946–2957. [CrossRef]

57. Cai, Y.H.; Chen, X.J.; Xie, K.; Xing, Q.K.; Wu, Y.W.; Li, J.; Du, C.H.; Sun, Z.X.; Guo, Z.J. Dlf1, a WRKY transcription factor, is involved in the control of flowering time and plant height in rice. *PLoS ONE* **2014**, *9*, e102529. [CrossRef]

58. Chen, Z.H.; Yuan, Y.; Fu, D.; Shen, C.J.; Yang, Y.J. Identification and expression profiling of the auxin response factors in Dendrobium officinale under abiotic stresses. *Int. J. Mol. Sci.* **2017**, *18*, 927. [CrossRef] [PubMed]

59. Porco, S.; Larrieu, A.; Du, Y.J.; Gaudinier, A.; Goh, T.; Swarup, K.; Swarup, R.; Kuempers, B.; Bishopp, A.; Lavenus, J.; et al. Lateral root emergence in Arabidopsis is dependent on transcription factor LBD29 regulation of auxin influx carrier LAX3. *Development* **2016**, *143*, 3340–3349. [CrossRef] [PubMed]

60. Dinesh, D.C.; Villalobos, L.I.A.C.; Abel, S. Structural biology of nuclear auxin action. *Trends Plant Sci.* **2016**, *21*, 302–316. [CrossRef] [PubMed]

61. Jung, H.R.; Lee, D.K.; Do Choi, Y.; Kim, J.K. OsIAA6, a member of the rice Aux/IAA gene family, is involved in drought tolerance and tiller outgrowth. *Plant Sci.* **2015**, *236*, 304–312. [CrossRef]

62. Asghar, M.A.; Li, Y.; Jiang, H.K.; Sun, X.; Ahmad, B.; Imran, S.; Yu, L.; Liu, C.Y.; Yang, W.Y.; Du, J.B. Crosstalk between Abscisic Acid and Auxin under Osmotic Stress. *Agron. J.* **2019**, *111*, 2157–2162. [CrossRef]

63. Kelley, K.B.; Riechers, D.E. Recent developments in auxin biology and new opportunities for auxinic herbicide research. *Pestic. Biochem. Phys.* **2007**, *89*, 1–11. [CrossRef]

64. Han, S.; Hwang, I. Integration of multiple signaling pathways shapes the auxin response. *J. Exp. Bot.* **2017**, *69*, 189–200. [CrossRef] [PubMed]

65. Luo, J.; Zhou, J.J.; Zhang, J.Z. Aux/IAA gene family in plants: Molecular structure, regulation, and function. *Int. J. Mol. Sci.* **2018**, *19*, 259. [CrossRef] [PubMed]

66. Bai, Q.S.; Hou, D.; Li, L.; Cheng, Z.C.; Ge, W.; Liu, J.; Li, X.P.; Mu, S.H.; Gao, J. Genome-wide analysis and expression characteristics of small auxin-up RNA (SAUR) genes in moso bamboo (*Phyllostachys edulis*). *Genome* **2016**, *60*, 325–336. [CrossRef]

67. Ren, H.; Gray, W.M. SAUR proteins as effectors of hormonal and environmental signals in plant growth. *Mol. Plant* **2015**, *8*, 1153–1164. [CrossRef]

68. Han, D.G.; Zhang, Z.Y.; Ding, H.B.; Chai, L.J.; Liu, W.; Li, H.X.; Yang, G.H. Isolation and characterization of MbWRKY2 gene involved in enhanced drought tolerance in transgenic tobacco. *J. Plant Interact.* **2018**, *13*, 163–172. [CrossRef]

69. Ahmad, N.; Malagoli, M.; Wirtz, M.; Hell, R. Drought Stress in Maize Causes Differential Acclimation Responses of Glutathione and Sulfur Metabolism in Leaves and Roots. *BMC Plant Biol.* **2016**, *16*, 247. [CrossRef]

70. Zhang, X.; Tao, L.; Qiao, S.; Du, B.H.; Guo, C.H. Roles of Glutathione S-transferase in Plant Tolerance to Abiotic Stresses. *J. Chin. Biotechnol.* **2017**, *37*, 92–98.

71. Kong, X.X.; Li, B.Z.; Yang, J.S. Research progress in microalgae resistance to cadmium stress. *Microbiol. Chin.* **2017**, *44*, 1980–1987.

72. George, S.; Venkataraman, G.; Parida, A.A. chloroplast-localized and auxin-induced glutathione S-transferase from phreatophyte Prosopis juliflora confer drought tolerance on tobacco. *J. Plant Physiol.* **2010**, *167*, 311–318. [CrossRef]

73. Nakamura, A.; Umemura, I.; Gomi, K.; Hasegawa, Y.; Kitano, H.; Sazuka, T.; Matsuoka, M. Production and characterization of auxin-insensitive rice by overexpression of a mutagenized rice IAA protein. *Plant J.* **2006**, *46*, 297–306. [CrossRef] [PubMed]

74. Gu, H.H.; Yang, Y.; Xing, M.H.; Yue, C.P.; Wei, F.; Zhang, Y.J.; Zhao, W.E.; Huang, J.Y. Physiological and transcriptome analyses of *Opisthopappus taihangensis* in response to drought stress. *Cell Biosci.* **2019**, *9*, 56. [CrossRef] [PubMed]

10

Silicon and the Association with an Arbuscular-Mycorrhizal Fungus (*Rhizophagus clarus*) Mitigate the Adverse Effects of Drought Stress on Strawberry

Narges Moradtalab [1,*], **Roghieh Hajiboland** [1], **Nasser Aliasgharzad** [2], **Tobias E. Hartmann** [3] and **Günter Neumann** [3]

[1]   Department of Plant Science, University of Tabriz, Tabriz 51666-16471, Iran; ehsan@tabrizu.ac.ir
[2]   Department of Soil Science, University of Tabriz, Tabriz 51666-16471, Iran; n-aliasghar@tabrizu.ac.ir
[3]   Institute of Crop Science, University of Hohenheim, 70593 Stuttgart, Germany;
       tobias.hartmann@uni-hohenheim.de (T.E.H.); guenter.neumann@uni-hohenheim.de (G.N.)
*   Correspondence: moradtalabnarges@gmail.com

Abstract: Silicon (Si) is a beneficial element that alleviates the effects of stress factors including drought (D). Strawberry is a Si-accumulator species sensitive to D; however, the function of Si in this species is obscure. This study was conducted to examine the effect of Si and inoculation with an arbuscular mycorrhizal fungus (AMF) on physiological and biochemical responses of strawberry plants under D. Plants were grown for six weeks in perlite and irrigated with a nutrient solution. The effect of Si (3 mmol L^{-1}), AMF (*Rhizophagus clarus*) and D (mild and severe D) was studied on growth, water relations, mycorrhization, antioxidative defense, osmolytes concentration, and micronutrients status. Si and AMF significantly enhanced plant biomass production by increasing photosynthesis rate, water content and use efficiency, antioxidant enzyme defense, and the nutritional status of particularly Zn. In contrast to the roots, osmotic adjustment did not contribute to the increase of leaf water content suggesting a different strategy of both Si and AMF for improving water status in the leaves and roots. Our results demonstrated a synergistic effect of AMF and Si on improving the growth of strawberry not only under D but also under control conditions.

Keywords: silicon; strawberry; total antioxidants; drought; stress responses; arbuscular mycorrhizal fungus (AMF); *Rhizophagus clarus*

1. Introduction

Although silicon (Si) is not considered an essential element for higher plants, numerous studies have demonstrated that Si is a beneficial element that alleviates abiotic and biotic stresses in plants [1–3]. Si is a quasi-essential element for the growth of rice, wheat, sorghum, potato, cucumber, zucchini, and soybean, under various biotic and abiotic stress conditions [4]. According to the Si tissue concentration, plants are classified into Si-accumulators and non-accumulators. The differences in Si accumulation among species can be attributed to the differential ability of roots to take up Si [2].

Drought (D) adversely influences several features of plant growth and development, and a prolonged D severely diminishes plant productivity [5]. Water loss through transpiration is reduced by stomatal closure as an immediate response of plants upon being exposed to D; however, it reduces also nutrient uptake and limits plant ability for dry matter production. In addition, reduced intercellular CO_2 concentration leads to an excess excitation energy that causes enhanced leakage of electrons to molecular oxygen and increases the production of reactive oxygen species (ROS) [6,7]. These cytotoxic ROS destroy normal metabolism through oxidative damage to lipids, proteins, and nucleic acids [8].

Plants have developed complex physiological and biochemical adjustments to tolerate D, including the activation of antioxidative enzymes, maintenance of cell turgor, and water status through the accumulation of organic osmolytes such as soluble carbohydrates and free amino acids, particularly proline [9,10].

Si supplementation of plants alleviates D stress. Several mechanisms including the activation of photosynthetic enzymes [11], the activation of enzymatic antioxidant defense systems, increased water use efficiency [12,13], nutrient uptake [14], root growth and hydraulic conductance [15], and the accumulation of organic osmolytes [16] are involved in Si-mediated growth improvement under D [11,17].

The association of roots with arbuscular mycorrhizal fungi (AMF) is the most abundant symbiosis in the plant kingdom [18]. The colonization of roots by AMF enhances the plant growth by increasing nutrient uptake and plant tolerance to stress [19,20]. Several studies evaluated the effects of AMF-inoculation in horticultural plants such as citrus, apple, and strawberry [21–23]. AMF symbiosis increased the rate of photosynthesis, stomatal conductance, and leaf water potential in colonized plants under D [24]. Moreover, AMF had a significant direct contribution to the uptake of phosphorus (P), zinc (Zn), and copper (Cu) under water stress [25].

Strawberry (*Fragaria x ananasa* Duch.) plants are extremely sensitive to drought because of a shallow root system, large leaf area, and high-water content of fruits. When the strawberry plants are not sufficiently irrigated, both yield and fruit size are reduced [22]. As a Si-accumulating species [26,27], strawberry has both functional influx (Lsi1) and efflux (Lsi2) transporters for Si uptake, and under a constant soluble Si application can absorb 3% Si per dry weight [26]. However, to the best of our knowledge, there is no study on the effect of Si on strawberry under abiotic stresses including D. Another obscure aspect in this regard is Si effect on the association of roots with AMF in this species. Therefore, given the potential of both Si and AMF for mitigation of drought stress effects, the objectives of the present study are (1) to elucidate the influence of Si on photosynthesis, water status, and activity of antioxidative defense system in strawberry plants under D conditions and (2) to investigate the Si effect on the response of mycorrhizal plants when exposed to D stress. We hypothesized the existence of a synergistic effect of Si and AMF on the protection against D in strawberry plants.

2. Materials and Methods

2.1. Preparation of Plant and Fungus Materials

The first-generation strawberry (*Fragaria × ananassa* var. Paros) plantlets of genetically different individuals originating from a strawberry field were prepared as donor mother plants. Second-generation strawberry plantlets from 10 cm stolons of these genetically different mother plants were propagated in a growth chamber. Four independent biological replicates were used per treatment. The offset plants were grown in a standard peat–perlite (1:1) mixture for one week to allow root development.

Inoculum of *Rhizophagus clarus* (Walker & Schüßler; isolated in symbiosis with *Poa annua* L. in a grassland in Cuba) (MUCL 46238–GINCO–BEL; Synonymy: *Glomus clarum* Nicolson & Schenck; [28]) was provided by the Department of Soil Science, University of Tabriz, Iran. Originally, fungi were obtained from Pal Axel Lab, Lund University, Sweden. *R. clarus* was propagated with *Trifolium repens* L. plants in 3.5 L pots containing sterile sandy loam soil. Rorison's nutrient solution, prepared with deionized water [29] with 50% strength of phosphorus, was added to the pots twice a week to bring the soil moisture to water holding capacity (WHC). The pots were incubated in a greenhouse with 28/20 °C day/night and 16/8 h light/dark periods. After four months, the tops of the plants were excised and the pot materials containing soil and mycorrhizal roots were thoroughly mixed and used as fungal inoculum.

2.2. Plant Treatments

The experiment was conducted using a completely randomized design with three factors including irrigation regimes (three levels), Si treatments (two levels), and AMF inoculation (two levels). Each treatment combination was represented by four independent pots as four replicates.

One-week-old strawberry seedlings were transferred to 3 L pots (one plant per pot) filled with washed perlite and containing 60 g autoclaved and non-autoclaved AMF inoculum in −AMF and +AMF treatments, respectively. The pots were irrigated daily with water or Hoagland nutrient solution at WHC of the perlite after weighing. The total volume of nutrient solution applied to the plants was 200 mL pot^{-1} week^{-1}. To avoid the accumulation of salts in the substrate, electric conductivity in the perlite was measured in samples taken weekly from the bottom of the pots. Si as sodium silicate (Na_2SiO_3, Sigma–Aldrich, Munich, Germany) prepared as the solution (0.6 mM, pH = 6.1) was added to the pots weekly by irrigation leading to a concentration of 3 mmol L^{-1} perlite (~84 mg L^{-1} perlite) at the end of the experiment after 6 weeks. One week after starting the Si application, the different irrigation regimes (IR) included well-watered (WW, 90% WHC), mild drought (MD, 75% WHC), and severe drought (SD, 35% WHC) and were assigned randomly to the pots, and watering was omitted from D treatments until they reached the respective WHC. This was achieved 4 and 6 days after starting a different IR for the MD and SD treatments, respectively. Well-watered and D plants received the same amount of nutrient solution, and the respective WHC was achieved by adjusting the volume of water used for irrigation.

In order to determine the possible effect of Na as the accompanying ion in the Si salt applied to the plants, an experiment was conducted parallel to the main experiment with an additional control (without the addition of salt or Si) and 6 mmol L^{-1} NaCl containing an equivalent Na with 3 mmol L^{-1} Na_2SiO_3. The dry weight (g plant^{-1}) of plants under control (0.48 ± 0.05) and 6 mmol L^{-1} salt (0.51 ± 0.04) was not significantly different (Tukey test, $p < 0.001$).

Plants were grown under controlled environmental conditions with a temperature regime of 25 °C/18 °C day/night, 14/10 h light/dark periods, a relative humidity of 30%, and at a photon flux density of about 400 μmol m^{-2} s^{-1}.

2.3. Plant Harvest

Six-week-old plants (five weeks after starting Si treatments and four weeks after reaching the respective WHC) were harvested. Shoots and roots were separated, washed with distilled water, and blotted dry on filter paper. After determination of the fresh weight (FW), the dry weight (DW) was determined after drying at 60 °C for 48 h. Subsamples were taken for biochemical analyses before drying. Before harvest, the gas exchange parameters were determined in attached leaves.

For evaluation of the AMF colonization, the fine roots (1 g FW) were cleared in 10% (v/v) KOH and stained with 0.05% (v/v) trypan blue in lacto–glycerin. The colonization rate of the roots (%) was estimated by counting the proportion of root length containing fungal structures (arbuscules, vesicles and hyphae) using the gridline intersect method [30,31]. In brief, stained root segments were spread out evenly in a 10 cm diameter Petri dish. A grid of lines was marked on the bottom of the dish to form 0.5 cm^2. Vertical and horizontal gridlines were observed with a binocular device, and the presence or absence of fungal structures was recorded at each point where the roots intersected a line. Three sets of observations were made recording all the root-gridline intersects. Each of the three replicate records was made on a fresh rearrangement of the same root segments [30,31].

2.4. Leaf Osmotic Potential and Relative Water Content

The leaf osmotic potential (ψ_s) was determined in the second leaves harvested 1 h after the light was turned on in the growth chamber. The leaves were homogenized in a prechilled mortar and pestle and centrifuged at 4000 g for 20 min at 4 °C. The osmotic pressure of the samples was measured by an osmometer (Micro–Osmometer, Herman Roebling Messtechnik, Germany), and the milliosmol data were recalculated to MPa. For the determination of the relative water content (RWC%), the leaf disks (5 mm diameter) were prepared, and after the determination of the fresh weight (FW), they were submerged for 20 h in distilled water; thereafter, they were blotted dry gently on a paper towel, and the turgid weight (TW) was determined. The dry weight (DW) of the samples was determined after drying in an oven at 70 °C for 24 h, and the RWC% was calculated according to the formula $(FW - DW)/(TW - DW) \times 100$.

2.5. Measurements of Photosynthetic Gas Exchange

Before the harvest gas exchange parameters were determined with the attached leaves. The net CO_2 fixation rate (μmol m^{-2} s^{-1}), transpiration rate (mmol m^{-2} s^{-1}), and stomatal conductance (mol m^{-2} s^{-1}) were determined with a calibrated portable gas exchange system (LCA–4, ADC Bioscientific Ltd., Hoddesdon, UK). Water use efficiency (WUE) was calculated as the ratio of photosynthesis/transpiration (μmol mmol^{-1}).

2.6. Biochemical Determinations

For the determination of carbohydrates, leaf and root samples (100 mg) were homogenized in a 100 mM potassium phosphate buffer (pH 7.5) at 4 °C. After centrifugation at 12,000 g for 15 min, the supernatant was used for the determination of total soluble sugars. An aliquot of the supernatant was mixed with an anthrone–sulfuric acid reagent and incubated for 10 min at 100 °C. After cooling, the absorbance was determined at 625 nm. The standard curve was created using glucose (Sigma–Aldrich, Munich, Germany) [32]. The total soluble protein was determined by the Bradford (1976) method using a commercial reagent (Roti®Quant, Roth GmbH, Karlsruhe, Germany) and bovine serum albumin (BSA) as standard. Total free α-amino acids were assayed using a ninhydrin colorimetric method. Glycine (Sigma–Aldrich, Munich, Germany) was used to produce a standard curve [33]. For the determination of proline, samples were homogenized with 3% (v/v) sulfosalicylic acid and the homogenate was centrifuged at 3000 g for 20 min. The supernatant was treated with acetic acid and acid ninhydrin and boiled for 1 h, and then the absorbance was determined at 520 nm. Proline (Sigma–Aldrich, Munich, Germany) was used to produce a standard curve [34].

2.7. Determination of Enzyme Activities and Concentration Of Oxidants

Fresh leaf samples (100 mg) were ground in liquid nitrogen using a mortar and pestle. Each enzyme assay was tested for linearity between the volume of crude extract and the measured activity. All measurements were undertaken through spectrophotometry (Specord 200, Analytical Jena AG, Jena, Germany) according to optimized protocols described elsewhere [35]. The activity of ascorbate peroxidase (APX, EC 1.11.1.11) was measured by determining the ascorbic acid oxidation; one unit of APX oxidizes ascorbic acid at a rate of 1 μmol min^{-1} at 25 °C. The catalase (CAT, EC 1.11.1.6) activity was assayed by monitoring the decrease in absorbance of H_2O_2 at 240 nm; unit activity was taken as the amount of enzyme which decomposes 1 μmol of H_2O_2 in one min. Peroxidase (POD, EC 1.11.1.7) activity was assayed using the guaiacol test. The enzyme unit was calculated as the enzyme protein required for the formation of 1 μmol tetra–guaiacol for 1 min. The total superoxide dismutase (SOD, EC 1.15.1.1) activity was determined using the mono–formazan formation test. One unit of SOD was defined as the amount of enzyme required to induce a 50% inhibition of nitro blue tetrazolium (NBT) reduction as measured at 560 nm compared with control samples without enzyme aliquot. The concentration of H_2O_2 was determined using KI at 508 nm. Lipid peroxidation was estimated from the amount of malondialdehyde (MDA) formed in a reaction mixture containing thio-barbituric acid (Sigma–Aldrich, Munich, Germany) at 532 nm. The MDA levels were calculated from a 1,1,3,3–tetraethoxypropane (Sigma–Aldrich, Munich, Germany) standard curve [35].

2.8. Mineral Nutrient Analysis

For the determination of the plant nutritional status, 250 mg of dried leaf material was ashed in a muffle furnace at 500 °C for 5 h. After cooling, the samples were extracted twice with 2.5 mL of 3.4 M HNO_3 until dryness to precipitate SiO_2. The ash was dissolved in 2.5 mL of 4 M HCl, subsequently diluted ten times with hot deionized water, and boiled for 2 min. After the addition of a 0.1 mL cesium chloride/lanthanum chloride buffer to the 4.9 mL ash solution, Fe, Mn, and Zn concentrations were measured by atomic absorption spectrometry (AAS, UNICAM 939, Offenbach/Main, Germany) [36].

2.9. Silicon Determination

Dry leaf material (0.2 g) was microwave digested with 3 mL concentrated HNO_3 + 2 mL H_2O_2 for 1 h. Samples were diluted with circa 15 mL deionized H_2O and transferred into 25 mL plastic flasks; 1 mL concentrated Hydrofluoric acid was added and left overnight. After the addition of 2.5 mL 2% (w/v) H_3BO_3, the flask volume was adjusted to 25 mL with deionized H_2O, and Si was determined by ICP–OES (Vista–PRO, Varian Inc., Palo Alto, USA) [36].

2.10. Statistical Analyses

A primary statistical analysis was carried out using the Sigma Plot 11.0 software Systat Software Inc. San Jose, USA. Experimental data were checked for normality using the Shapiro–Wilk test. Where necessary, data were transformed through standard methods to meet the requirements of statistical analysis. In a second analytical step, a so-called insert-and-absorb algorithm was used to truthfully present all relevant significant differences for the main factors and interactions between the main factors. The algorithm was implemented using the SAS 9.4 macro% (Multi factors)based on the work of Piepho, 2012 [37]. The %MULT macro uses output generated from the MIXED, GLIMMIX, or GENMOD procedures. It allows up to three by-variables for factorial experiments but can process the least squares means for one effect only. If Least Squares Means (LSMEANS) are needed for several effects, the linear model procedure must be run several times, each time using only one LSMEANS statement with only one effect. It means each level of one main factor (e.g. IR) was compared separately for each level of the remaining two factors (e.g. AMF and Si) as pairwise comparisons. In our three-factorial analyses (IR, Si and AMF factors), the main effects of the experiment (IR, AMF, Si, IR×AMF, IR×Si, Si×AMF, IR×Si×AMF) were compared using a proc mixed model (MIXED procedures) in the SAS environment at a significance level of $\alpha = 0.05$. LSMEANS of the main and interaction effects were determined.

3. Results

3.1. Effect of Si and Inoculation with AMF on Plants Biomass And Root Colonization

Mycorrhization or Si as single treatments did not significantly affect shoot biomass under well-watered (WW) conditions while the combination of both treatments resulted in a higher shoot biomass suggesting a synergistic effect between AMF and Si. In the plants exposed to mild drought (MD) and severe drought (SD) stresses, in contrast, Si and AMF as single treatments increased the shoot biomass; however, the effect of AMF was not significant in SD plants (Figure 1A). Root biomass was increased by the AMF treatment under WW conditions. Under MD and SD, in comparison, the effect of both Si and AMF as single treatments was significant on root biomass; the effect of AMF was much higher than Si particularly under MD (Figure 1B).

The relative water content decreased with the severity of D. Under WW conditions, there was no significant effect of Si or AMF as single treatments on RWC while the combination of both treatments resulted in higher RWC. In MD and SD plants, in contrast, the effect of single treatments was mainly significant (Figure 1C). The osmotic potential of the leaves and roots was affected by an inverse trend of RWC (Figure 1D). There was a significant interaction among the three main factors including IR, Si, and AMF on the shoot and root biomass, RWC, and osmotic potential where all decreased with the severity of D (IR factor) but were modified by Si and AMF applications (Figure 1).

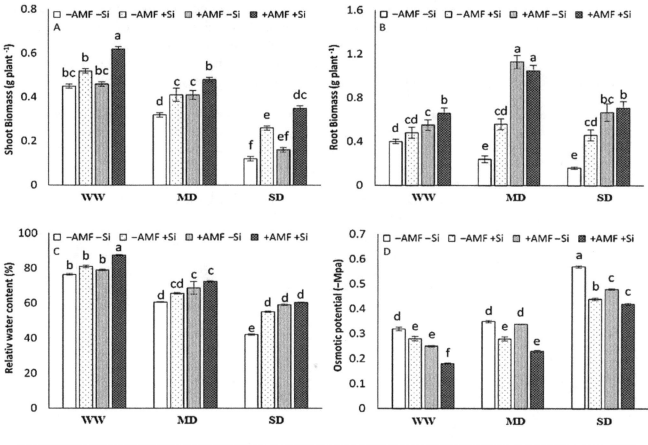

F		Shoot Biomass	Root Biomass	Leaf RWC	Leaf Osmotic Potential
				p	
IR	WW MD SD	<0.001 ***	<0.001 ***	<0.001 ***	<0.001 ***
AMF	−AMF +AMF	<0.001 ***	<0.001 ***	<0.001 ***	<0.001 ***
Si	−Si +Si	<0.001 ***	<0.001 ***	<0.001 ***	<0.001 ***
IR×AMF		0.47 ns	<0.001 ***	<0.001 ***	<0.001 ***
IR×Si		<0.001 ***	0.62 ns	0.21 ns	0.002 **
AMF×Si		0.01 *	<0.001 ***	0.02 *	0.04 *
IR×AMF×Si		0.01 *	0.01 *	<0.001 ***	<0.001 ***

Figure 1. The biomass of shoot (**A**) and root (**B**), the leaf relative water content (RWC) (**C**), and the osmotic potential (**D**) in strawberry plants at harvest after six experimental weeks under three irrigation regimes (IR): well-watered (WW), mild drought (MD), and severe drought (SD) without (−AMF) or with inoculation with arbuscular mycorrhizal fungus (+AMF) *Rhizophagus clarus* (Walker & Schüßler), in the absence (−Si) or presence of silicon (+Si, 3 mmol L^{-1} Na_2SiO_3). The bars show the treatment means (4 replicates) ±SE of the mean. The interactions among the main factors are in the table (**F**); *** $p < 0.001$, ** $p < 0.01$, * $p < 0.05$, and ns is not significant (Tukey test, alpha = 0.05).

There was a low colonization percentage detectable even in −AMF plants, which might be caused by carryover of some fungal populations from the field-grown mother plants or from the peat culture substrate used for the preculture (Table 1). Interestingly, D decreased the hyphal and arbuscular colonization rates (%) in the −AMF plants while not influencing them in the +AMF ones (Table 1). The pairwise comparison indicated that the hyphal colonization percentage of +AMF plants was increased by Si under all IR treatments. The same was true for the frequency of arbuscules that was significant even for the −AMF plants under MD and SD conditions. The frequency of vesicles increased in the −AMF plants under SD conditions. In the +AMF plants, a significant effect was observed under both MD and SD conditions. Si did not affect this parameter. Interestingly, inoculation with AMF decreased the frequency of vesicles in the WW while it increased in the MD and SD plants (Table 1).

Table 1. The root colonization rate (%) in strawberry plants at harvest after six experimental weeks under three irrigation regimes (IR): well-watered (WW), mild drought (MD), and severe drought (SD) without (−AMF) or with inoculation with arbuscular mycorrhizal fungus (+AMF) *Rhizophagus clarus* (Walker & Schüßler), in the absence (−Si) or presence of silicon (+Si, 3 mmol L^{-1} Na_2SiO_3). The numbers show the treatment means (4 replicates) ±SE of the mean. Means with the same letters are not significantly different. The interactions among the main factors include *** $p < 0.001$, ** $p < 0.01$, * $p < 0.05$, and ns as not significant (Tukey test, alpha = 0.05).

			Hyphae	Arbuscules	Vesicles
WW	−AMF	−Si	1.1 ± 0.1 [c]	2.0 ± 0.1 [cd]	0.4 ± 0.0 [c]
		+Si	1.1 ± 0.1 [c]	2.4 ± 0.1 [c]	0.4 ± 0.0 [c]
	+AMF	−Si	1.6 ± 0.1 [b]	4.9 ± 0.2 [b]	0.2 ± 0.1 [d]
		+Si	2.6 ± 0.1 [a]	7.3 ± 0.3 [a]	0.2 ± 0.1 [d]
MD	−AMF	−Si	0.1 ± 0.0 [d]	1.1 ± 0.1 [e]	0.4 ± 0.0 [c]
		+Si	0.1 ± 0.0 [d]	1.7 ± 0.1 [d]	0.4 ± 0.0 [c]
	+AMF	−Si	2.4 ± 0.1 [b]	6.3 ± 0.3 [b]	1.0 ± 0.1 [b]
		+Si	2.6 ± 0.1 [a]	8.1 ± 0.4 [a]	1.0 ± 0.1 [b]
SD	−AMF	−Si	0.1 ± 0.0 [d]	0.0 ± 0.0 [f]	0.9 ± 0.0 [b]
		+Si	0.1 ± 0.0 [d]	1.2 ± 0.1 [e]	0.9 ± 0.0 [b]
	+AMF	−Si	2.4 ± 0.1 [b]	2.8 ± 0.1 [c]	1.9 ± 0.1 [a]
		+Si	2.7 ± 0.1 [a]	3.8 ± 0.2 [b]	2.1 ± 0.1 [a]
				p	
IR	WW MD SD		0.72 [ns]	0.02 *	0.01 *
AMF	−AMF +AMF		<0.001 ***	<0.001 ***	0.04 *
Si	−Si +Si		0.20 [ns]	0.05 *	0.87 [ns]
IR×AMF			0.001 **	<0.001 ***	0.05 *
IR×Si			0.12 [ns]	<0.001 ***	0.06 [ns]
AMF×Si			0.06 [ns]	0.01 *	0.12 [ns]
IR×AMF×Si			0.14 [ns]	<0.001 ***	0.11 [ns]

3.2. Effect of Si and Inoculation with AMF on the Leaf Gas Exchange Parameters

The single application of Si or AMF did not influence the rate of photosynthesis under WW conditions. A significant effect of Si as the single treatment, however, was observed under the MD and SD conditions, and a significant AMF effect was observed under MD conditions. The combined application of Si and AMF, in contrast, increased the rate of photosynthesis under WW, MD, and SD conditions, and the highest photosynthesis rate was obtained under the combination of both treatments with a significant difference with each single treatment (Figure 2A). In the absence of AMF and Si treatments, SD decreased the transpiration rate. This parameter increased by AMF only under SD conditions and by Si under MD and SD conditions. Under WW conditions, in contrast, the rate of transpiration was only increased by the combined application of Si and AMF (Figure 2B). The stomatal conductance showed a similar pattern to the rate of photosynthesis (Figure 2C). The sater use efficiency (WUE) decreased by D irrespective of the AMF or Si treatments. Significant effects of the single treatments were observed in the SD plants for Si and in both MD and SD plants for AMF, and the highest value of WUE was obtained in the combination of both treatments (Figure 2D). There was a significant interaction among the three main factors on photosynthetic activity, transpiration rate, and stomatal conductance. There was not any three-way interaction evident for water use efficiency. Significant differences were observed in IR, AMF, Si, and IR×Si (Figure 2).

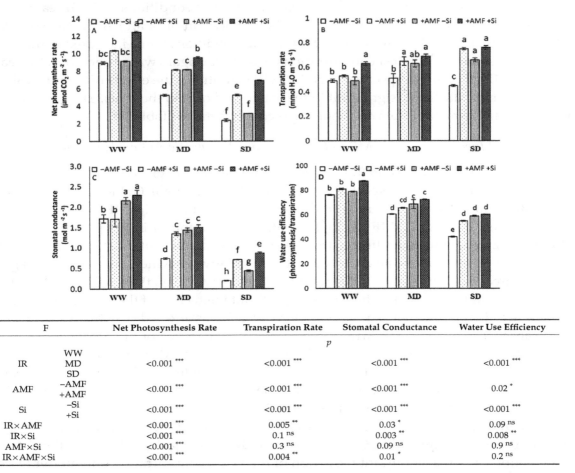

F		Net Photosynthesis Rate	Transpiration Rate	Stomatal Conductance	Water Use Efficiency
			p		
IR	WW MD SD	<0.001 ***	<0.001 ***	<0.001 ***	<0.001 ***
AMF	−AMF +AMF	<0.001 ***	<0.001 ***	<0.001 ***	0.02 *
Si	−Si +Si	<0.001 ***	<0.001 ***	<0.001 ***	<0.001 ***
IR×AMF		<0.001 ***	0.005 **	0.03 *	0.09 ns
IR×Si		<0.001 ***	0.1 ns	0.003 **	0.008 **
AMF×Si		<0.001 ***	0.3 ns	0.09 ns	0.9 ns
IR×AMF×Si		<0.001 ***	0.004 **	0.01 *	0.2 ns

Figure 2. The net photosynthesis rate (**A**), transpiration rate (**B**), stomatal conductance (**C**), and water use efficiency (**D**) of strawberry plants at harvest after six experimental weeks under three irrigation regimes (IR): well-watered (WW), mild drought (MD), and severe drought (SD) without (−AMF) or with inoculation with arbuscular mycorrhizal fungus (+AMF) *Rhizophagus clarus* (Walker & Schüßler), in the absence (−Si) or presence of silicon (+Si, 3 mmol L^{-1} Na$_2$SiO$_3$). The bars show the treatment means (4 replicates) ±SE of the mean. The interactions among the main factors are in the table (**F**); *** $p < 0.001$, ** $p < 0.01$, * $p < 0.05$, and ns is not significant (Tukey test, alpha = 0.05).

3.3. Effect of Si and Inoculation with AMF on the Concentrations of Osmolytes

Under WW conditions, there was no effect of either AMF or Si on the proline concentrations (Table 2). Under MD and SD, in comparison, both Si and AMF treatments decreased leaf proline concentrations; a synergistic effect, however, was observed only under SD conditions (Table 2). The opposite trend of the proline concentration was observed in the root under SD, which was increased by Si and AMF applications where the combined application was not significantly different from the single application. There was a significant interaction among the three main factors including IR, Si, and AMF on the leaf proline concentration (Table 2). D conditions decreased the concentration of proteins while increased the concentration of free amino acids (AA) in leaf and root tissues. The application of Si in the −AMF plants increased leaf protein concentrations under D (but not under WW) conditions while decreasing that of free AA. In the roots, in contrast, both protein and free AA concentrations increased by Si in the −AMF plants under SD conditions. Similar to Si as a single treatment, AMF application as a single treatment decreased the concentration of free AA in the leaves while increased that in the roots under SD conditions (Table 2). The total free AA concentration of the leaf was significantly affected by all two-way and three-way interactions, while there was not IR×AMF interaction regarding leaf protein concentration. For the roots, there was only an interaction of AMF and Si factors on protein concentrations and of IR and Si on free AA concentration (Table 2).

The concentration of soluble sugars increased under D conditions in both leaves and roots irrespective the AMF and Si treatments. Upon the application of Si and AMF, the soluble sugars concentration decreased in the leaves under MD and SD conditions while increased in the roots of SD plants. The lowest and the highest concentrations of soluble sugars was observed in the leaves and roots in the +AMF+Si plants, respectively. Under WW conditions, the effects of Si and AMF as single treatments were not statistically significant in the leaves and of Si in the roots. There was a three-way interaction among the main factors on shoot sugar concentrations (Table 2).

3.4. Effect of Si and Inoculation with AMF on the Function of Enzymatic Antioxidant Defense

The activity of CAT, SOD, and POD in the leaves and the activity of CAT and SOD in the roots were increased under D conditions irrespective the Si and AMF treatments (Table 3). The highest activity of antioxidative enzymes was observed in the combination of Si and AMF treatments (+AMF+Si). A significant effect of Si and AMF as single treatments was found in SD plants for all analyzed antioxidative enzymes while this effect in the leaves was not significant for POD in MD and for SOD and POD in WW plants. Among all analyzed leaf antioxidative enzymes, only SOD was significantly affected by a three-way interaction. In the roots, the effect of AMF on the CAT and SOD activity was higher than Si as single treatments. There was a significant interaction of the three main factors in CAT but not SOD activities of the root. The activity of root SOD was affected only by IR×AMF (Table 3).

In the absence of Si and AMF, MDA concentration as an indicator of damage to the membrane increased with increasing severity of D. Both Si and AMF treatments decreased the concentration of the leaf MDA that was observed only in D plants. AMF was more effective than Si as single treatment on the reduction of MDA concentrations; the lowest value was observed in +AMF+Si plants. A significant three-way interaction affected the leaf MDA (Table 3).

D treatment led to the accumulation of H_2O_2 in the roots that increased with increasing severity of stress. Si treatment decreased H_2O_2 concentration that was significant only in the D treatments. AMF inoculation caused a significant reduction of the H_2O_2 concentration only under D treatment that was significant in SD plants. The H_2O_2 concentration of the root was decreased by a significant interaction among IR, Si, and AMF factors (Table 3).

3.5. Effect of Si and Inoculation with AMF on the Leaf Concentrations of Nutrients and Si

The Si concentration significantly decreased in SD plants and increased by Si application in the presence or absence of AMF under WW and D conditions (Table 4). The effect of AMF on Si concentration was significant only in +Si plants under WW and in −Si ones under SD conditions (Table 4). The interaction effects between two (IR×AMF, IR×Si, and AMF×Si) and among three main factors (IR×AMF×Si) on Si concentration were significant (Table 4).

A significant effect of D on the leaf concentrations of Mn, Fe, and Cu was observed only in the SD treatment while leaf Zn concentration decreased under both MD and SD conditions (Table 4). Si and AMF treatments alone or in combination did not influence the concentrations of Mn, Fe, and Cu. However, Si and AMF significantly increased the leaf Zn concentration under MD conditions (Table 4). Furthermore, significant two-way (IR×Si) and three-way (IR×AMF×Si) interactions were observed for the leaf Zn concentration (Table 4).

Table 2. The concentrations of proline ($\mu g\ g^{-1}$ FW), total free amino acids (AA, $\mu g\ g^{-1}$ FW), total soluble proteins (mg g^{-1} FW), and soluble sugars (mg g^{-1} FW) in the leaf and roots of strawberry plants at harvest after six experimental weeks under three irrigation regimes (IR): well-watered (WW), mild drought (MD), and severe drought (SD) without (−AMF) or with inoculation with arbuscular mycorrhizal fungus (+AMF) *Rhizophagus clarus* (Walker & Schüßler), in the absence (−Si) or presence of silicon (+Si, 3 mmol L^{-1} Na_2SiO_3). The numbers show the treatment means (4 replicates) ±SE of the mean. Means with the same letters are not significantly different. Interactions among the main factors are indicated as *** $p < 0.001$, ** $p < 0.01$, and * $p < 0.05$ and ns as not significant (Tukey test, alpha = 0.05).

IR	AMF	Si	Leaf Proline	Leaf Free AA	Leaf Protein	Leaf Soluble Sugars	Root Proline	Root Free AA	Root Protein	Root Soluble Sugars
WW	−AMF	−Si	1.4 ± 0.0 ^{e}	2.6 ± 0.1 ^{g}	9.0 ± 0.3 ^{bc}	5.1 ± 0.0 ^{d}	2.9 ± 0.4 ^{ab}	1.9 ± 0.3 ^{df}	2.9 ± 0.3 ^{ab}	1.5 ± 0.0 ^{f}
		+Si	1.3 ± 0.0 ^{e}	2.5 ± 0.0 ^{g}	10.3 ± 0.3 ^{b}	4.5 ± 0.0 ^{d}	3.0 ± 0.3 ^{ab}	1.4 ± 0.4 ^{f}	3.0 ± 0.3 ^{ab}	2.8 ± 0.1 ^{ef}
	+AMF	−Si	1.4 ± 0.0 ^{e}	2.5 ± 0.0 ^{g}	9.1 ± 0.2 ^{bc}	4.3 ± 0.1 ^{d}	3.4 ± 0.4 ^{ab}	1.4 ± 0.4 ^{f}	3.4 ± 0.3 ^{ab}	2.3 ± 0.2 ^{e}
		+Si	1.3 ± 0.0 ^{e}	2.9 ± 0.0 ^{g}	12.5 ± 0.3 ^{a}	2.0 ± 0.0 ^{e}	3.7 ± 0.2 ^{a}	1.4 ± 0.3 ^{f}	3.7 ± 0.1 ^{a}	2.6 ± 0.1 ^{e}
MD	−AMF	−Si	39.5 ± 0.0 ^{b}	8.2 ± 0.1 ^{c}	6.4 ± 0.1 ^{d}	9.8 ± 0.0 ^{b}	1.7 ± 0.1 ^{bc}	4.8 ± 0.5 ^{d}	1.7 ± 0.0 ^{bc}	3.7 ± 0.3 ^{ed}
		+Si	7.4 ± 0.2 ^{d}	5.1 ± 0.1 ^{e}	8.2 ± 0.5 ^{c}	6.5 ± 0.0 ^{c}	2.4 ± 0.3 ^{b}	5.8 ± 0.3 ^{cd}	2.4 ± 0.0 ^{b}	5.9 ± 0.1 ^{d}
	+AMF	−Si	7.7 ± 0.1 ^{d}	4.8 ± 0.1 ^{e}	8.2 ± 0.4 ^{c}	6.3 ± 0.0 ^{c}	2.3 ± 0.1 ^{b}	6.2 ± 0.3 ^{cd}	2.3 ± 0.2 ^{b}	4.5 ± 0.1 ^{d}
		+Si	5.1 ± 0.0 ^{de}	3.3 ± 0.1 ^{f}	9.5 ± 0.1 ^{b}	6.0 ± 0.0 ^{cd}	2.0 ± 0.2 ^{b}	8.9 ± 0.8 ^{c}	2.0 ± 0.2 ^{b}	5.7 ± 0.4 ^{d}
SD	−AMF	−Si	62.1 ± 2.9 ^{a}	12.3 ± 0.1 ^{a}	2.4 ± 0.2 ^{f}	14.8 ± 0.1 ^{a}	1.3 ± 0.2 ^{c}	11.0 ± 0.8 ^{c}	1.3 ± 0.3 ^{c}	8.2 ± 0.4 ^{c}
		+Si	16.2 ± 0.1 ^{c}	9.1 ± 0.0 ^{b}	5.3 ± 0.1 ^{e}	9.2 ± 0.0 ^{b}	2.8 ± 0.1 ^{ab}	20.6 ± 1.1 ^{a}	2.8 ± 0.2 ^{ab}	12.7 ± 0.5 ^{b}
	+AMF	−Si	18.1 ± 0.0 ^{c}	6.3 ± 0.0 ^{d}	3.2 ± 0.1 ^{f}	10.1 ± 0.0 ^{b}	2.6 ± 0.2 ^{ab}	15.7 ± 0.6 ^{b}	2.6 ± 0.4 ^{ab}	12.0 ± 0.4 ^{b}
		+Si	9.2 ± 0.0 ^{d}	6.1 ± 0.0 ^{d}	7.0 ± 0.2 ^{d}	7.2 ± 0.1 ^{c}	2.9 ± 0.3 ^{ab}	23.3 ± 1.6 ^{a}	2.9 ± 0.2 ^{ab}	15.1 ± 0.6 ^{a}
p										
IR (WW, MD, SD)			<0.001 ***	<0.001 ***	<0.001 ***	<0.001 ***	<0.001 ***	<0.001 ***	<0.001 ***	<0.001 ***
AMF (−AMF, +AMF)			<0.001 ***	<0.001 ***	<0.001 ***	<0.001 ***	0.004 **	0.151 ns	0.001 **	<0.001 ***
Si (−Si, +Si)			<0.001 ***	<0.001 ***	<0.001 ***	<0.001 ***	0.006 **	0.061 ns	0.002 **	<0.001 ***
IR×AMF			<0.001 ***	<0.001 ***	0.5 ns	<0.001 ***	0.2 ns	0.1 ns	0.2 ns	0.01 *
IR×Si			<0.001 ***	<0.001 ***	<0.001 ***	<0.001 ***	0.1 ns	0.02 *	0.09 ns	0.6 ns
AMF×Si			0.05 ns	0.01 *	0.01 *	<0.001 ***	0.05 ns	0.4 ns	0.03 *	<0.001 ***
IR×AMF×Si			<0.001 ***	<0.001 ***	0.01 *	<0.001 ***	0.1 ns	0.3 ns	0.08 ns	0.04 *

Table 3. The activity of catalase (CAT, μmol mg^{-1} protein min^{-1}), superoxide dismutase (SOD, Unit mg^{-1} protein min^{-1}), and peroxidase (POD, (μmol tetra guaicol mg^{-1} protein min^{-1}) and the concentration of malondialdehyde (MDA, nmol g^{-1} FW) in the leaf and the activity of CAT and SOD and the concentration of hydrogen peroxide (H_2O_2, μmol g^{-1} FW) in the roots of strawberry plants at harvest after six experimental weeks under three irrigation regimes (IR): well-watered (WW), mild drought (MD), and severe drought (SD) without ($-$AMF) or with inoculation with arbuscular mycorrhizal fungus (+AMF) *Rhizophagus clarus* (Walker & Schüßler), in the absence ($-$Si) or presence of silicon (+Si, 3 mmol L^{-1} Na$_2$SiO$_3$). The numbers show the treatment means (4 replicates) ±SE of the mean. Means with the same letters are not significantly different. Interactions among the main factors are indicated as *** $p < 0.001$, ** $p < 0.01$, and * $p < 0.05$ and ns as not significant (Tukey test, alpha = 0.05).

			Leaf				Root		
			CAT	SOD	POD	MDA	CAT	SOD	H_2O_2
WW	$-$AMF	$-$Si	36.7 ± 1.7 ^{g}	4.5 ± 0.5 ^{d}	1.7 ± 0.3 ^{f}	0.2 ± 0.0 ^{i}	2.8 ± 0.1 ^{e}	3.7 ± 0.2 ^{e}	0.5 ± 0.0 ^{e}
		+Si	50.0 ± 4.3 ^{f}	8.1 ± 0.4 ^{d}	3.8 ± 0.2 ^{f}	0.2 ± 0.0 ^{i}	3.8 ± 0.3 ^{e}	4.5 ± 0.3 ^{e}	0.3 ± 0.0 ^{e}
	+AMF	$-$Si	54.1 ± 1.6 ^{f}	5.2 ± 0.5 ^{d}	3.8 ± 0.4 ^{f}	0.1 ± 0.0 ^{i}	4.0 ± 0.1 ^{e}	4.7 ± 0.4 ^{e}	0.5 ± 0.0 ^{e}
		+Si	61.2 ± 0.9 ^{df}	9.2 ± 0.7 ^{d}	4.5 ± 0.6 ^{df}	0.1 ± 0.0 ^{i}	4.5 ± 0.1 ^{ed}	5.1 ± 0.4 ^{ed}	0.3 ± 0.0 ^{e}
MD	$-$AMF	$-$Si	66.1 ± 1.2 ^{e}	18.0 ± 0.8 ^{c}	7.1 ± 0.3 ^{d}	25.2 ± 0.1 ^{b}	7.2 ± 0.2 ^{d}	8.2 ± 0.4 ^{de}	2.5 ± 0.2 ^{c}
		+Si	84.1 ± 1.9 ^{d}	34.2 ± 2.8 ^{b}	9.8 ± 0.9 ^{d}	10.2 ± 0.0 ^{e}	8.4 ± 0.2 ^{cd}	9.4 ± 0.5 ^{dc}	1.5 ± 0.1 ^{d}
	+AMF	$-$Si	86.0 ± 1.0 ^{d}	35.4 ± 3.1 ^{b}	9.9 ± 1.0 ^{d}	7.2 ± 0.0 ^{g}	14.1 ± 0.7 ^{c}	13.5 ± 0.4 ^{c}	2.1 ± 0.0 ^{c}
		+Si	111.6 ± 4.1 ^{c}	37.1 ± 3.6 ^{b}	10.8 ± 0.7 ^{d}	6.3 ± 0.1 ^{h}	17.1 ± 1.3 ^{c}	17.3 ± 0.6 ^{c}	1.0 ± 0.0 ^{d}
SD	$-$AMF	$-$Si	113.1 ± 1.2 ^{c}	23.0 ± 1.7 ^{c}	30.0 ± 3.5 ^{c}	49.5 ± 0.2 ^{a}	18.3 ± 1.1 ^{c}	19.0 ± 1.1 ^{c}	3.9 ± 0.1 ^{a}
		+Si	132.9 ± 1.0 ^{b}	45.1 ± 1.6 ^{a}	37.8 ± 6.1 ^{b}	17.5 ± 0.1 ^{c}	26.3 ± 2.0 ^{b}	26.3 ± 3.0 ^{b}	2.7 ± 0.1 ^{b}
	+AMF	$-$Si	126.0 ± 2.0 ^{b}	43.0 ± 1.7 ^{b}	37.1 ± 4.6 ^{b}	12.3 ± 0.2 ^{d}	36.7 ± 2.0 ^{a}	36.7 ± 4.0 ^{a}	2.2 ± 0.1 ^{bc}
		+Si	151.0 ± 1.3 ^{a}	50.5 ± 2.4 ^{a}	41.3 ± 1.8 ^{a}	9.2 ± 0.0 ^{f}	38.0 ± 2.5 ^{a}	38.8 ± 4.0 ^{a}	2.5 ± 0.2 ^{b}
						p			
IR	WW MD SD		<0.001 ***	<0.001 ***	<0.001 ***	<0.001 ***	<0.001 ***	<0.001 ***	<0.001 ***
AMF	$-$AMF +AMF		0.006 **	<0.001 ***	0.5 ns	<0.001 ***	0.8 ns	<0.001 ***	<0.001 ***
Si	$-$Si +Si		0.6 ns	<0.001 ***	0.7 ns	<0.001 ***	0.02 *	<0.001 ***	0.008 **
IR×AMF			<0.001 ***	0.5 ns	0.08 ns	<0.001 ***	<0.001 ***	0.002 **	0.06 ns
IR×Si			<0.001 ***	<0.001 ***	0.2 ns	<0.001 ***	<0.001 ***	0.9 ns	<0.001 ***
AMF×Si			<0.001 ***	<0.001 ***	0.08 ns	<0.001 ***	0.03 *	0.4 ns	0.5 ns
IR×AMF×Si			0.2 ns	<0.001 ***	0.06 ns	0.002 **	<0.001 ***	0.3 ns	0.008 **

Table 4. The concentrations of Si (%), Zn ($\mu g\ g^{-1}$ DW), Mn ($\mu g\ g^{-1}$ DW), Fe ($\mu g\ g^{-1}$ DW), and Cu ($\mu g\ g^{-1}$ DW) in the leaf of strawberry plants at harvest after six experimental weeks under three irrigation regimes (IR): well-watered (WW), mild drought (MD), and severe drought (SD) without (−AMF) or with inoculation with arbuscular mycorrhizal fungus (+AMF) *Rhizophagus clarus* (Walker & Schüßler), in the absence (−Si) or presence of silicon (+Si, 3 mmol L^{-1} Na_2SiO_3). The numbers show the treatment means (4 replicates) ±SE of the mean. Means with the same letters are not significantly different. Interactions among the main factors are indicated as *** $p < 0.001$, ** $p < 0.01$, and * $p < 0.05$ and ns as not significant (Tukey test, alpha = 0.05).

			Si	Zn	Mn	Fe	Cu
WW	−AMF	−Si	0.3 ± 0.0 c	70.6 ± 4.0 a	63.1 ± 9.4 a	80.6 ± 14.9 a	7.1 ± 0.4 a
		+Si	1.4 ± 0.2 b	75.3 ± 5.0 a	65.1 ± 5.1 a	102.5 ± 18.9 a	8.0 ± 0.4 a
	+AMF	−Si	0.4 ± 0.1 bc	78.0 ± 2.8 a	74.2 ± 6.3 a	92.2 ± 15.0 a	7.8 ± 0.9 a
		+Si	1.9 ± 0.1 a	79.4 ± 7.0 a	79.4 ± 8.0 a	99.4 ± 21.3 a	8.7 ± 0.3 a
MD	−AMF	−Si	0.2 ± 0.0 c	31.5 ± 4.3 c	37.5 ± 5.9 ab	32.8 ± 13.0 ab	3.9 ± 0.9 ab
		+Si	1.1 ± 0.1 b	58.1 ± 2.7 b	57.9 ± 6.4 a	47.6 ± 6.3 ab	5.8 ± 0.3 a
	+AMF	−Si	0.5 ± 0.0 bc	47.5 ± 2.5 bc	57.5 ± 6.3 a	50.0 ± 4.1 a	6.0 ± 0.9 a
		+Si	0.8 ± 0.1 b	61.7 ± 1.7 b	64.8 ± 4.9 a	69.8 ± 8.2 a	6.2 ± 0.2 a
SD	−AMF	−Si	0.1 ± 0.0 d	13.4 ± 3.5 d	32.2 ± 8.6 b	12.2 ± 2.3 b	3.5 ± 0.5 ab
		+Si	0.5 ± 0.1 bc	16.3 ± 4.0 dc	26.8 ± 7.3 b	19.3 ± 3.5 b	3.4 ± 0.6 b
	+AMF	−Si	0.5 ± 0.0 bc	21.0 ± 8.4 dc	38.8 ± 6.6 b	23.8 ± 5.5 b	3.4 ± 0.6 b
		+Si	1.0 ± 0.1 b	15.5 ± 8.3 dc	35.0 ± 2.9 b	22.5 ± 6.3 b	4.8 ± 0.8 ab
		p					
IR	WW MD SD		<0.001 ***	<0.001 ***	<0.001 ***	<0.001 ***	<0.001 ***
AMF	−AMF +AMF		<0.001 ***	0.04 *	<0.007 **	<0.01 **	0.02 *
Si	−Si +Si		<0.001 ***	0.001 ***	0.3 ns	0.03 **	<0.02 *
IR×AMF			0.03 *	0.5 ns	0.8 ns	0.4 ns	0.7 ns
IR×Si			<0.001 ***	0.02 *	0.2 ns	0.3 ns	0.9 ns
AMF×Si			0.03 *	0.5 ns	0.7 ns	0.4 ns	0.9 ns
IR×AMF×Si			<0.001 ***	0.002 **	0.6 ns	0.6 ns	0.2 ns

4. Discussion

Our results showed that the application of Si and AMF in strawberry might alleviate the adverse effects of D stress in a synergistic manner. Different mechanisms could be involved in this synergistic effect, including Si-mediated improvement of the carbon supply for fungi and likely an increase in the formation of arbuscules. Our results also provide evidence for the effect of Si and AMF on the improvement of strawberry growth under optimum growth conditions through an elevated photosynthesis and water use efficiency.

4.1. Effect of Si and AMF on Growth and Photosynthesis of Plants under Water Stress

Biomass production, water content, and photosynthetic activity of leaves decreased under D conditions in the strawberry plants of this work. Both the Si and AMF treatments alleviated the effects of D and increased leaf water content and photosynthesis rate, leading to a higher biomass production. The observations of gas exchange parameters indicated a D-induced decrease in CO_2 assimilation caused by the closure of stomata. The application of Si and AMF increased net photosynthesis rate through an elevation of stomatal conductance. Our results on the effect of Si are in agreement with those of Ma, 2004 [38] for cucumber, Chen et al. 2011 [14] for rice, and Pilon et al. 2013 [39] for potato. Further research has shown that AMF significantly increased leaf area, carboxylation efficiency, chlorophyll content, net photosynthetic rate, and the photochemical efficiency of PS II under water stress [40,41]. Although an improved stomatal conductance upon Si and AMF treatments resulted also in a higher transpiration rate, a greater stimulation of photosynthetic capacity than water loss led to higher water use efficiency in +AMF and +Si plants.

Despite lower photosynthesis activity, soluble sugars accumulated in the leaves of D plants following an impaired growth. It has been stated that water stress triggers sugar accumulation and leads to an adjustment of the rate of photosynthesis [42]. This accumulation of soluble sugars under water stress, in turn, causes an impaired plant metabolism by changing either the composition or the translocation of sugars in the leaves [43]. In the leaves, the concentration of soluble sugars decreased by AMF and Si treatments most likely because of the growth resumption and consumption of carbohydrates for biomass production. Thus, Si and AMF may modulate the accumulation of soluble sugars in water-stressed leaves in a negative feedback mechanism of biochemical limitations.

The same effect of D on the soluble sugars concentration was observed in the roots. However, in contrast to the leaves, the soluble sugars concentration in the roots increased by AMF and Si treatments. This increase may be resulted from an improved net CO_2 assimilation and/or allocation of photosynthates to the roots and may, in turn, contribute to the stimulation of root growth under these conditions. Considering the osmotic effect of soluble carbohydrates, elevated soluble sugars pool may also improve root water uptake capacity from a dry substrate (see below).

4.2. Effect of Si and AMF on the Water Status and Concentration of Organic Osmolytes

The accumulation of organic osmolytes leading to an osmotic gradient with the environment, as a common response in plants under water stress [44], was observed in the strawberry plants in this work for proline, free AA, and soluble sugars, concomitant with the reduction of osmotic potential. The alleviating effect of AMF and Si, however, was not mediated by an osmo-adjustment, and the concentration of organic osmolytes rather decreased in the leaves of +AMF and +Si plants. These results suggest that the Si-mediated increase in leaf water uptake was not due to an increase in the osmotic driving force in strawberry plants under water stress. An increase in the leaf RWC was achieved apparently by an increased capacity for water uptake that in turn hindered triggering the stomatal closure and allowed the maintenance of a high photosynthetic capacity for supporting growth and dry matter production. Increasing levels of organic compounds under osmotic stress are usually thought to adversely affect growth because of the cost associated with their synthesis [45]. Thus, the method of stress alleviation of AMF and Si for an increase in water uptake capacity may be less expensive than the strategy of osmo-adjustment. This result is in contrast with our previous observation on tobacco plants showing a Si-mediated improvement of plant water status through the leaf accumulation of organic osmolytes including soluble sugars, free amino acids, and proline [13].

In contrast to the leaves, the root concentration of organic osmolytes increased by AMF and Si treatments, suggesting a different strategy for the adjustment of the water economy triggered by AMF and Si in the roots than in the leaves of strawberry. In tomatos, water stress did not change the root osmotic potential in Si-treated plants [46], and in cucumbers, the role of the osmotic driving force in the Si-mediated enhancement of water uptake was genotype-dependent [47]. Collectively, these results suggest different strategies for the improvement of water content and uptake capacity under osmotic stress in Si-treated plants depending on plant organ, species, and genotypes. There are reports on the increased root hydraulic conductance by Si, and the increase was attributed to the Si-mediated upregulation of transcription of some aquaporin genes [48].

Under D conditions, proline accumulated in the leaves while the application of AMF and Si reduced leaf proline concentrations. The accumulation of proline in the leaves under water stress is a well-documented phenomenon, but the role of proline in osmotolerance remains controversial. In some studies, the accumulation of proline has been correlated with stress tolerance [49], but other researchers suggest that proline accumulation is a symptom of stress impairment rather than stress tolerance [50]. Our results support the view that proline accumulation under stress is a symptom of stress and, thus, the Si-mediated reduction of proline concentrations is a sign of stress alleviation.

Similarly, the AMF-mediated reduction of the proline concentration suggests that the AMF colonization of plants, to an extent, mitigated the effects of drought stress and reduced proline concentrations in leaves. These results are in agreement with a previous report [51].

An inhibited formation of proteins from amino acids, which could be judged by the accumulation of free AA concomitant with a reduced protein concentration, was observed under water stress of leaves. Both AMF and Si treatments caused the reduction of the free AA pool associated with an increase in soluble proteins. The accumulation of proteins helps the plant to maintain the water-status of leaves, reduce negative effects from active and reactive oxygen species [52] under severe and long-term drought, and maintain the water-status of leaves [10].

4.3. Effect of Si and AMF on the Antioxidative Defense System

Water stress caused the activation of antioxidative defense enzymes in the leaves and roots. However, this activation was not obviously sufficient for the protection of the plants against ROS that was reflected well in the increasing MDA concentrations in the D plants. The application of AMF and Si to the D plants similarly increased the activity of antioxidative defense enzymes (particularly of SOD). However, compared to the stress-induced activation of enzymes, it led to a decline of stress metabolites (MDA, H_2O_2). It may be suggested that AMF and Si contributed to the alleviation of oxidative damage not only by an elevated capacity of defense system but also through less production of the stress metabolites. It has been frequently shown that plants with higher root colonization with AMF exhibit greater enzymatic and non-enzymatic antioxidative defense systems activity [21,35] than non-inoculated plants. A clear biochemical link between Si and antioxidative capacity in stressed plants, however, has not yet been found. It has been argued that the biochemical enhancement of antioxidant defense mechanisms is a beneficial, physical result of Si-deposition in the cell membrane [4]. Several investigators argue that the Si-induced increases in the activity of antioxidant enzymes and the levels of non-enzymatic antioxidative substances in plants exposed to abiotic stress lead to an implication of Si in the plant metabolism [46,47,53]. According to Ma and Yamaji, 2006 [54], the Si-mediated increase in antioxidant defense abilities is a beneficial result of Si rather than a direct effect.

4.4. Effect of Si and AMF on Plants Nutrients Uptake

Water stress reduced the nutritional status of plants, causing deficiencies in Zn, Mn, Cu, and Fe, particularly in more severely stressed plants, but was already partially detectable under MD conditions. The application of AMF and Si led to an improved micronutrient status, equaling or even exceeding the critical deficiency thresholds of Fe, Zn, and Mn. Maksimovi´c et al., 2012 [55] and Pavlovic et al., 2013 [56] found that Si application increased the uptake of Zn and Fe at low concentrations on the rhizoplane. In this work, the effect of Si on nutrient acquisition under D stress was more pronouncedly observed for Zn than other micronutrients. This effect is likely mediated by stimulation of root growth [57] that increases the spatial availability of Zn for plants [58] or by an enhanced concentrations of low molecular weight organic compounds by Si (e.g., citrate) that might contribute to metal uptake and transport from root to shoot, thereby diminishing deficiency symptoms [59]. The higher Zn uptake after the application of Si under D conditions is also likely to result from the effect of the Si on Zn transporters. It has been observed that Si increases the expression levels of the Fe transporters (IRT1 and IRT2) [56] belonging to the ZIP (Zrt/IRT-like protein) family that include also Zn transporters. A limited Zn/Mn availability in the D plants of this work disbalanced Zn/Mn-dependent ROS detoxification systems produced excessive ROS accumulation and caused oxidative damage. The excessive production of ROS can promote oxidative degradation of indoleacetic acid, as was demonstrated in Zn-deficient maize plants under cold stress, which is restored by the Si application [60]. Auxin deficiency is an important factor for growth limitation in Zn-deficient plants [60]. Regarding the role of AMF, plants with a higher root colonization by AMF are more efficient in the uptake and translocation of macro- and micronutrients to the shoot than non-inoculated plants [61,62].

4.5. A Synergistic Effect of Si and AMF

The synergistic effects of Si and AMF as a combined treatment (+AMF+Si) on the low-Si medium used as growth substrate in this work may partly be related to the contribution of AMF to Si uptake observed in this work and in other works [63–66], the Si-induced stimulation in root growth that in turn promotes AMF colonization in the combination treatment, and the effect of Si on an increase in the root soluble sugars pool, which is important for supporting AMF entry, and further establishment in the roots are other probable mechanisms. The mycorrhizal association is completely dependent on the organic carbon supply from their photosynthetic partner since 4 to 20% of the C fixed through photosynthesis is transferred to the AM fungi [67]. Similarly, the Si-induced increase in the percentage of arbuscules formation observed in this work may result from the improved root growth, the enhancement of nutrients uptake and transfer within the plant, and the induced photosynthesis rate that provides more carbon sources for the fungi partner. A significant increase in the percentage of arbuscule formation in response to Si added to a sand substrate has been reported for Banana [65]. In contrast to our results, in a report on the effect of Si on mycorrhizal chickpea [66], an increase was observed in the salinity tolerance by both Si and AMF, but a synergistic effect was not detected.

Another possible explanation for the synergistic effect of AMF and Si is a Si-induced alteration of the AMF-hosts metabolism. In another report, the authors reported an enhanced metabolism of phenolic compounds (flavonoid-type phenolics) influenced by Si [68]. Phenolic compounds such as flavonoids may play a role in facilitating the interactions between fungus and host [69] and have some positive effects on fungal growth parameters, e.g., hyphal growth and branching, germination of spores [70], and formation of secondary spores. Moreover, they play a role during the fungal invasion and arbuscule formation inside the root [71]. The recent identification of strigolactones as host-recognition signals for AM fungi, however, raises the question about the role of flavonoids as general signaling molecules in AMF-plant interactions [72].

4.6. Effect of Si and AMF on Plants Growth in the Absence of Stress

In the well-watered (WW) strawberry plants grown as unstressed controls, Si treatment caused a significant increase in the shoot growth, where the highest biomass production of the shoot was observed in the +AMF+Si treatment. This Si effect under WW conditions disagrees with some of the previous reports [4] describing the beneficial effects of Si on plant growth only under stress conditions. The Si application has been frequently related to the stimulation of enzymatic defense strategies involved in the detoxification of ROS [12]. However, the lower growth of –Si plants under WW conditions in our experiment was not associated with significant changes in the physiological stress indicators, such as MDA and proline. Furthermore, the positive effects of Si on plant growth under WW conditions could also not be attributed to the increased concentrations of the micronutrients. Even in –Si control plants, the nutritional status exceeded the critical levels reported for the respective micronutrient deficiencies. The unexpected positive effects of Si supplementation on the growth of WW plants may be attributed to a significant improvement of the leaf photosynthesis and water content. Considering a higher leaf area in the +Si plants, it is expected that the photosynthesis of the whole strawberry plants is considerably higher than the –Si ones. Improved Si supply may increase the physical stability of the leaves, leading to a more horizontal orientation of the leaves and thereby improving photosynthetic efficiency as previously reported for cucumber [73]. A recent unified model, so-called apoplastic obstruction hypothesis (74), argued for a fundamental role of Si as an extracellular prophylactic agent as opposed to an active cellular agent. In this model, Si, rather than being involved directly in the regulation of gene expression and metabolism, regulates plant metabolism through a cascading effect [74]. Here in our work, the highest growth improvement was observed in the WW plants under the combination of Si with AMF treatments because a Si-induced shoot growth was associated with an AMF-mediated increase in the root growth. The soil-free culture systems that are based on perlite or vermiculite and are being widely used in horticultural practices and are characterized by low plant availability of Si [75]. Thus, the significant effect of Si supplementation in plants cultivated on these potting substrates, in contrast to the soil-grown plants, could be related to supply of plants with Si and meeting their requirement at least in the accumulator species.

5. Conclusions

The findings of the present study suggest that the major factors determining the sensitivity of strawberry plants to D stress are a reduction of micronutrients uptake, particularly Zn, a reduced photosynthesis rate and protein level, a ROS overproduction, and the consequent membrane damage. In this context, the protective effects of Si and AMF treatments seems to be related to an improved micronutrients status, an increased expression of the enzymatic antioxidative defense system, and an elevated water uptake capacity and use efficiency. Our results indicate that Si and AMF alleviated water stress in a synergistic manner. The AMF colonization and formation of fungal structures were increased by Si, and, in turn, Si uptake was increased upon mycorrhization. Other probable interactions at the metabolic levels need to be elucidated. A conceptual model of these proposed roles of Si and AMF, mediating D tolerance in strawberry plants is presented in Figure 3. Our results provide a theoretical basis for the application of Si fertilizers and AMF in water-conserving irrigation systems for strawberry cultivation under field conditions and for greenhouse production, particularly in the soil-free culture systems.

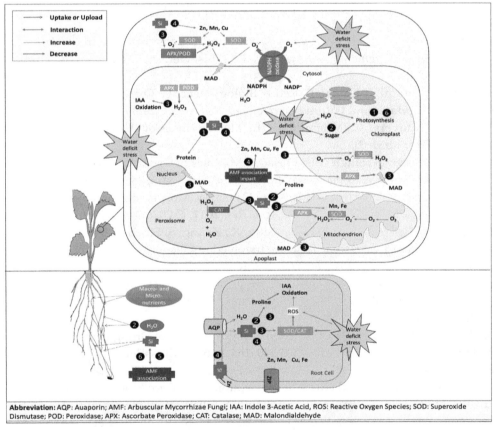

Abbreviation: AQP: Auaporin; AMF: Arbuscular Mycorrhizae Fungi; IAA: Indole 3-Acetic Acid, ROS: Reactive Oxygen Species; SOD: Superoxide Dismutase; POD: Peroxidase; APX: Ascorbate Peroxidase; CAT: Catalase; MAD: Malondialdehyde

Figure 3. A conceptual model representing the effect of Si and AMF in drought-stressed strawberry plants reverting plant performance to well-watered conditions. Si and AMF 1) enhanced growth and photosynthesis of plants, 2) regulated the water status and concentration of organic osmolytes, 3) promoted the antioxidative defense system, 4) increased plants nutrients uptake, 5) had synergistic effects, and 6) enhanced plant growth even in the absence of stress. Abbreviations: AQP: Aquaporin, AMF: Arbuscular Mycorrhizae Fungi, IAA: Indole 3-Acetic Acid, ROS: Reactive Oxygen Species, SOD: Superoxide Dismutase, POD: Peroxidase, APX: Ascorbate Peroxidase, CAT: Catalase, MAD: Malondialdehyde, NADPH: nicotinamide adenine dinucleotide phosphate hydrogen, NADP$^+$: Nicotinamide adenine dinucleotide phosphate.

Author Contributions: R.H. and N.M. conceived and designed the experiments. N.M. conducted the experiments, performed the analyses, and collected the data. N.A., R.H., and G.N. provided the facilities and advised on the preparation of materials. N.M. wrote the manuscript. N.M. and T.E.H. did the statistics evaluations. G.N. and R.H. read and edited the manuscript. All authors approved the final manuscript.

Acknowledgments: The University of Tabriz, Iran, is greatly appreciated for its financial support. Thanks to Zarrin Eshaghi (Payame Noor University, Mashhad) for giving support to the analytical facilities and to Filippo Capezzone (Hohenheim University, Biostatistics department) for the support with SAS and statistical analyses. Very special thanks to Hans Lambers (University of Western Australia) for reviewing the manuscript.

References

1.   Etesami, H.; Jeong, B.R. Silicon (Si) Review and Future Prospects on the Action Mechanisms in Alleviating Biotic and Abiotic Stresses in Plants. *Ecotoxicol. Environ. Saf.* **2018**, *147*, 881–896. [CrossRef] [PubMed]

2.   Broadley, M.; Brown, P.; Cakmak, I.; Ma, J.F.; Rengel, Z.; Zhao, F. Chapter 8—Beneficial Elements Marschner, Petra BT. In *Marschner's Mineral. Nutrition of Higher Plants*, 3rd ed.; Academic Press: San Diego, CA, USA, 2012; pp. 249–269. ISBN 9780123849052.

3.   Liang, Y.; Sun, W.; Zhu, Y.G.; Christie, P. Mechanisms of Silicon–mediated Alleviation of Abiotic Stresses in Higher Plants: A Review. *Environ. Pollut.* **2007**, *147*, 4228. [CrossRef]

4.   Savvas, D.; Ntatsi, G. Biostimulant Activity of Silicon in Horticulture. *Sci. Hort.* **2015**, *196*, 66–81. [CrossRef]

5.   Basu, S.; Ramegoda, V.; Kumar, A.; Pereira, A. Plant Adaptation to Drought Stress. *F1000Res* **2016**, *5*, F1000 Faculty Rev-1554. [CrossRef] [PubMed]

6.   Hajiboland, R. Chapter 1—Reactive Oxygen Species and Photosynthesis. In *Oxidative Damage to Plants, Antioxidant Networks and Signaling*; Ahmad, P., Ed.; Elsevier: San Diego, CA, USA, 2014; pp. 1–63, ISBN 978-0-12-799963-0.

7.   Singh, R.; Parihar, P.; Singh, S.; Kumar, R. Redox Biology Reactive Oxygen Species Signaling and Stomatal Movement: Current Updates and Future Perspectives. *Redox Biol.* **2017**, *11*, 213–218. [CrossRef] [PubMed]

8.   Noctor, G.; Lelarge–Trouverie, C.; Mhamdi, A. The Metabolomics of Oxidative Stress. *Phytochemistry* **2014**, *112*, 33–53. [CrossRef] [PubMed]

9.   Singh, M.; Kumar, J.; Singh, S.; Singh, V.P.; Prasad, S.M. Roles of Osmoprotectants in Improving Salinity and Drought Tolerance in Plants: A Review. *Rev. Environ. Sci. Biol.* **2015**, *14*, 407–426. [CrossRef]

10.   Anjum, S.A.; Xie, X.; Wang, L.; Saleem, M.F.; Man, C.; Lei, W. Morphological, Physiological and Biochemical Responses of Plants to Drought Stress. *Afr. J. Agric Res.* **2011**, *6*, 2026–2032. [CrossRef]

11.   Yin, L.; Wang, S.; Liu, P.; Wang, W.; Cao, D.; Deng, X.; Zhang, S. Silicon-mediated Changes in Polyamine and 1-aminocyclopropane-1-carboxylic Acid are Involved in Silicon-induced Drought Resistance in *Sorghum bicolor* L. *Plant. Physiol. Biochem.* **2014**, *80*, 268–277. [CrossRef] [PubMed]

12.   Shen, X.; Zhou, Y.; Duan, L.; Li, Z.; Eneji, E.; Li, J. Silicon Effects on Photosynthesis and Antioxidant Parameters of Soybean Seedlings under Drought and Ultraviolet-B Radiation. *J. Plant Physiol.* **2010**, *167*, 1248–1252. [CrossRef]

13.   Hajiboland, R.; Cheraghvareh, L.; Poschenrieder, C. Improvement of Drought Tolerance in Tobacco (*Nicotiana rustica* L.) Plants by Silicon. *J. Plant Nutr.* **2017**, *40*, 1661–1676. [CrossRef]

14.   Chen, W.; Yao, X.; Cai, K.; Chen, J. Silicon Alleviates Drought Stress of Rice Plants by Improving Plant Water Status, Photosynthesis and Mineral Nutrient Absorption. *Biol. Trace Elem. Res.* **2011**, *142*, 67–76. [CrossRef]

15.   Lux, A.; Luxová, M.; Hattori, T.; Inanaga, S.; Sugimoto, Y. Silicification in Sorghum (*Sorghum bicolor*) Cultivars with Different Drought Tolerance. *J. Plant Physiol.* **2002**, *115*, 87–92. [CrossRef]

16.   Ming, D.F.; Pei, Z.F.; Naeem, M.S.; Gong, H.J.; Zhou, W.J. Silicon alleviates PEG-induced Water-deficit Stress in Upland Rice Seedlings by Enhancing Osmotic Adjustment. *J. Agron. Crop. Sci.* **2012**, *198*, 14–26. [CrossRef]

17.   Rizwan, M.; Ali, S.; Rizwan, M.; Ali, S.; Ibrahim, M.; Farid, M. Mechanisms of Silicon–mediated Alleviation of Drought and Salt Stress in Plants: A Review. *Environ. Sci. Pollut. Res. Int.* **2015**, *22*, 15416–15431. [CrossRef]

18.   Willis, A.; Rodrigues, B.F.; Harris, P.J.C. The Ecology of Arbuscular Mycorrhizal Fungi. *CRC Crit. Rev. Plant Sci.* **2013**, *32*, 1–20. [CrossRef]

19.   Abdel, A.; Abdel, H.; Hashem, A.; Rasool, S.; Fathi, E.; Allah, A. Arbuscular Mycorrhizal Symbiosis and Abiotic Stress in Plants: A Review. *J. Plant Biol.* **2016**, *59*, 407. [CrossRef]

20. Hajiboland, R.; Bahrami-Rad, S.; Bastani, S. Phenolics Metabolism in Boron Deficient Tea (*Camellia sinensis* (L.) O. Kuntze) Plants. *Acta Biol. Hung.* **2013**, *64*, 196–206. [CrossRef]
21. Wu, Q.S.; Srivastava, A.K.; Zou, Y.N. AMF-induced Tolerance to Drought Stress in Citrus: A Review. *Sci. Hort.* **2013**, *164*, 77–87. [CrossRef]
22. Krishna, H.; Das, B.; Attri, B.L.; Grover, M.; Ahmed, N. Suppression of Botryosphaeria Canker of Apple by Arbuscular Mycorrhizal Fungi. *Crop. Prot.* **2010**, *29*, 1049–1054. [CrossRef]
23. Boyer, L.R.; Brain, P.; Xu, X.M.; Jeffries, P. Inoculation of Drought-stressed Strawberry with a Mixed Inoculum of Two Arbuscular Mycorrhizal Fungi: Effects on Population Dynamics of Fungal Species in Roots and Consequential Plant Tolerance to Water Deficiency. *Mycorrhiza* **2015**, *25*, 215–227. [CrossRef] [PubMed]
24. Augé, R.M.; Toler, H.D.; Saxton, A.M. Mycorrhizal Stimulation of Leaf Gas Exchange in Relation to Root Colonization, Shoot Size, Leaf Phosphorus and Nitrogen: A Quantitative Analysis of the Literature Using Meta-Regression. *Front. Plant Sci.* **2016**, *7*, 1084. [CrossRef] [PubMed]
25. Smith, F.A.; Smith, S.E. What is the Significance of the Arbuscular Mycorrhizal Colonization of Many Economically Important Crop Plants? *Plant Soil* **2011**, *348*, 63–79. [CrossRef]
26. Ouellette, S.; Goyette, M.H.; Labbé, C.; Laur, J.; Gaudreau, L.; Gosselin, A.; Dorais, M.; Deshmukh, R.K.; Bélanger, R.R. Silicon Transporters and Effects of Silicon Amendments in Strawberry under High Tunnel and Field Conditions. *Front. Plant Sci.* **2017**, *8*, 1–11. [CrossRef] [PubMed]
27. Wang, S.Y.; Galletta, G.J. Foliar Application of Potassium Silicate Induces Metabolic Changes in Strawberry Plants. *J. Plant Nutr.* **1998**, *21*, 157–167. [CrossRef]
28. Nicolson, T.H.; Schenck, N.C. Endogonaceous Mycorrhizal. Endophytes in Florida. *Mycologia* **1979**, *71*, 178–198. [CrossRef]
29. Merryweather, J.W.; Fitter, A.H. A Modified Method for Elucidating the Structure of the Fungal Partner in a Vesicular Arbuscular Mycorrhiza. *Mycol. Res.* **1991**, *95*, 1435–1437. [CrossRef]
30. Giovanetti, M.; Mosse, B. An Evaluation of Techniques for Measuring Vesicular Arbuscular Mycorrhizal Infection in Roots. *New Phytol.* **1980**, *84*, 489–500. [CrossRef]
31. McGonigle, T.P.; Miller, M.H.; Evans, D.G.; Fairchild, G.L.; Swan, J.A. A New Method which Gives an Objective Measure of Colonization of Roots by Vesicular Arbuscular Mycorrhizal Fungi. *New Phytol.* **1990**, *115*, 495–501. [CrossRef]
32. Yemm, E.W.; Willis, A.J. The Estimation of Carbohydrates in Plant Extracts by Anthrone. *Biochem. J.* **1954**, *57*, 508–514. [CrossRef]
33. Yemm, E.W.; Cocking, E.C. The Determination of Amino Acids with Ninhydrin. *Analyst* **1955**, *80*, 209–213. [CrossRef]
34. Bates, L.S.; Waldren, R.P.; Teare, I.D. Rapid Determination of Free Proline for Water–stress Studies. *Plant Soil* **1973**, *39*, 205–207. [CrossRef]
35. Hajiboland, R.; Aliasgharzadeh, N.; Laiegh, S.F.; Poschenrieder, C. Colonization with Arbuscular Mycorrhizal Fungi Improves Salinity Tolerance of Tomato (*Solanum lycopersicum* L.) Plants. *Plant Soil* **2010**, *1*, 313–327. [CrossRef]
36. *VDLUFA Method Book VII Environmental Analysis*; VDLUVA-Verlag: Darmstadt, Germany, 2011; p. 690, ISBN 978-3-941273-10-8.
37. Piepho, H.P. A SAS Macro for Generating Letter Displays of Pairwise Mean Comparisons. *Com. Biom. Crop. Sci.* **2012**, *7*, 4–13.
38. Ma, J.F. Role of Silicon in Enhancing the Resistance of Plants to Biotic and Abiotic Stresses. *J. Soil Sci. Plant Nutr.* **2004**, *50*, 11–18. [CrossRef]
39. Pilon, C.; Soratto, R.P.; Moreno, L.A. Effects of Soil and Foliar Spplication of Soluble Silicon on Mineral Nutrition, Gas Exchange, and Growth of Potato Plants. *J. Crop. Sci.* **2013**, *53*, 1605–1614. [CrossRef]
40. Wu, Q.S.; Xia, R.X. Arbuscular Mycorrhizal Fungi Influence Growth, Osmotic Adjustment and Photosynthesis of Citrus under Well–watered and Water Stress Conditions. *J. Plant Physiol.* **2006**, *163*, 417–425. [CrossRef]
41. Zhu, X.Q.; Wang, C.Y.; Chen, H.; Tang, M. Effects of Arbuscular Mycorrhizal Fungi on Photosynthesis, Carbon Content, and Calorific Value of Black Locust Seedlings. *Photosynthetica* **2014**, *52*, 247–252. [CrossRef]
42. McCormick, A.J.; Cramer, M.D.; Watt, D.A. Regulation of Photosynthesis by Sugars in Sugarcane Leaves. *J. Plant Physiol.* **2008**, *165*, 1817.e29. [CrossRef]

43. Silva, E.N.; Ribeir, R.V.; Ferreira-Silva, S.L.; Vieira, S.; Ponte, L.F.; Silveira, J.G. Coordinate Changes in Photosynthesis, Sugar Accumulation and Antioxidative Enzymes Improve the Performance of *Jatropha curcas* Plants under Drought Stress. *Biomass Bioenergy* **2012**, *45*, 270–279. [CrossRef]

44. Ashraf, M.; Akram, N.A.; Foolad, M.R. Drought Tolerance: Roles of Organic Osmolyts, Growth Regulators, and Mineral Nutrients. *Adv. Agron.* **2011**, *111*, 249–296. [CrossRef]

45. Munns, R. Comparative Physiology of Salt and Water Stress. *Plant Cell Environ.* **2002**, *25*, 239–250. [CrossRef] [PubMed]

46. Shi, Y.; Zhang, Y.; Han, W.; Feng, R.; Hu, Y.; Guo, J.; Gong, H. Silicon Enhances Water Stress Tolerance by Improving Root Hydraulic Conductance in *Solanum lycopersicum* L. *Front. Plant Sci.* **2016**, *7*, 196. [CrossRef]

47. Zhu, Y.X.; Xu, X.B.; Hu, Y.H.; Han, W.H.; Yin, J.L.; Li, H.L.; Gong, H.J. Silicon Improves Salt Tolerance by Increasing Root Water Uptake in *Cucumis sativus* L. *Plant Cell Rep.* **2015**, *34*, 1629–1646. [CrossRef] [PubMed]

48. Liu, P.; Yin, L.N.; Deng, X.P.; Wang, S.W.; Tanaka, K.; Zhang, S.Q. Aquaporin-mediated Increase in Root Hydraulic Conductance is Involved in Silicon-induced Improved Root Water Uptake under Osmotic Stress in *Sorghum bicolor* L. *J. Exp. Bot.* **2014**, *65*, 4747–4756. [CrossRef] [PubMed]

49. Zou, Y.N.; Wu, Q.S.; Huang, Y.M.; Ni, Q.D.; He, X.H. Mycorrhizal Mediated Lower Proline Accumulation in *Poncirus trifoliata* under Drought Derives from the Integration of Inhibition of Proline Synthesis with Increase of Proline Degradation. *PLoS ONE* **2013**, *8*, e80568. [CrossRef] [PubMed]

50. Crusciol, C.C.; Pulz, A.L.; Lemos, L.B.; Soratto, R.P.; Lima, G.P.P. Effects of Silicon and Drought Stress on Tuber Yield and Leaf Biochemical Characteristics in Potato. *Crop. Sci.* **2009**, *49*, 949–954. [CrossRef]

51. Porcel, R.; Aroca, R.; Ruiz-Lozano, J.M. Salinity Stress Alleviation Using Arbuscular Mycorrhizal Fungi. *Agron. Sustain. Dev.* **2012**, *32*, 181–200. [CrossRef]

52. Martinelli, T.; Whittaker, A.; Bochicchio, A.; Vazzana, C.; Suzuki, A.; Masclaux-Daubresse, C. Amino acid Pattern and Glutamate Metabolism during Dehydration Stress in the "Resurrection" Plant *Sporobolus stapfianus*: A Comparison Between Desiccation-sensitive and Desiccation-tolerant Leaves. *J. Exp. Bot.* **2007**, *58*, 3037–3046. [CrossRef] [PubMed]

53. Zhu, Y.; Gong, H. Beneficial Effects of Silicon on Salt and Drought Tolerance in Plants. *Agron. Sustain. Dev* **2014**, *34*, 455–472. [CrossRef]

54. Ma, J.F.; Yamaji, N. Silicon Uptake and Accumulation in Higher Plants. *Trends Plant Sci.* **2006**, *11*, 392–397. [CrossRef] [PubMed]

55. Maksimović, J.D.; Mojović, M.; Maksimović, V.; Römheld, V.; Nikolic, M. Silicon Ameliorates Manganese Toxicity in Cucumber by Decreasing Hydroxylradical Accumulation in the Leaf Apoplast. *J. Exp. Bot.* **2012**, *63*, 2411–2420. [CrossRef] [PubMed]

56. Pavlovic, J.; Samardzic, J.; Maksimović, V.; Timotijevic, G.; Stevic, N.; Laursen, K.H.; Hansen, T.H.; Husted, S.; Schjoerring, J.K.; Liang, Y.; et al. Silicon Alleviates Iron Deficiency in Cucumber by Promoting Mobilization of Iron in the Root Apoplast. *New Phytol.* **2013**, *198*, 1096–1107. [CrossRef] [PubMed]

57. Hattori, T.; Inanaga, S.; Tanimoto, E.; Lux, A.; Luxová, M.; Sugimoto, Y. Silicon-induced Changes in Viscoelastic Properties of Sorghum Root Cell Walls. *Plant Cell Physiol.* **2003**, *44*, 743–749. [CrossRef] [PubMed]

58. Rengel, Z. Availability of Mn, Zn and Fe in the Rhizosphere. *J. Soil. Sci. Plant Nutr.* **2015**, *15*, 397–409. [CrossRef]

59. Hernandez-apaolaza, L. Can Silicon Partially Alleviate Micronutrient Deficiency in Plants? A Review. *Planta* **2014**, *240*, 447–458. [CrossRef] [PubMed]

60. Moradtalab, N.; Weinmann, M.; Walker, F.; Höglinger, B.; Ludewig, U.; Neumann, G. Silicon Improves Chilling Tolerance during Early Growth of Maize by Effects on Micronutrient Homeostasis and Hormonal Balances. *Front. Plant Sci.* **2018**, *9*, 420. [CrossRef]

61. Cakmak, I.; Marschner, H.; Bangerth, F. Effect of Zinc Nutritional Status on Growth, Protein Metabolism and Levels of Indole-3-acetic Acid and Other Phytohormones in Bean (*Phaseolus vulgaris* L.). *J. Exp. Bot.* **1989**, *40*, 405–412. [CrossRef]

62. Rouphael, Y.; Franken, P.; Schneider, C.; Schwarz, D.; Giovannetti, M.; Agnolucci, M.; De Pascale, S.; Bonini, P.; Colla, G. Arbuscular Mycorrhizal Fungi Act as Biostimulants in Horticultural Crops. *Sci. Hortic.* **2015**, *196*, 91–108. [CrossRef]

63. Singh, L.P.; Gill, S.S.; Tuteja, N. Unraveling the Role of Fungal Symbionts in Plant Abiotic Stress Tolerance. *Plant Signal. Behav.* **2011**, *6*, 175–19164. [CrossRef]

64. Clark, R.B.; Zeto, S.K. Mineral Acquisition by Arbuscular Mycorrhizal Plants. *J. Plant Nutr.* **2000**, *23*, 867–902. [CrossRef]

65. Anda, O.C.C.; Opfergelt, S.; Declerck, S. Silicon Acquisition by Bananas (c.V. Grande Naine) is Increased in Presence of the Arbuscular Mycorrhizal Fungus *Rhizophagus irregularis* MUCL 41833. *Plant Soil* **2016**, *409*, 77–85. [CrossRef]

66. Garg, N.; Bhandari, P. Silicon Nutrition and Mycorrhizal Inoculations Improve Growth, Nutrient Status, K+/Na+ Ratio and Yield of *Cicer arietinum* L. Genotypes under Salinity Stress. *Plant Growth Regul.* **2016**, *78*, 371–387. [CrossRef]

67. Smith, S.E.; Read, D.J. *Mycorrhizal Symbiosis*, 3rd ed.; Academic Press: London, UK, 2008; p. 800, ISBN 9780123705266.

68. Rodrigues, F.A.; McNally, D.J.; Datnoff, L.E.; Jones, J.B.; Labbé, C.; Benhamou, N.; Menzies, J.G.; Bélanger, R.R. Silicon Enhances the Accumulation of Diterpenoid Phytoalexins in Rice: A Potential Mechanism for Blast Resistance. *Phytopathology* **2004**, *94*, 177–183. [CrossRef]

69. Mandal, S.M.; Chakraborty, D.; Dey, S. Phenolic Acids Act as Signalling Molecules in Plant-microbe Symbioses. *Plant Signal. Behav.* **2010**, *5*, 359–368. [CrossRef] [PubMed]

70. Steinkellner, S.; Lendzemo, V.; Langer, I.; Schweiger, P.; Khaosaad, T.; Toussaint, J.P.; Vierheilig, H. Flavonoids and Strigolactones in Root Exudates as Signals in Symbiotic and Pathogenic Plant–Fungus Interactions. *Molecules* **2007**, *12*, 1290–1306. [CrossRef]

71. Hassan, S.; Mathesius, U. The Tole of Flavonoids in Rootrhizosphere Signalling: Opportunities and Challenges for Improving Plant-microbe Interactions. *J. Exp. Bot.* **2012**, *63*, 3429–3444. [CrossRef]

72. Abdel-Lateif, K.; Bogusz, D.; Hocher, V. The Role of Flavonoids in the Establishment of Plant Roots Endosymbioses with Arbuscular Mycorrhiza Fungi, Rhizobia and Frankia Bacteria. *Plant Signal. Behav.* **2012**, *7*, 636–641. [CrossRef]

73. Botta, A.; Rodrigues, F.A.; Sierras, N.; Marin, C.; Cerda, J.M.; Brossa, R. Evaluation of Armurox® (cComplex of Peptides with Soluble Silicon) on Mechanical and Biotic Stresses in Gramineae. In Proceedings of the 6th International Conference on Silicon in Agriculture, Stockholm, Sweden, 26–30 August 2014.

74. Coskun, D.; Deshmukh, R.; Sonah, H.; Menzies, J.G.; Reynolds, O.; Ma, J.F.; Kronzucker, H.J.; Bélanger, R.R. The Controversies of Silicon's Role in Plant Biology. *New Phytol.* **2019**, *221*, 67–85. [CrossRef]

75. Reddy, S. Time to Say Sí to Silicon—And Bring Back the Missing Element in Soilless Growing. Available online: http://www.sungro.com/time-say-si-silicon-bring-back-missing-element-soilless-growing (accessed on 12 May 2014).

11

Phytotoxic Effects of Three Natural Compounds: Pelargonic Acid, Carvacrol and Cinnamic Aldehyde, against Problematic Weeds in Mediterranean Crops

Marta Muñoz [1,2], Natalia Torres-Pagán [1], Rosa Peiró [3], Rubén Guijarro [2], Adela M. Sánchez-Moreiras [4,5] and Mercedes Verdeguer [1,*]

[1]   Instituto Agroforestal Mediterráneo (IAM), Universitat Politècnica de València, Camino de Vera s/n, 46022 Valencia, Spain; mmunoz@seipasa.com (M.M.); natorpa@etsiamn.upv.es (N.T.-P.)

[2]   SEIPASA S.A. C/Ciudad Darío, Polígono Industrial La Creu naves 1-3-5, L'Alcudia, 46250 Valencia, Spain; rguijarro@seipasa.com

[3]   Centro de Conservación y Mejora de la Agrodiversidad Valenciana (COMAV), Universitat Politècnica de València, Camino de Vera s/n, 46022 Valencia, Spain; ropeibar@btc.upv.es

[4]   Department of Plant Biology and Soil Science, Faculty of Biology, University of Vigo, Campus Lagoas-Marcosende s/n, 36310 Vigo, Spain; adela@uvigo.es

[5]   CITACA. Agri-Food Research and Transfer Cluster, Campus da Auga. University of Vigo, 32004 Ourense, Spain

*   Correspondence: merversa@doctor.upv.es

Abstract: Weeds and herbicides are important stress factors for crops. Weeds are responsible for great losses in crop yields, more than 50% in some crops if left uncontrolled. Herbicides have been used as the main method for weed control since their development after the Second World War. It is necessary to find alternatives to synthetic herbicides that can be incorporated in an Integrated Weed Management Program, to produce crops subjected to less stress in a more sustainable way. In this work, three natural products: pelargonic acid (PA), carvacrol (CV), and cinnamic aldehyde (CA) were evaluated, under greenhouse conditions in postemergence assays, against problematic weeds in Mediterranean crops *Amaranthus retroflexus*, *Avena fatua*, *Portulaca oleracea*, and *Erigeron bonariensis*, to determine their phytotoxic potential. The three products showed a potent herbicidal activity, reaching high efficacy (plant death) and damage level in all species, being PA the most effective at all doses applied, followed by CA and CV. These products could be good candidates for bioherbicides formulations.

Keywords: weeds; abiotic stress; natural herbicides; secondary metabolites; postemergence; phytotoxicity

1. Introduction

One of the main challenges for the agriculture in this 21st century is to be capable to feed the increasing world population in a sustainable way, because natural resources are becoming even more scarce [1]. Crop protection measures can prevent yield losses due to pests [2]. Herbicides have been the most used method to control weeds since their development, at the end of the Second World War because they are effective and economical [3,4].

Herbicides cause stress in crops and can make them more susceptible to other pests [5]. Other problems derived from the overuse of herbicides are environmental pollution, toxicity for nontarget organisms, and the development of herbicide-resistant weed biotypes [6]. In the latest 10 years, integrated weed management (IWM) strategies have been promoted worldwide [7,8] to control weeds. They consist of a combination of methods: cultural, mechanical, physical, biological, biotechnological, and chemical. In Europe, IWM has been promoted through the European Union Directive 2009/128/EC [8].

The society is demanding new solutions for weed control and "greener" weed management products. The use of natural products as bioherbicides could be one alternative to reduce the stress that synthetic herbicides promote in crops and all their negative impacts aforementioned. Bioherbicides could be incorporated in IPM programs as an innovative weed control method. They are less persistent than synthetic herbicides and are potentially more environmentally friendly and safe [9] and also, they have different modes of action, which can prevent the development of herbicide-resistant weed biotypes [10].

Bailey [11] defined bioherbicides as products of natural origin for weed control. The EPA (USA Environmental Protection Agency), considers three categories of biopesticides: (1) biochemical pesticides, which include naturally occurring substances that control pests; (2) microbial pesticides or biocontrol agents, which are microorganisms that control pests; and (3) plant-incorporated protectants, or PIPs, which are pesticide substances produced by plants that contain added genetic material) [10]. In recent years, the search for natural substances that can act as bioherbicides has been very extensive.

The weeds selected for this study were *Amaranthus retroflexus* L., *Avena fatua* L., *Portulaca oleracea* L., and *Erigeron bonariensis* L. because of their importance in many crops worldwide and their difficult management. *A. fatua* is a very important weed mainly in cereals and also in other crops around the world [12], and this weed is on the fourth position in resistance to herbicides worldwide, having developed resistance to nine different modes of action [13]. *A. retroflexus* is a serious and aggressive weed in summer crops, with cosmopolite distribution [14]. It has developed resistance to five modes of action and is on the eight position worldwide in resistance to herbicides [13]. *E. bonariensis*, which can be found both in summer or winter crops, especially with no-tillage practices [15], is on the ninth position in resistance to herbicides worldwide, with resistance to four modes of action. *P. oleracea*, which is a summer weed difficult to control in Mediterranean crops [16], has developed resistance only to two modes of action [13]. *A. fatua* and *E. bonariensis* have developed resistance to glyphosate, which is the herbicide most commonly used around the world [13,17].

There are several examples of natural products that have been tested as potential bioherbicides to control *A. fatua*, *A. retroflexus*, *E. bonariensis*, and *P. oleracea*, mainly essential oils (EOs) [14,18–26], or extracts from plants with different solvents [27–29], or their isolated compounds [30,31]. Most studies have been carried out only in in vitro conditions. Of the weeds considered, *A. retroflexus* has been the most tested. In vitro studies with EOs from *Artemisia vulgaris*, *Mentha spicata*, *Ocimum basilicum*, *Salvia officinalis*, and *Thymbra spicata* from Turkey demonstrated high phytotoxic effects on seed germination and seedling growth of *A. retroflexus*, with stronger effects with higher doses [18]. EOs from *Tanacetum* species growing in Turkey, rich in oxygenated monoterpenes, inhibited completely *A. retroflexus* germination in in vitro assays [19]. In addition, EOs from *Nepeta meyeri*, with high content in oxygenated monoterpenes controlled completely *A. retroflexus* germination [20]. The phytotoxic potential of 12 EOs was studied in vitro against *A. retroflexus* and *A. fatua*, and the most phytotoxic EOs were those constituted mainly by oxygenated monoterpenes [21]. Other EOs which showed strong herbicidal potential against *A. retroflexus* seed germination and seedling growth were *Rosmarinus officinalis*, *Satureja hortensis*, and *Laurus nobilis* [14], and a nanoemulsion of *S. hortensis* EO was tested against *A. retroflexus* in greenhouse conditions killing the weed at 4000 μL/mL dose [22]. *P. oleracea* germination was completely inhibited by *Eucalyptus camaldulensis* EO in in vitro conditions [23]. The application of leaf extracts (obtained using water, methanol, and ethanol as solvents) of cultivated *Cynara cardunculus* in in vitro bioassays inhibited seed germination and germination time in *A. retroflexus* and *P. oleracea* [27].

Different natural compounds have demonstrated herbicidal potential against the germination and seedling growth of *A. fatua*, such as EOs from *Artemisia herba-alba* [24] and *Eucalyptus citriodora* EOs [25] and extracts from *Sapindus mukorossi*, which inhibited *A. fatua* and *A. retroflexus* growth in vitro and in pots [28] or from *Iris sibirica* rhizomes [29].

EOs from *Thymbra capitata*, *Mentha piperita*, *Eucalyptus camaldulensis*, and *Santolina chamaecyparissus* were tested in vivo against *E. bonariensis*. *T. capitata* EO, with high content in carvacrol, was the most effective to control *E. bonariensis*, showing an excellent potential to develop bioherbicide formulations [26].

Some studies carried out in recent years relate the herbicidal activity of plant extracts or EOs to their composition in monoterpenes, and these substances are postulated as the future of natural herbicide components [32–35]. For example, eugenol, a monoterpene that can be found in many EOs as the major compound, like in *Syzygium aromaticum* EO, has shown strong phytotoxic potential against *A. retroflexus* [30] and *A. fatua* [31]. In *A. fatua*, eugenol inhibited its seedling growth, affecting more the roots than the coleoptiles. In addition, sesquiterpenes, secondary metabolites in plants, present in some EOs, have demonstrated strong herbicidal activity [36,37].

The natural products studied on this work for their potential as bioherbicides were pelargonic acid, trans-cinnamaldehyde and carvacrol. Pelargonic acid (PA) ($CH_3(CH_2)_7CO_2H$, n-nonanoic acid), which is present as esters in the EO of *Pelargonium* spp., is a saturated fatty acid with nine carbons in its structure [28–40]. PA and its salts are used like active ingredients in bioherbicide formulations for garden and professional uses worldwide. They are applied as burndown herbicides, which in a short time, attack cell membranes, causing cell leakage, followed by breakdown of membrane acyl lipids [41], and finally causing visible effects of desiccation of green areas of the weeds [38]. All the symptoms caused by PA on weeds involve extreme phytotoxicity for the plants and their cells, which rapidly begin to oxidize, causing necrotic lesions on aerial parts of plants [42,43].

Herbicidal fatty acids have been used for a long time in weed management, and some of them are used as natural herbicides. Still, the high dosage and the high cost are some of the drawbacks of its practical application in the current agriculture. In 2015, the bioherbicide Beloukha® was authorized as plant protection product to be marketed in Europe [44]. It is derived from oleic acid from different origin. Actually, it is authorized also for markets in USA and Canada. This work aims to find an optimal formulation of PA capable to be effective at reduced doses compared to the existing products in the market.

Trans-cinnamaldehyde (CA) (C_9H_8O) is one of the major components of two different cinnamon species (*Cinnamomum zeylanicum* and *Cinnamomum cassia*) and their EOs [45–48]. This compound has shown strong antioxidant properties and is responsible for various observed biological activities of cinnamon like bactericidal, fungicidal, or acaricidal [49–52]. The antimicrobial activity of CA is well known, however, its potential as bioherbicide has been less studied. Despite that, recent research demonstrated the herbicidal activity of CA against *Echinochloa crus-galli* by reducing the fresh weight and growth of this important weed [53]. To our knowledge, the mode of action of CA on weeds has not been elucidated.

The third natural compound evaluated was carvacrol (CV), a phenolic monoterpene frequently present on EOs obtained from many species belonging to Lamiaceae family like *Thymus* spp., *Thymbra* spp., and *Origanum* spp. [34]. CV presents antimicrobial properties that make it helpful for controlling diseases in crop protection [54–58]. In relation to its mode of action, CV exhibited membrane-disrupting activity that was dependent on long exposure at high concentration [33]. Postemergence exposure of plants to high concentrations of CV causes severe phytotoxicity. One of the effects associated with the mode of action of CV is the reduction of weed growth [22,41,54].

This work is a collaboration between the Universitat Politècnica de València (UPV) and the company Seipasa S.A., which develops and commercializes biopesticides, with the purpose to manage agricultural ecosystems in a more sustainable way. The objective of the present study was to evaluate the herbicidal potential of the natural compounds pelargonic acid, trans-cinnamaldehyde, and carvacrol against important cosmopolite weeds (*Amaranthus retroflexus* L., *Portulaca oleracea* L., *Erigeron bonariensis* L., and *Avena fatua* L.) as an alternative to synthetic herbicides to reduce the abiotic stress that they cause on crops. Effective compounds were formulated as emulsifiable concentrates (ECs) by Seipasa S.A., and evaluated for their postemergence herbicidal activity in greenhouse conditions in the UPV (Spain).

2. Materials and Methods

2.1. Postemergence Herbicidal Assays against Targeted Weed Species

2.1.1. Weeds

Seeds of *Amaranthus retroflexus* L., *Portulaca oleracea* L., and *Avena fatua* L. purchased from Herbiseed (Reading, UK) (year of collection 2017), which have been previously tested in a plant growth chamber EGCHS series from Equitec (Madrid, Spain) (30 ± 0.1 °C, 16 h light and 20 ± 0.1 °C, 8 h dark for *A. retroflexus* and *P. oleracea*; 23.0 ± 0.1 °C, 8 h light and 18.0 ± 0.1 °C 16 h dark for *A. fatua*) to assure their germination viability, were sown in pots (8 × 8 × 7 cm) filled with 2 cm of perlite and 5 cm of soil collected from a citrus orchard nontreated with herbicides. In Figure 1, the location (39°37′24.8″ N, 0°17′25.6″ W Puzol, Valencia province, Spain) and a view of the citrus orchard (0.4 ha) from which the soil was collected is reported. Table 1 shows the main physical characteristics of the soil used for the experiments.

Figure 1. Location (**A**) and view (**B**) of the citrus orchard where the soil for the herbicidal tests was collected.

Table 1. Physical properties of the soil used for the experiments [59].

Soil Properties
Clay 21.85%
Silt 47.55%
Sand 30.60%

Erigeron bonariensis L. seeds were collected from an ecological weed management persimmon orchard located in Carlet (Valencia province, Spain) in July 2018. They were previously tested in the plant growth chamber described before (30 ± 0.1 °C, 16 h light and 20 ± 0.1 °C 8 h dark) to assure their germination capability and after that, sown in plastic pots filled with a mix of three-fourth peat and one-fourth perlite instead of soil because it was very difficult to germinate the seeds on the soil, as *E. bonariensis* germinates better in lighter soils [60] and, therefore, the properties of the soil collected from the citrus orchard (Table 1) did not fit the needs for their germination.

All weeds were irrigated by capillarity from trays (43 cm × 28 cm × 65 cm) placed under the pots and filled with water, until the plants were ready for the herbicidal experiments.

2.1.2. Treatments

Ten pots were prepared for each treatment, described in Table 2. The treatments were applied when plants reached the phenological stage of 2-3-true leaves, corresponding to stage 12-13 BBCH (Biologische Bundesanstalt, Bundessortenamt und Chemische Industrie) scale for the monocotyledonous *A. fatua*, and 3-4-true leaves, corresponding to stage 13-14 BBCH scale for the dicotyledonous *A. retroflexus* and *P. oleracea* and in rosette stage for *E. bonariensis*, stage 14-15 BBCH scale (Figure 2). Pelargonic acid, cinnamic aldehyde and carvacrol were provided formulated as emulsifiable concentrates (ECs) by the company Seipasa S.A. (L'Alcudia, Valencia province, Spain). Beloukha® was purchased from Ferlasa (Museros, Valencia province, Spain) and Roundup® Ultra Plus was purchased from Cooperativa Agrícola Nuestra Señora del Oreto (CANSO, L'Alcudia, Valencia province, Spain).

Table 2. Treatments tested.

	Treatments	**Abbreviations**
T1	Control treated with water	CW
T2	Pelargonic acid 5%	PA5
T3	Pelargonic acid 8%	PA8
T4	Pelargonic acid 10%	PA10
T5	Cinnamic aldehyde 6%	AC6
T6	Cinnamic aldehyde 12%	AC12
T7	Cinnamic aldehyde 24%	AC24
T8	Carvacrol 8%	CV8
T9	Carvacrol 16%	CV16
T10	Carvacrol 32%	CV32
T11	Bioherbicide reference: pelargonic acid (Beloukha® 8%)	BE
T12	Chemical reference: glyphosate (Roundup® Ultra Plus 10%)	GL

Figure 2. Pots ready for the postemergence treatments. (**A**) *A. fatua*, (**B**) *A. retroflexus*, (**C**) *P. oleracea*, and (**D**) *E. bonariensis*.

In Table 3, the dates of the herbicidal tests and the greenhouse conditions during the experimental periods are reported. Data were registered using a HOBO U23 Pro v2 data logger (Onset Computer Corporation, Bourne, MA, USA).

Table 3. Greenhouse conditions during the herbicidal tests.

Species	**Starting-End Date**	**Temperature (°C)**			**Relative Humidity (%)**		
		Mean	**Max.**	**Min.**	**Mean**	**Max.**	**Min.**
P. oleracea	August 9, 2018–September 9, 2018	28.03	38.39	22.87	68.04	87.03	37.18
A. retroflexus	September 2, 2018–October 2, 2018	26.38	35.42	19.82	70.91	85.88	31.14
A. fatua	December 3, 2018–January 3, 2019	18.57	25.72	12.75	57.87	75.56	29.84
E. bonariensis	February 15, 2019–March 15, 2019	22.62	27.16	17.99	45.88	50.26	40.40

2.2. Evaluation of the Herbicidal Activity of Each Natural Product

During the experiments, images from the plants were taken 24 h and 3, 7, 15, and 30 days after the treatments application to be processed with Digimizer v.4.6.1 software (MedCalc Software, Ostend, Belgium, 2005–2016).

To evaluate the herbicidal activity, two variables were measured for each plant: the efficacy, which was scored 0 if the plant was alive and 100 if the plant was dead, and the damage level, which was assessed between 0 and 4 as reported in Table 4 and Figure 3. The efficacy and damage level for each treatment were calculated as the mean of the 10 treated plants.

Table 4. Damage level assessment.

Level of Damage	
0	Undamaged plant
1	Plant with slight damage
2	Plant with severe damage
3	Dead plant
4	Regrown plant

Figure 3. Damage scale for each species: (**A**) *A. fatua*, (**B**) *P. oleracea*, (**C**) *A. retroflexus*, and (**D**) *E. bonariensis*.

2.3. Statistical Analyses

Data were processed using Statgraphics® Centurion XVII (StatPoint Technologies Inc., Warrenton, VA, USA) software. A multifactor analysis of variance (ANOVA) was performed on efficacy and damage level including species, treatments, time after treatments application, and their double significant interactions as effects, followed by Fisher's multiple comparison test (LSD intervals, least significant difference, at $p \leq 0.05$) for the separation of the means.

3. Results and Discussion

3.1. Efficacy of Pelargonic Acid, Cinnamic Aldehyde, and Carvacrol against Target Weeds

A. retroflexus was the weed species most susceptible to the treatments tested, with 73.50 efficacy (Table 5). No significant differences were observed between the other species, which showed around 55 efficacies. The fact that all species tested were susceptible to all treatments with natural products assayed confirm that they could be a more sustainable alternative to synthetic herbicides, and they also offer new modes of action to control weeds that have developed resistant biotypes to many herbicides.

Table 5. Efficacy according to the species, time, and treatment.

Species	Efficacy
Portulaca oleracea	56.17 ± 1.11 b
Amaranthus retroflexus	73.50 ± 1.11 a
Avena fatua	54.83 ± 1.11 b
Erigeron bonariensis	55.67 ± 1.11 b
Time (Days after application)	**Efficacy**
1	41.67 ± 1.24 c
3	81.88 ± 1.24 b
7	87.08 ± 1.24 a
15	89.58 ± 1.24 a
Treatment	**Efficacy**
Control treated with water	4.00 ± 1.92 g
Pelargonic acid 5%	70.50 ± 1.92 b
Pelargonic acid 8%	73.50 ± 1.92 ab
Pelargonic acid 10%	74.50 ± 1.92 ab
Cinnamic aldehyde 6%	53.50 ± 1.92 e
Cinnamic aldehyde 12%	70.00 ± 1.92 bc
Cinnamic aldehyde 24%	70.00 ± 1.92 bc
Carvacrol 8%	60.50 ± 1.92 d
Carvacrol 16%	64.50 ± 1.92 d
Carvacrol 32%	65.00 ± 1.92 cd
Bioherbicide reference: pelargonic acid (Beloukha® 8%)	78.50 ± 1.92 a
Chemical reference: glyphosate (Roundup® Ultra Plus 10%)	36.00 ± 1.92 f

Values are efficacy ± standard error. Means followed by different letters in the same column differ significantly ($p \leq 0.05$).

Efficacy increased with time after treatments application, with values close to 90 between 7 and 15 days (Table 5). This happened because PA, at all doses applied, and the higher doses of CA and CV acted very quickly in the treated species, causing the death of all plants between 24 h and 3 days after application of treatment (Figures 4–7, Tables S1–S4). The same happened for the bioherbicide reference BE (as PA was also the active compound on it), while GL acted more slowly, depending on the species against which it was applied; it killed *A. retroflexus* plants after 3 days, *A. fatua* and *P. oleracea* after 15 days, and *E. bonariensis* after 30 days (Figures 4–7, Tables S1-S4). It has been reported that weed damage caused by PA can be observed visually few hours after application [61]. Thymol, *trans*-cinnamaldehyde, eugenol, farnesol, and nerolidol were tested in postemergence in *E. crus-galli* applied at two-leaf stage, and significantly reduced the shoot growth and the fresh and dry weight 2 days after the foliar treatments with 0.5%, 1.0%, and 2.0% concentrations. All treatments except thymol controlled the weed completely when applied at 1.0% and 2.0% [52]. The concentrations of CA used in this work were higher, and this could explain the quicker toxic effect observed on weeds. It is also remarkable that weed species displayed different sensitivity to low doses of CA; *E. bonariensis* and *P. oleracea* showed more resistance to this compound than the other weeds tested (Figures 4–7, Tables S1-S4), as the lowest concentration (6%) used took more time (15 days) to kill all the plants in *E. bonariensis* than in *A. retroflexus* (24 h) or *A. fatua* (3 days), whereas in *P. oleracea*, this dose reached 50 efficacy, i.e., only 50% of plants were dead at the end of the experiment (30 days). Previous studies also confirmed the rapid activity of carvacrol in plants; in a greenhouse experiment, a nanoemulsion (NE) of *Satureja hortensis* L. EO, rich in carvacrol (55.6%), was applied against *A. retroflexus* and *C. album*, and after 30 min, the weeds were exhibiting injury symptoms, reaching the maximum lethality within 24 h of treatment application. The lethality percentage was dependent on the doses applied and the species against which NE was applied [21]. As observed with CA, also weed species showed different sensibility to CV application, especially at the lower dose, which took more time to control the weeds (Figures 4–7, Tables S1–S4): *A. retroflexus* was the more sensitive species, being controlled by all doses 24 h after application of treatment (Figure 4, Table S1), whereas in *A. fatua* and *E. bonariensis*, the lowest dose took 7 and 15 days, respectively, to reach 100 efficacy (Figures 5 and 6, Tables S2 and S3), being again *P. oleracea* the most resistant weed species, 7 days after treatment application, all plants were killed in all CV treatments, although then some regrew 15 and 30 days after treatments application (Figure 7, Table S4).

All the treatments managed to control the weed species tested, and the results of the treatments were statistically significant compared to CW (Table 5). The most effective treatment was the PA formulation at 10%, achieving 74.50 efficacy. This treatment did not show significant differences compared to the results obtained by the commercial product used as biological reference, also containing PA as active ingredient, which obtained an efficacy of 78.50. Moreover, there were no significant statistical differences in the efficacy between the three doses of the PA-based formulations (5%, 8%, and 10%). The next most effective treatment was the CA-based formulation, which exhibited the same efficacy values for the two higher doses applied (12% and 24%), while the lowest dose (6%) had significant less efficacy. This can be explained by the different sensitivity of the weed species to low doses of CA, as commented above. Finally, the treatments with carvacrol did not show significant differences in efficacy between doses, but with the control, and were also very effective, reaching an efficacy between 60.50 and 65.00 (Table 5).

All treatments tested with natural products showed higher efficacy for the control of weeds than GL, which showed efficacy values of 36. This was because of its slower activity. Mechanism of action of GL is by affecting the enzyme 5-enolpyruvlyshikimate-3-phosphate synthase (EPSPS), and it is the only herbicide with this mode of action. The inhibition of EPSPS reduces levels of amino acids needed for the synthesis of proteins, cell walls, and secondary plant products. In addition, the inhibition of EPSPS causes deregulation of the shikimic acid pathway, promoting the disruption of plant carbon metabolism [62]. GL is translocated in plants and differential responses of weed species may be caused by differences in herbicide translocation, i.e., weeds capable to translocate GL more efficiently are more severely damaged [63]. In field experiments conducted for 2 years, it was verified that GL controlled more effectively *A. retroflexus* than other species [64], which supports our results. Decreased herbicide translocation to the meristem causes reduced glyphosate efficacy [65]. The necessity of being translocated explains the slow effect of GL compared with the natural compounds, as ^{14}C translocation throughout the plant demonstrated that glyphosate took 3 days to reach and accumulate in the meristematic tips of the roots and shoots [66]."

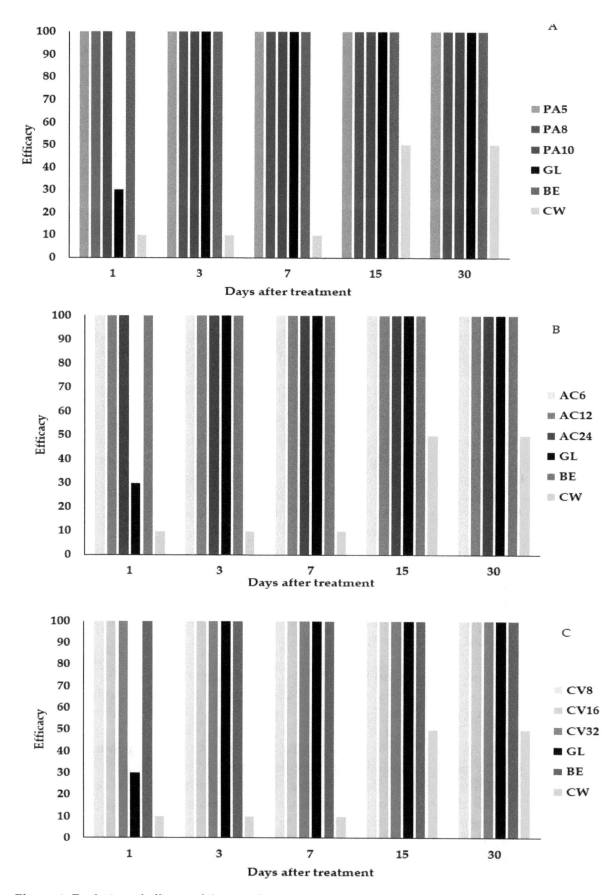

Figure 4. Evolution of efficacy of the tested treatments (**A**) pelargonic acid, (**B**) cinnamic aldehyde and (**C**) carvacrol in *A. retroflexus* during 30 days after application.

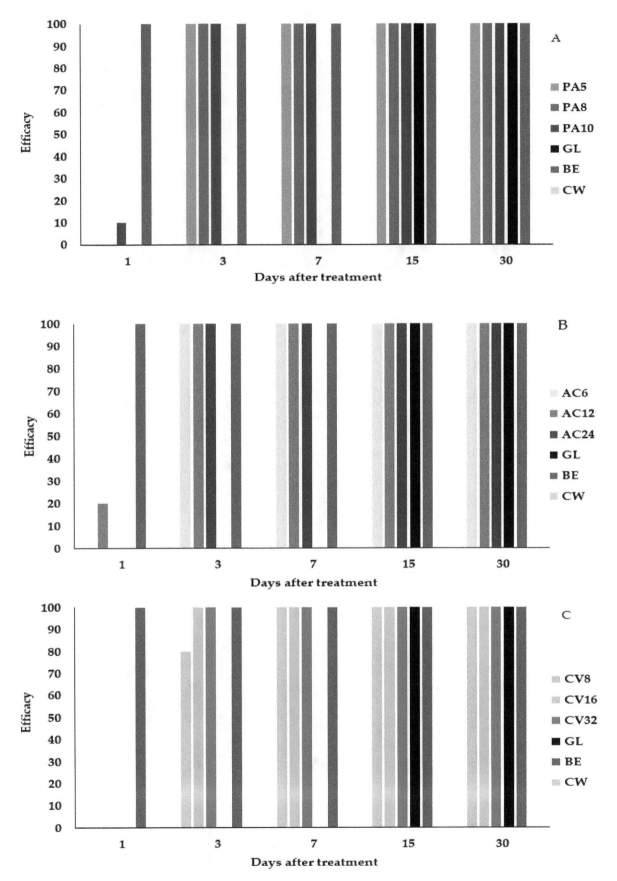

Figure 5. Evolution of efficacy of the tested treatments (**A**) pelargonic acid, (**B**) cinnamic aldehyde, and (**C**) carvacrol in *A. fatua* during 30 days after their application.

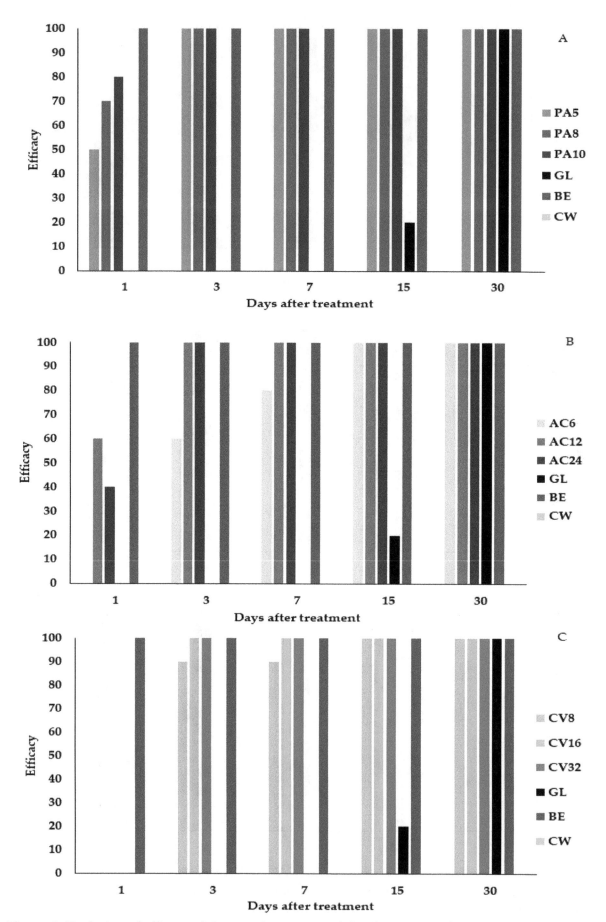

Figure 6. Evolution of efficacy of the tested treatments (**A**) pelargonic acid, (**B**) cinnamic aldehyde, and (**C**) carvacrol in *E. bonariensis* during 30 days after their application.

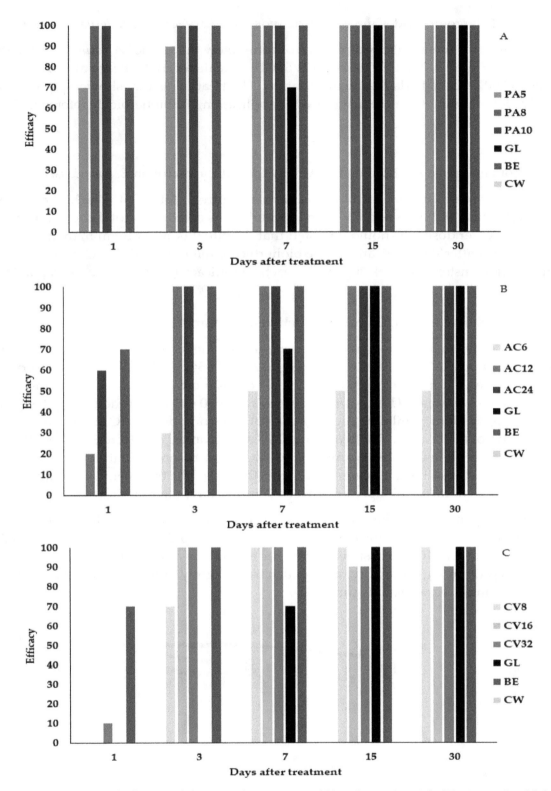

Figure 7. Evolution of efficacy of the tested treatments (**A**) pelargonic acid, (**B**) cinnamic aldehyde, and (**C**) carvacrol in *P. oleracea* during 30 days after their application.

3.1.1. Efficacy of Pelargonic Acid, Cinnamic Aldehyde, and Carvacrol on *A. retroflexus*

In the species *A. retroflexus* (Figure 4, Table S1) all the treatments tested obtained 100 efficacy (all treated plants were dead) one day after the application of the treatment, except for the chemical reference. The treatment with GL managed to control the species on the third day after its application. In this trial, there was a relevant percentage of mortality in the CW, especially at the end of the trial.

3.1.2. Efficacy of Pelargonic Acid, Cinnamic Aldehyde, and Carvacrol on *A. fatua*

All the tested treatments managed to control completely the species *A. fatua* from the third day after application (Figure 5, Table S2), except CV6, which achieved 100 efficacy after 7 days, and GL, which reached 100 efficacy 15 days after application. The treatments that showed phytotoxic effects more quickly were, starting from the first day after application, the bioherbicide reference (BE), AC12, and PA10.

3.1.3. Efficacy of Pelargonic Acid, Cinnamic Aldehyde, and Carvacrol on *E. bonariensis*

All treatments were able to control *E. bonariensis* (Figure 6, Table S3). The higher doses of the treatments performed with CA- and CV-based formulations achieved a total control of this species faster than their lower doses. It should be noted that despite this, all of them managed to control it completely 15 days after the application. The bioherbicide reference (BE) reached 100 efficacy 24 h after its application, instead GL took 30 days to reach 100 efficacy (death of all treated plants).

3.1.4. Efficacy of Pelargonic Acid, Cinnamic Aldehyde, and Carvacrol on *P. oleracea*

The most effective treatments to control *P. oleracea* were the three treatments carried out with the PA-based formulation (PA5, PA8, and PA10) (Figure 7, Table S4). A dose effect was observed in this species for the tested natural products, being higher doses more effective and showing phytotoxic effects faster than lower ones. The treatment AC6 reached 50 efficacy at the end of the experiment (30 days after application), while the higher doses of this compound (AC12 and AC24) killed all plants after 3 days of application. The treatments CV8, CV16, and CV32 decreased their efficacy from day 7, when some of the evaluated plants regrew. It should be noted that the treatment with the chemical reference, GL, exhibited a slower action than the rest of the treatments with natural products, showing phytotoxic effects on this species between 7 and 15 days after application.

When analyzing the effect of the interaction between species and time after treatments with respect to efficacy, the species that showed the highest sensitivity most rapidly was *A. retroflexus*. On the other hand, the species that took longer to show phytotoxic effects was *A. fatua*. However, at the end of the trials, all species showed high mortality rates, which were slightly higher in *A. retroflexus* and *A. fatua* than in *P. oleracea* and *E. bonariensis* (Figure 8).

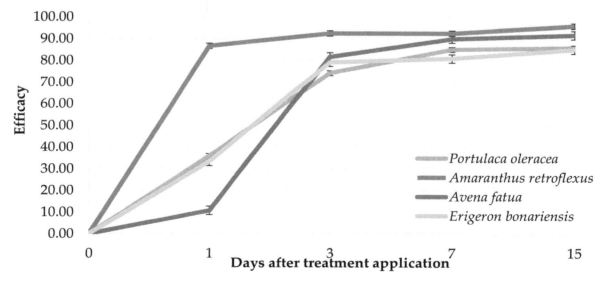

Figure 8. Effect of the interaction between treatment and days after treatment application in the efficacy per species.

3.2. Damage Level of Pelargonic Acid, Cinnamic Aldehyde, and Carvacrol against Target Weeds

A. retroflexus was the species which presented higher damage level, followed by *P. oleracea* and *A. fatua* (without significant differences between them), and finally *E. bonariensis* (Table 6). All species exhibited damage level near 2 or higher, which means severe damage (Table 4). It is important to consider the damage level caused by the treatments on the weed species in addition to their efficacy because it represents the state of the plants that were not killed. If the plants remaining alive were more damaged, it would mean that in field conditions, they would be less competitive with crops, causing less stress to them.

Table 6. Damage level depending on the species, time after application, and treatment.

Species	Level of Damage
Portulaca oleracea	1.98 ± 0.02 b
Amaranthus retroflexus	2.24 ± 0.02 a
Avena fatua	1.96 ± 0.02 bc
Erigeron bonariensis	1.92 ± 0.02 c
Time (Days after Application)	**Level of Damage**
0	0.00 ± 0.02 e
1	2.08 ± 0.02 d
3	2.59 ± 0.02 c
7	2.68 ± 0.02 b
15	2.78 ± 0.02 a
Treatment	**Damage level**
Control treated with water	0.16 ± 0.04 g
Pelargonic acid 5%	2.31 ± 0.04 abc
Pelargonic acid 8%	2.34 ± 0.04 ab
Pelargonic acid 10%	2.35 ± 0.04 ab
Cinnamic aldehyde 6%	2.13 ± 0.04 e
Cinnamic aldehyde 12%	2.30 ± 0.04 abc
Cinnamic aldehyde 24%	2.30 ± 0.04 abc
Carvacrol 8%	2.18 ± 0.04 de
Carvacrol 16%	2.23 ± 0.04 cd
Carvacrol 32%	2.25 ± 0.04 bcd
Bioherbicide reference: pelargonic acid (Beloukha 8%)	2.39 ± 0.04 a
Chemical reference: glyphosate (Roundup Ultra Plus 10%)	1.40 ± 0.03 f

Values are mean of damage level ± standard error (ten replicates). Different letters in the same column indicate significant differences ($p \leq 0.05$).

Throughout time, more severe levels of damage were reached as more days after treatment applications passed, with significant differences in the damage level assessment between different days after the applications (Table 6). All the treatments tested successfully controlled the weed species inducing a high level of damage compared with CW. The treatments that showed the strongest phytotoxicity on weeds were PA10 and BE, with no significant differences between them. PA10 showed no significant differences with the other two doses of PA-based formulations tested (PA5 and PA8), neither with the two highest doses of CA based formulations tested (CA12 and CA24) nor with the highest doses of CV tested (CV32) (Table 6).

The damage level increased in all species with time after treatments (Figure 9). *A. retroflexus* was confirmed as the most susceptible species to the treatments, as it showed a higher level of damage than the other species 24 h after the treatments were administrated. No differences between species were observed 15 days after treatment, as all showed similar levels of damage.

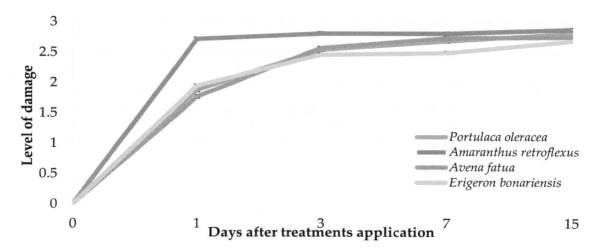

Figure 9. Effect of treatment and time after treatment interaction on damage level.

The effects induced by the different treatments on *E. bonariensis* 24 h after their administration are presented in Figure 10. This species is shown because of its intermediate response to all treatments as compared with *A. retroflexus* that was more sensitive or *P. oleracea*, which was more resistant and because phytotoxic effects can be better visualized in it than in *A. fatua*. The intermediate concentration tested for PA, CV, and CA is shown to be representative of the effects of the other concentrations tested. All the natural compounds tested caused more severe plant damage than the synthetic herbicide GL 1 day after treatment. The effects of 8% PA were very similar to those induced by the positive bioherbicide control Beloukha (also containing PA as active compound). Probably due to the effect of PA, the cuticles exhibited alteration on membrane permeability and peroxidation of thylakoid membranes [67] and leaves appeared desiccated, with reduced photosynthetic pigments but without punctual damages on the leaves, which resulted in a stoppage of growth and development of the whole plant. In contrast, CV-treated leaves showed signs of dehydration, resulting in curling and punctual damages on the leaves with increased necrotic spots related to application spots, which could be due to the disruption of cell membranes [68]. Finally, CA treatment resulted in growth reduction and loss of photosynthetic pigments, which could be related to oxidative damage induced by this compound. This oxidative damage has to be further investigated as no mode of action of CA has been reported in the literature up to now.

Bioherbicides are new products on the international markets and consequently, the processes for obtaining natural raw materials are not yet very efficient or the final cost of its extraction is elevated compared to synthetics. This fact affects the final cost of these formulated products, making them more expensive in some cases than conventional herbicides for farmers. Nevertheless, it is important to evaluate the cost–benefit factor of bioherbicides, including sustainability, reduction of soil and water contamination, or the absence of residues on crops. In line with legal framework, policies, and global sustainability objectives, the higher price of bioherbicides justifies the benefits that can be achieved with their implementation [69]. On the other hand, the rapid action, broad spectrum, and eco-friendly profile make bioherbicides molecules more attractive to the pesticide market, which is increasingly concerned with the sustainability of treatments applied in agriculture. Herbicide market is expected to reach a value of $37.99 billion by 2025 [38]. Improving the efficiency of raw material extraction, decreasing the applied doses per hectare using improved formulations, as well as combining active substances in search of synergies may be the future of new sustainable herbicides.

The natural products tested, PA, CV, and CA, performed strong herbicidal activity in all the treated weeds, causing high lethality and damage levels; hence, they demonstrated that they could be good candidates for bioherbicides formulations. Further investigations should focus on determining the dose–response of different weed species to these compounds in order to find the optimal doses, which is very important in the context of integrated weed management and sustainable agriculture.

Another key point is to find out the optimum phenological stage in which the products should be applied to weeds and crops, to achieve the maximum phytotoxic effect on weeds minimizing their phytotoxic effects and consequent stress on crops. A better understanding of their mode of action could lead to a more efficient administration. Finally, different combinations between these natural products could be a powerful tool for weed management. Their synergies and antagonisms must be also considered and studied.

Figure 10. Images of *Erigeron bonariensis* plants 24 h after treatment applications.

4. Conclusions

The natural products PA, CV, and CA showed great herbicidal activity against the weeds *A. retroflexus*, *A. fatua*, *E. bonariensis*, and *P. oleracea* and could be good candidates for bioherbicides formulations. *A. retroflexus* was the most sensitive weed to all the applied treatments. For CV and CA, the higher doses applied exhibited greater and quicker phytotoxic effects than the lowest, with different responses in the weed species, while there were no significant differences in the herbicidal activity between the tested doses of PA. This study demonstrates that natural products could be sustainable as well as effective alternatives to synthetic herbicides, and they contribute to integrated weed management.

Supplementary Materials
Table S1. Efficacy of the tested treatments on *A. retroflexus* after 1, 3, 7, 15 and 30 days of application. Table S2. Efficacy of the tested treatments on *A. fatua* after 1, 3, 7, 15 and 30 days of application. Table S3. Efficacy of the tested treatments on *E. bonariensis* after 1, 3, 7, 15 and 30 days of application. Table S4. Efficacy of the tested treatments on *P. oleracea* after 1, 3, 7, 15 and 30 days of application.

Author Contributions: Conceptualization, M.V., M.M., and A.M.S.-M.; methodology M.V., M.M., N.T.-P., and R.G.; formal analysis, M.V., M.M., and N.T.-P.; investigation, M.V., M.M., A.M.S.-M., R.P., N.T.-P., and R.G.; resources, M.V., M.M., and R.G.; data curation, N.T., M.M., R.G. and R.P.; writing—original draft preparation, M.V., M.M., and N.T.-P.; writing—review and editing, M.V., A.M.S.-M., and R.P.; visualization, M.V., M.M., N.T.-P., A.M.S.-M., R.P., and R.G.; supervision, M.V., M.M., and A.M.S.-M.; project administration, M.V.; and funding acquisition, M.V. All authors have read and agreed to the published version of the manuscript.

Acknowledgments: Thanks to Vicente Estornell Campos and the Library staff from Polytechnic University of Valencia that assisted us to get some helpful references.

References

1. Vos, R.; Bellù, L.G. Global trends and challenges to food and agriculture into the 21st century. In *Sustainable Food and Agriculture: An Integrated Approach*; Campanhola, C., Pandey, S., Eds.; Academic Press: London, UK, 2019; pp. 11–30. [CrossRef]
2. Oerke, E.C. Crop losses to pests. *J. Agric. Sci.* **2006**, *144*, 31–43. [CrossRef]
3. Vats, S. Herbicides: History, classification and genetic manipulation of plants for herbicide resistance. In *Sustainable Agriculture Reviews*; Lichtfouse, E., Ed.; Springer: Cham, Switzerland, 2015; Volume 15. [CrossRef]
4. Heap, I.M. Global perspective of herbicide-resistant weeds. *Pest Manag. Sci.* **2014**, *70*, 1306–1315. [CrossRef] [PubMed]
5. Bagavathiannan, M.; Singh, V.; Govindasamy, P.; Abugho, S.B.; Liu, R. Impact of concurrent weed or herbicide stress with other biotic and abiotic stressors on crop production. In *Plant Tolerance to Individual and Concurrent Stresses*; Senthil-Kumar, M., Ed.; Springer: New Delhi, India, 2017; pp. 33–45. [CrossRef]
6. Abbas, T.; Zahir, Z.A.; Naveed, M.; Kremer, R.J. Limitations of existing weed control practices necessitate development of alternative techniques based on biological approaches. In *Advances in Agronomy*; Sparks, D.L., Ed.; Academic Press: Cambridge, MA, USA, 2018; Volume 147, pp. 239–280. [CrossRef]
7. World Health Organization; Food and Agriculture Organization of the United Nations. The International Code of Conduct on Pesticide Management. Rome. 2014. Available online: http://www.fao.org/agriculture/crops/thematic-sitemap/theme/pests/code/en/ (accessed on 22 April 2020).
8. Villa, F.; Cappitelli, F.; Cortesi, P.; Kunova, A. Fungal biofilms: Targets for the development of novel strategies in plant disease management. *Front. Microbiol.* **2017**, *8*, 654–664. [CrossRef] [PubMed]
9. De Mastro, G.; Fracchiolla, M.; Verdini, L.; Montemurro, P. Oregano and its potential use as bioherbicide. *Acta Hortic.* **2006**, *723*, 335–346. [CrossRef]
10. Seiber, J.N.; Coats, J.; Duke, S.O.; Gross, A.D. Biopesticides: State of the art and future opportunities. *J. Agric. Food Chem.* **2014**, *62*, 11613–11619. [CrossRef] [PubMed]
11. Bailey, K.L. The bioherbicide approach to weed control using plant pathogens. In *Integrated Pest Management*; Abrol, D.P., Ed.; Academic Press, Elsevier: San Diego, CA, USA, 2014; pp. 245–266.
12. Dahiya, A.; Sharma, R.; Sindhu, S.; Sindhu, S.S. Resource partitioning in the rhizosphere by inoculated *Bacilluss* pp. towards growth stimulation of wheat and suppression of wild oat (*Avena fatua* L.) weed. *Physiol. Mol. Biol. Plants* **2019**, *25*, 1483–1495. [CrossRef]
13. Heap, I. The International Herbicide-Resistant Weed Database. Available online: www.weedscience.org (accessed on 25 April 2020).
14. Hazrati, H.; Saharkhiz, M.J.; Moein, M.; Khoshghalb, H. Phytotoxic effects of several essential oils on two weed species and tomato. *Biocatal. Agric. Biotechnol.* **2018**, *13*, 204–212. [CrossRef]
15. Bajwa, A.A.; Sadia, S.; Ali, H.H.; Jabran, K.; Peerzada, A.M.; Chauhan, B.B. Biology and management of two important Conyza weeds: A global review. *Environ. Sci. Pollut. Res.* **2016**, *23*, 24694–24710. [CrossRef]
16. Graziani, F.; Onofri, A.; Pannacci, E.; Tei, F.; Guiducci, M. Size and composition of weed seedbank in long-term organic and conventional low-input cropping systems. *Eur. J. Agron.* **2012**, *39*, 52–56. [CrossRef]
17. Benbrook, C.M. Trends in glyphosate herbicide use in the United States and globally. *Environ. Sci. Eur.* **2016**, *28*, 3–18. [CrossRef]
18. Önen, H.; Özer, Z.; Telci, I. Bioherbicidal effects of some plant essential oils on different weed species. *J. Plant Dis. Prot.* **2002**, *18*, 597–605.
19. Salamci, E.; Kordali, S.; Kotan, R.; Cakir, A.; Kaya, Y. Chemical compositions, antimicrobial and herbicidal effects of essential oils isolated from Turkish *Tanacetum aucheranum* and *Tanacetum chiliophyllum* var. *chiliophyllum*. *Biochem. Syst. Ecol.* **2007**, *35*, 569–581. [CrossRef]
20. Mutlu, S.; Atici, Ö.; Esim, N. Bioherbicidal effects of essential oils of *nepeta meyeri* benth. On weed spp. *Allelopath. J.* **2010**, *26*, 291–300.
21. Synowiec, A.; Kalemba, D.; Drozdek, E.; Bocianowski, J. Phytotoxic potential of essential oils from temperate climate plants against the germination of selected weeds and crops. *J. Pest Sci.* **2017**, *90*, 407–419. [CrossRef]
22. Hazrati, H.; Saharkhiz, M.J.; Niakousari, M.; Moein, M. Natural herbicide activity of *Satureja hortensis* L. essential oil nanoemulsion on the seed germination and morphophysiological features of two important weed species. *Ecotoxicol. Environ. Saf.* **2017**. [CrossRef]

23. Verdeguer, M.; Blazquez, M.A.; Boira, H. Phytotoxic effects of *Lantana camara*, Eucalyptus camaldulensis and *Eriocephalus africanus* essential oils in weeds of Mediterranean summer crops. *Biochem. Syst. Ecol.* **2009**, *37*, 362–369. [CrossRef]

24. Benarab, H.; Fenni, M.; Louadj, Y.; Boukhabti, H.; Ramdani, M. Allelopathic activity of essential oil extracts from *Artemisia herba-alba* Asso. on seed and seedling germination of weed and wheat crops. *Acta Sci. Nat.* **2020**, *7*, 86–97. [CrossRef]

25. Benchaa, S.; Hazzit, M.; Abdelkrim, H. Allelopathic Effect of *Eucalyptus citriodora* essential oil and its potential use as bioherbicide. *Chem. Biodivers.* **2018**, *15*, e1800202. [CrossRef]

26. Verdeguer, M.; Castañeda, L.G.; Torres-Pagan, N.; Llorens-Molina, J.A.; Carrubba, A. Control of Erigeron bonariensis with *Thymbra capitata*, *Mentha piperita*, *Eucalyptus camaldulensis*, and *Santolina chamaecyparissus* essential oils. *Molecules* **2020**, *25*, 562. [CrossRef]

27. Scavo, A.; Pandino, G.; Restuccia, A.; Mauromicale, G. Leaf extracts of cultivated cardoon as potential bioherbicide. *Sci. Hortic.* **2020**, 109024. [CrossRef]

28. Ma, S.; Fu, L.; He, S.; Lu, X.; Wu, Y.; Ma, Z.; Zhang, X. Potent herbicidal activity of *Sapindus mukorossi* Gaertn. against *Avena fatua* L. and *Amaranthus retroflexus* L. *Ind. Crops Prod.* **2018**, *122*, 1–6. [CrossRef]

29. Pacanoski, Z.; Mehmeti, A. Allelopathic effect of Siberian iris (*Iris sibirica*) on the early growth of wild oat (*Avena fatua*) and Canada thistle (*Cirsium arvense*). *J. Cent. Eur. Agric.* **2019**, *20*, 1179–1187. [CrossRef]

30. Bainard, L.D.; Isman, M.B.; Upadhyaya, M.K. Phytotoxicity of clove oil and its primary constituent eugenol and the role of leaf epicuticular wax in the susceptibility to these essential oils. *Weed Sci.* **2006**, *54*, 833–837. [CrossRef]

31. Ahuja, N.; Singh, H.P.; Batish, D.R.; Kohli, R.K. Eugenol-inhibited root growth in *Avena fatua* involves ROS-mediated oxidative damage. *Pestic. Biochem. Phys.* **2015**, *118*, 64–70. [CrossRef]

32. Vaughn, S.F.; Spencer, G.F. Volatile Monoterpenes as Potential Parent Structures for New Herbicides[l]. *Weed Sci.* **1993**, *41*, 114–119. [CrossRef]

33. Chaimovitsh, D.; Shachter, A.; Abu-Abied, M.; Rubin, B.; Sadot, E.; Dudai, N. Herbicidal Activity of Monoterpenes is associated with disruption of microtubule functionality and membrane integrity. *Weed Sci.* **2017**, *65*, 19–30. [CrossRef]

34. Amri, I.; Lamia, H.; Mohsen, H.; Bassem, J. Review on the phytotoxic effects of essential oils and their individual components: News approach for weed management. *Int. J. Appl. Biol. Pharm. Technol.* **2013**, *4*, 96–114.

35. Verdeguer, M.; García-Rellán, D.; Boira, H.; Pérez, E.; Gandolfo, S.; Blázquez, M.A. Herbicidal activity of Peumus boldus and Drimys winterii essential oils from Chile. *Molecules* **2011**, *16*, 403–411. [CrossRef]

36. Saad, M.M.G.; Abdelgaleil, S.A.M.; Suganuma, T. Herbicidal potential of pseudoguaninolide sesquiterpenes on wild oat, *Avena fatua* L. *Biochem. Syst. Ecol.* **2012**, *44*, 333–337. [CrossRef]

37. Araniti, F.; Sánchez-Moreiras, A.M.; Graña, E.; Reigosa, M.J.; Abenavoli, M.R. Terpenoid trans-caryophyllene inhibits weed germination and induces plant water status alteration and oxidative damage in adult Arabidopsis. *Plant Biol. (Stuttg)* **2017**, *19*, 79–89. [CrossRef]

38. Ciriminna, R.; Fidalgo, A.; Ilharco, L.M.; Pagliaro, M. Herbicides based on pelargonic acid: Herbicides of the bioeconomy. *Biofuels Bioprod. Biorefining* **2019**, *13*, 1476–1482. [CrossRef]

39. Coleman, R.; Penner, D. Organic acid enhancement of pelargonic acid. *Weed Technol.* **2008**, *22*, 38–41. [CrossRef]

40. Crmaric, I.; Keller, M.; Krauss, J.; Delabays, N. Efficacy of natural fatty acid based herbicides on mixed weed stands. *Jul. Kühn Arch.* **2018**, *458*, 327–332. [CrossRef]

41. Dayan, F.E.; Duke, S.O. Natural compounds as next-generation herbicides. *Plant Physiol.* **2014**, *166*, 1090–1105. [CrossRef]

42. Croteau, R.; Kutchan, T.M.; Lewis, N.G. Natural products (secondary metabolites). In *Biochemistry & Molecular Biology of Plants*, 2nd ed.; Buchanan, B.B., Gruissem, W., Jones, R.L., Eds.; Wiley: Rockville, MD, USA, 2000; pp. 1250–1318.

43. Lebecque, S.; Lins, L.; Dayan, F.E.; Fauconnier, M.L.; Deleu, M. Interactions between natural herbicides and lipid bilayers mimicking the plant plasma membrane. *Front. Plant Sci.* **2019**, *10*, 329–340. [CrossRef]

44. Cordeau, S.; Triolet, M.; Wayman, S.; Steinberg, C.; Guillemin, J.P. Bioherbicides: Dead in the water? A review of the existing products for integrated weed management. *Crop Prot.* **2016**, *87*, 44–49. [CrossRef]

45. Gruenwald, J.; Freder, J.; Armbruester, N. Cinnamon and health. *Crit. Rev. Food Sci. Nutr.* **2010**, *50*, 822–834. [CrossRef]

46. Ranasinghe, P.; Pigera, S.; Premakumara, G.S.; Galappaththy, P.; Constantine, G.R.; Katulanda, P. Medicinal properties of "true" cinnamon (*Cinnamomum zeylanicum*): A systematic review. *BMC Complement. Altern. Med.* **2013**, *13*, 275. [CrossRef]

47. Doyle, A.A.; Krämer, T.; Kavanagh, K.; Stephens, J.C. Cinnamaldehydes: Synthesis, antibacterial evaluation, and the effect of molecular structure on antibacterial activity. *Results Chem.* **2019**, *1*, 100013–100018. [CrossRef]

48. Chericoni, S.; Prieto, J.M.; Iacopini, P.; Cioni, P.; Morelli, I. In vitro activity of the essential oil of *Cinnamomum zeylanicum* and eugenol in peroxynitrite-induced oxidative processes. *J. Agric. Food Chem.* **2005**, *53*, 4762–4765. [CrossRef]

49. Viazis, S.; Akhtar, M.; Feirtag, J.; Diez-Gonzalez, F. Reduction of Escherichia coli O157:H7 viability on leafy green vegetables by treatment with a bacteriophage mixture and trans-cinnamaldehyde. *Food Microbiol.* **2011**, *28*, 149–157. [CrossRef]

50. Kwon, J.A.; Yu, C.B.; Park, H.D. Bacteriocidal effects and inhibition of cell separation of cinnamic aldehyde on *Bacillus cereus*. *Lett. Appl. Microbiol.* **2003**, *37*, 61–65. [CrossRef]

51. Friedman, M. Chemistry, antimicrobial mechanisms, and antibiotic activities of cinnamaldehyde against pathogenic bacteria in animal feeds and human foods. *J. Agric. Food. Chem.* **2017**, *65*, 10406–10423. [CrossRef]

52. Kim, H.K.; Kim, J.R.; Ahn, Y.J. Acaricidal activity of cinnamaldehyde and its congeners against *Tyrophagus putrescentiae* (Acari: Acaridae). *J. Stored Prod. Res.* **2004**, *40*, 55–63. [CrossRef]

53. Saad, M.M.G.; Gouda, N.A.A.; Abdelgaleil, S.A.M. Bioherbicidal activity of terpenes and phenylpropenes against *Echinochloa crus-galli*. *J. Environ. Sci. Health B* **2019**, *54*, 954–963. [CrossRef]

54. Roselló, J.; Sempere, F.; Sanz-Berzosa, I.; Chiralt, A.; Santamarina, M.P. Antifungal activity and potential use of essential oils against *Fusarium culmorum* and *Fusarium verticillioides*. *J. Essent. Oil Bear Plants* **2015**, *18*, 359–367. [CrossRef]

55. Santamarina, M.; Ibáñez, M.; Marqués, M.; Roselló, J.; Giménez, S.; Blázquez, M. Bioactivity of essential oils in phytopathogenic and post-harvest fungi control. *Nat. Prod. Res.* **2017**, *31*, 2675–2679. [CrossRef]

56. Krepker, M.; Shemesh, R.; Danin Poleg, Y.; Kashi, Y.; Vaxman, A.; Segal, E. Active food packaging films with synergistic antimicrobial activity. *Food Control* **2017**, *76*, 117–126. [CrossRef]

57. Ye, H.; Shen, S.; Xu, J.; Lin, S.; Yuan, Y.; Jones, G.S. Synergistic interactions of cinnamaldehyde in combination with carvacrol against food-borne bacteria. *Food Control* **2013**, *34*, 619–623. [CrossRef]

58. De Sousa, J.P.; de Azerêdo, G.A.; de Araújo Torres, R.; da Silva Vasconcelos, M.A.; da Conceição, M.L.; de Souza, E.L. Synergies of carvacrol and 1,8-cineole to inhibit bacteria associated with minimally processed vegetables. *Int. J. Food Microbiol.* **2012**, *154*, 145–151. [CrossRef]

59. Oddo, M. Effects of Different weed Control Practices on Soil Quality in Mediterranean Crops. Ph.D. Thesis, Universita' degli Studi di Palermo, Palermo, Italy, Polytechnic University of Valencia, Valencia, Spain, 2 October 2017.

60. Wu, H.; Walker, S.; Rollin, M.J.; Tan, D.K.Y.; Robinson, G.; Werth, J. Germination, persistence, and emergence of flaxleaf fleabane (*Conyza bonariensis* [L.] Cronquist). *Weed Biol. Manag.* **2007**, *7*, 192–199. [CrossRef]

61. Webber, C.L.; Shrefler, J.W. Pelargonic acid weed control parameters. *HortScience* **2006**, *41*, 1034. [CrossRef]

62. Velini, E.D.; Duke, S.O.; Trindade, M.B.; Meschede, D.K.; Carbonari, C.A. Modo de acao do Glyphosate (Mode of Action of Glyphosate in Portuguese). In *Glyphosate*; Velini, E.D., Meschede, D.K., Carbonari, C.A., Trindade, M.L.B., Eds.; Fundaçã de Estudos e Pesquisas Agricoloas e Florestais: Botucato-SP, Brazil, 2009; pp. 113–133.

63. Hoss, N.; Al-Khatib, K.; Peterson, D.; Loughin, T. Efficacy of glyphosate, glufosinate, and imazethapyr on selected weed species. *Weed Sci.* **2003**, *51*, 110–117. [CrossRef]

64. Jordan, D.; York, A.; Griffin, J.; Clay, P.; Vidrine, P.; Reynolds, D. Influence of Application Variables on Efficacy of Glyphosate. *Weed Technol.* **1997**, *11*, 354–362. [CrossRef]

65. Mithila, J.; Swanton, C.J.; Blackshaw, R.E.; Cathcart, R.J.; Hall, J.C. Physiological basis for reduced glyphosate efficacy on weeds grown under low soil nitrogen. *Weed Sci.* **2008**, *56*, 12–17. [CrossRef]

66. Sandberg, C.L.; Meggitt, W.F.; Penner, D. Absorption, translocation and metabolism of ^{14}C-glyphosate in several weed species. *Weed Res.* **1980**, *20*, 195–200. [CrossRef]

67. Lederer, B.; Fujimori, T.; Tsujino, Y.; Wakabayashi, K.; Bögera, P. Phytotoxic activity of middle-chain fatty acids II: Peroxidation and membrane effects. *Pestic. Biochem. Physiol.* **2004**, *80*, 151–156. [CrossRef]

68. Albuquerque, C.C.; Camara, T.R.; Sant'ana, A.E.G.; Ulisses, C.; Willadino, L.; Marcelino Júnior, C. Effects of the essential oil of *Lippia gracilis* Schauer on caulinary shoots of heliconia cultivated in vitro. *Rev. Bras. Plantas Med.* **2012**, *14*, 26–33. [CrossRef]

69. Hasanuzzaman, M.; Mohsin, S.M.; Borhannuddin Bhuyan, M.H.M.; Farha Bhuiyan, T.; Anee, T.I.; Awal, A.; Masud, C.; Nahar, K. Phytotoxicity, environmental and health hazards of herbicides: Challenges and ways forward. In *Agrochemicals Detection, Treatment and Remediation. Pesticides and Chemical Fertilizers*; Vara Prasad, M.N., Ed.; Butterworth Heinemann: Hyderabad, India, 2020; pp. 55–99. [CrossRef]

12

Treatment of Sweet Pepper with Stress Tolerance-Inducing Compounds Alleviates Salinity Stress Oxidative Damage by Mediating the Physio-Biochemical Activities and Antioxidant Systems

Khaled A. Abdelaal [1], Lamiaa M. EL-Maghraby [2], Hosam Elansary [3,4], Yaser M. Hafez [1], Eid I. Ibrahim [5], Mostafa El-Banna [6], Mohamed El-Esawi [7,8] and Amr Elkelish [9,*]

[1] Plant Pathology and Biotechnology Lab., Excellence Center (EPCRS), Faculty of Agriculture, Kafrelsheikh University, Kafrelsheikh 33516, Egypt; khaled_elhaies@yahoo.com (K.A.A.); hafezyasser@gmail.com (Y.M.H.)

[2] Agricultural Biochemistry Department, Faculty of Agriculture, Zagazig University, Zagazig 44511, Egypt; dr_lamiaa222@yahoo.com

[3] Plant Production Department, College of Food and Agricultural Sciences, King Saud University, P.O. Box 2455, Riyadh 11451, Saudi Arabia; helansary@ksu.edu.sa

[4] Floriculture, Ornamental Horticulture and Garden Design Department, Faculty of Agriculture, Alexandria University, Alexandria 21526, Egypt

[5] Rice Biotechnology Lab., Rice Research Dep., Field Crops Research Institute, Sakha, Kafr El-Sheikh 33717, ARC, Egypt; Eid.ibrahim@gmail.com

[6] Agricultural Botany Department, Faculty of Agriculture, Mansoura University, Mansoura 35516, Egypt; el-banna@mans.edu.eg

[7] Botany Department, Faculty of Science, Tanta University, Tanta 31527, Egypt; mohamed.elesawi@science.tanta.edu.eg

[8] Sainsbury Laboratory, University of Cambridge, Cambridge CB2 1LR, UK

[9] Botany Department, Faculty of Science, Suez Canal University, Ismailia 41522, Egypt

* Correspondence: amr.elkelish@science.suez.edu.eg

Abstract: Salinity stress occurs due to the accumulation of high levels of salts in soil, which ultimately leads to the impairment of plant growth and crop loss. Stress tolerance-inducing compounds have a remarkable ability to improve growth and minimize the effects of salinity stress without negatively affecting the environment by controlling the physiological and molecular activities in plants. Two pot experiments were carried out in 2017 and 2018 to study the influence of salicylic acid (1 mM), yeast extract (6 g L^{-1}), and proline (10 mM) on the physiological and biochemical parameters of sweet pepper plants under saline conditions (2000 and 4000 ppm). The results showed that salt stress led to decreasing the chlorophyll content, relative water content, and fruit yields, whereas electrolyte leakage, malondialdehyde (MDA), proline concentration, reactive oxygen species (ROS), and the activities of antioxidant enzymes increased in salt-stressed plants. The application of salicylic acid (1 mM), yeast extract (6 g L^{-1}), and proline (10 mM) markedly improved the physiological characteristics and fruit yields of salt-stressed plants compared with untreated stressed plants. A significant reduction in electrolyte leakage, MDA, and ROS was also recorded for all treatments. In conclusion, our results reveal the important role of proline, SA, and yeast extracts in enhancing sweet pepper growth and tolerance to salinity stress via modulation of the physiological parameters and antioxidants machinery. Interestingly, proline proved to be the best treatment.

Keywords: *Capsicum annuum* L.; salt stress; salicylic acid; yeast; proline

1. Introduction

Global food safety is seriously dependent on crops and their supplies, which require considerable increases for servicing the gap between production and demand [1]. The necessity of improving crop production has been much more emergent in the last few years due to the expanding population, which will exceed to 9.7 billion by 2050. Undoubtedly, increases in the population will exert pressure on crops and food resources [1]. Simultaneously, global warming, as well as various biotic and abiotic stresses, hinder the growth and yields of agricultural crops [2]. Among abiotic stresses, salinity is recognized as one of the main restricting factors affecting the growth and productivity of agricultural crops, especially in arid and semiarid regions [3]. Salinity stress causes a reduction in growth and biomass, chlorophyll degradation, water status modification, malfunctions in stomatal functions, modifications in transpiration and respiration, and disequilibria in ion ratios [4,5]. Furthermore, plants develop cytotoxic-activated oxygen under saline conditions, which might seriously interfere with healthy metabolisms as a result of the oxidative damage of lipids, proteins, and nucleic acids [6,7]. Salinization may additionally lead to the excessive intracellular generation of reactive oxygen species (ROS) such as hydroxyl radicals (OH) and superoxide radicals (O_2^-) [8]. Plants confront these sorts of oxidants by developing several defensive mechanisms, including antioxidant enzymes and molecules that eliminate potentially cytotoxic types of activated oxygen [9,10].

Sweet pepper (*Capsicum annuum* L.) is an important vegetable crop that is grown for local consumption, and which has a high economic value in the Egyptian agricultural market. Farmers started to utilize saline water to partially fulfil crop water demands. The pepper plant is not a salt-tolerant vegetable, and about 14% of fruit yield loss occurs as a result of each increase in salt level of 1.0 dS/m [11]. Previous investigations have been conducted to mitigate the harmful impact of salt stress on sweet pepper, but most have not been sufficient or broadly applicable. As a result, the search for cheaper, ecologically-friendly strategies for salinity amelioration which enhance the growth and productivity of sweet pepper has been very important to the agriculture sector [12].

Numerous studies have found that implementing exogenous chemicals improves salt stress tolerance in plants [13]; examples of such chemicals are phytohormones such as salicylic acid, sterols, and methyl jasmonate [2,14]. Other chemicals such as polyamines, melatonin, and sodium nitroprusside have also been used to enhance the tolerance of various crop plants to saline conditions [15].

Salicylic acid is an essential phenolic compound that regulates plant growth processes and responses to different environmental factors [16]. It is a stress tolerance inducer and an important signal in many physiological processes, such as proline metabolism and photosynthesis. It reduces oxidative stress in plants under environmental stress and enhances plant growth and productivity under salt- [17] and drought-stress conditions [18]. Foliar application of SA-enhanced growth characteristics of sweet pepper plants [6] has increased the chlorophyll concentrations and enzyme activities in barley plants, as well as counteracting the deleterious impacts of salinity on faba beans [19]. Yeast extracts are the main source of various important compounds, such as amino acids, phytohormones, and vitamins [20,21]. The use of active yeast extracts has been shown to decrease the damaging impact of drought conditions on pea plants, and enhanced the growth performance and yield of stressed plants [22]. Yeast extract applications have led to improvements in the growth characteristics of bean and corn plants, such as the dry weight of leaves, the leaf area, and the number of leaves under drought conditions [23]. The application of yeast and NPK fertilizers has significantly enhanced chlorophyll concentrations and root yields in sugar beet plants [22]. Seaweed extracts have also improved plant tolerance to abiotic stresses. For example, the application of *Ecklonia maxima* seaweed extract has been shown to enhance the tolerance of zucchini squash plants to salinity stress by improving plant performance, shoot biomass yield, fruit quality, leaf gas exchange rate, SPAD index, and leaf nutritional status under saline conditions [24]. Furthermore, proline has a positive impact on the activity of enzymes and osmotic adjustment under stress conditions, while protecting enzyme denaturation and modulating osmoregulation [25]. The application of proline-modulated antioxidant enzymes such as peroxidase (POX) and catalase (CAT) in tobacco plants under salinity conditions plays a significant role in protein synthesis and accumulation in plants under stress conditions like drought and salinity in order to enhance the growth characteristics and yield [26–30].

Considering the variable effectiveness levels of salicylic acid, proline, and yeast extract on plants, as well as the harmful impact of salinity stress on the growth and productivity of important crops, the present study aims to evaluate and compare the levels of effectiveness of the three stress tolerance inducers, i.e., salicylic acid (1 mM), yeast extract (6 g L^{-1}), and proline (10 mM), on the growth characteristics, antioxidants, physiological and biochemical parameters, and yield of sweet pepper plants (*Capsicum annuum* L.) grown under the same saline conditions in order to determine which stress tolerance inducer should be recommended for further enhancements of crop performance and tolerance.

2. Materials and Methods

2.1. Experiments Design and Treatments

Pot experiments were performed at Agricultural Botany Department, Faculty of Agriculture, Kafrelsheikh University, Egypt during the growing seasons of 2017 and 2018. Laboratory analyses were carried out at the Plant Pathology & Biotechnology Lab, and the EPECRS Excellence Center Kafrelsheikh University, Egypt. This research was conducted to study the impacts of salicylic acid (1 mM), yeast extract (6 g L^{-1}), and proline (10 mM) on the growth characteristics and biochemical and yield parameters of salt-stressed sweet pepper plants (*Capsicum annuum* L.). Irrigation water was artificially salinized by applying NaCl at concentrations of 2000 and 4000 ppm. The seeds of sweet pepper cv. California Wonder were obtained from Sun Seed Company in USA. Ten seeds were sown in the nursery using foam trays. Forty-two days after sowing, seedlings were transplanted into pots (30 cm diameter); each pot contained 8 kg soil and 2 plants. The physical and chemical soil characteristics were recorded, according to the methods described by Abdelaal et al. [21], as follows. pH: 8.2; N: 32.4 ppm; P: 10.5 ppm; K: 289 ppm; electrical conductivity: 1.8 dS m^{-1}, soil organic matter: 1.9%; sand: 17.3%; silt: 35.5%; and clay: 47.2%. Fertilizers were added in two equal doses as recommended (NPK, 135:40:35 kg/ha), plus essential micronutrients, whereas the first dose was added 15 days after transplanting and the second at the beginning of flowering stage [31]. The plants were treated twice (20 and 40 days after transplanting) with salicylic acid (1 mM), yeast (6 g L^{-1}) and proline (10 mM). The experiment was done in a completely randomized design with five replicates (five pots with two plants each), and the following measurements were recorded after collecting the plant samples.

2.2. Physiological and Biochemical Analysis

For physiological and biochemical analyses, the samples were collected at 90 days after transplantation for use in the following assays.

2.2.1. Chlorophyll a and b Determination

For chlorophyll a and b determination, 5 mL N-N Dimethyl formamid was added to 1 g sweet pepper fresh leaves and placed in a refrigerator for 24 h. Following the centrifugation at 4000 *g* for 15 min, the optical density was calculated using spectrophotometer at 647 and 664 nm, according to Moran [32].

2.2.2. Calculation of Leaves Relative Water Content (RWC %) and Electrolyte Leakage (EL %)

The relative water content (RWC) in leaves was recorded according to the formula of Sanchez et al. [33] as follows: RWC = (FW − DW)/ (TW − DW) × 100, where FW is fresh weight, DW is dry weight, and TW is turgid weight. Electrolyte leakage (EL %) was estimated using the formula of Dionisio-Sese and Tobita [34] as follows: EL (%) = Initial electrical conductivity/final electrical conductivity × 100.

2.2.3. Proline Content Determination

Proline was assayed according to the method described by Bates et al. [35] with minor modifications. In brief, a plant sample (0.6 g) was extracted in sulfosalicylic acid (5%) followed by centrifugation at 10000 g for 7 min. The supernatants were diluted with water, mixed with 2% ninhydrin, heated at 94 °C for 30 min, and then cooled. Toluene was then added to the mixture, and the upper aqueous phase was spectrophotometrically assayed at 520 nm.

2.2.4. Calculation of Lipid Peroxidation and Reactive Oxygen Species (Superoxide and Hydrogen Peroxide)

The lipid peroxidation as malondialdehyde (MDA) in plant samples was calculated according to the method described by Heath and Packer [36] with minor modifications. In brief, 0.6 g of plant sample was extracted in TCA (0.1%), followed by centrifugation at 13,000 g for 8 min. The supernatants were mixed with thiobarbituric acid (0.5%) and TCA, and heated at 92 °C for 35 min, followed by cooling and centrifugation at 12,000 g for 8 min. Next, the supernatants' absorbance was measured at 532 and 660 nm. Superoxide and hydrogen peroxide levels were also determined according to the method described by Badiani et al. [37].

2.2.5. Antioxidant Enzymes Activity (CAT and POX)

Plant samples (1.5 g) were extracted in Tris-HCl (100 mM, pH 7.5) containing Dithiothreitol (5 mM), $MgCl_2$ (10 mM), EDTA (1 mM), magnesium acetate (5 mm), PVP-40 (1.6%), aphenylmethanesulfonyl fluoride (1 mM), and aproptinin (1 µg mL^{-1}). The mixed solutions were filtered and centrifuged for 8 min at 13,000 rpm. The supernatants were utilized to record enzymes activities. The activity of CAT and POX of leafy samples was determined according to the method described by Aebi [38] and Hammerschmidt et al. [39]. The supernatant absorbance was shown spectrophotometrically to be 470 nm.

2.2.6. Fruit yields

At 120 days after transplanting, the number of fruits per plant, the fruit fresh weight per plant (g), and the total fruit yield (ton hectare^{-1}) were recorded.

2.3. Statistical Analysis

Data represent the mean ± SD (standard deviation). Two-way analysis of variance was performed using SPSS ver. 19 (SPSS Inc., Chicago, IL, USA). A Tukey's test was also carried out to determine whether a significant difference ($p < 0.05$) existed between mean values.

3. Results

3.1. Chlorophyll a and b Concentrations

According to our results in Figure 1, the concentrations of chlorophyll a and b were significantly decreased in sweet pepper plants under salt-stress conditions; the lowest values were recorded with 4000 ppm compared with 2000 ppm and control plants in the two growing seasons. However, the salt stressed plants treated with salicylic acid, yeast extract, and proline showed significant increases in chlorophyll a and chlorophyll b concentrations compared with stressed untreated plants in both seasons. Under salt stresses of 2000 and 4000 ppm, the maximum concentrations of chlorophyll a and b were recorded with proline treatment in both seasons.

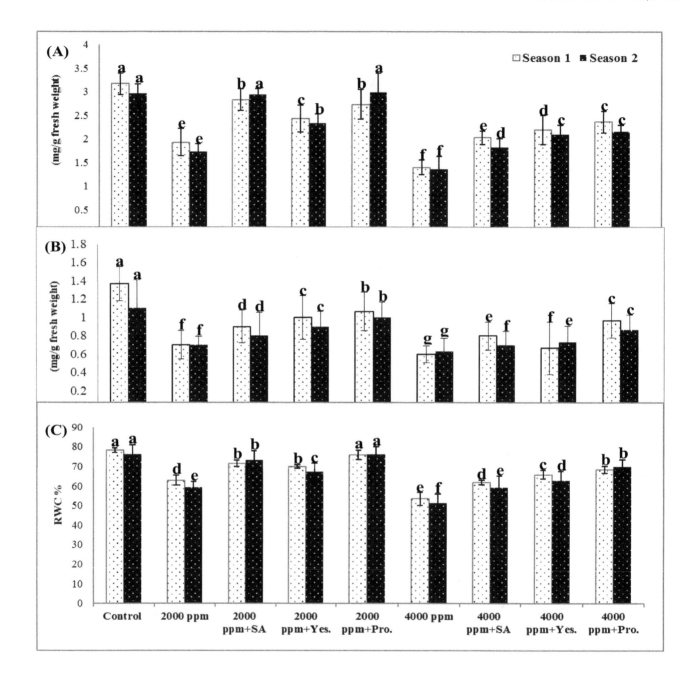

Figure 1. Effect of salinity stress (2000 and 4000 ppm NaCl) and supplementation of SA, yeast, and proline on the contents of (**A**) chlorophyll a, (**B**) chlorophyll b, (**C**) relative water content (RWC) in sweet pepper in the seasons of 2017 and 2018. Data is mean (±SE) of five replicates. Different letters in each Figure represent significant differences at $p < 0.05$.

3.2. Relative Water Content (RWC %)

Data obtained in Figure 1 showed that RWC decreased considerably in salt stressed plants; the greatest reduction was recorded in the plants exposed to salinity at 4000 ppm compared with control plants. The exogenous application of salicylic acid (1 mM), yeast (6 g L^{-1}), and proline (10 mM) caused a significant increase in RWC in salt stressed plants (2000 and 4000 ppm) compared with salt stressed untreated plants. Furthermore, the best treatments under salinity of 2000 ppm were salicylic acid and proline. Under salt treatment at 4000 ppm, the application of yeast extract (6 g L^{-1}) and proline (10 mM) showed the highest RWC in sweet pepper plants compared with SA treatment in stressed untreated plants in both seasons.

3.3. Electrolyte Leakage (EL %)

It may be noted from Figure 2 that salt stress at 2000 and 4000 ppm caused a significant increase in electrolyte leakage (EL); the maximum increase was recorded with a salinity level of 4000 ppm in both seasons. Interestingly, electrolyte leakage was significantly decreased upon the foliar application of salicylic acid (1 mM), yeast extract (6 g L^{-1}), and proline (10 mM) compared with control plants in both seasons. The best treatment was proline under a salt stress of 2000 ppm in both seasons (Figure 2).

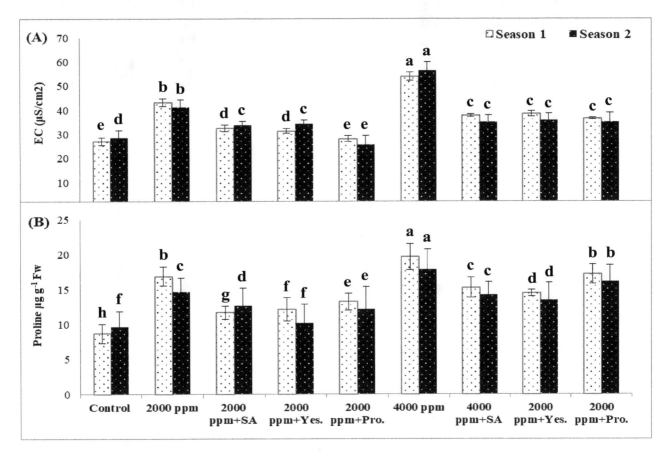

Figure 2. Effect of salinity stress (2000 and 4000 ppm NaCl) and supplementation of SA, yeast, and proline on the contents of (**A**) electrolyte leakage, (**B**) proline in sweet pepper in the seasons of 2017 and 2018. Data is mean (±SE) of five replicates. Different letters in each Figure represent significant differences at $p < 0.05$.

3.4. Proline Concentration

It is evident that proline had markedly accumulated in sweet pepper plants; the highest concentration was recorded with a salinity at 4000 ppm in comparison to the control plants (Figure 2). Intriguingly, the application of salicylic acid, yeast extract, and proline resulted in enhanced proline concentration under all salinity levels; the greatest result was observed with proline (10 mM).

3.5. Lipid Peroxidation (MDA) and Reactive Oxygen Species (Superoxide and Hydrogen Peroxide).

The results showed that lipid peroxidation (i.e., malondialdehyde or MDA), superoxide, and hydrogen peroxide were significantly increased under salt conditions compared with control plants in both seasons (Figure 3). The maximum levels of MDA, superoxide, and hydrogen peroxide were recorded at a salinity level of 4000 ppm, followed by 2000 ppm, in both seasons. On the other hand, the application of salicylic acid, yeast extract, and proline significantly reduced MDA, O_2^-, and H_2O_2 concentrations under all salinity levels compared to the stressed untreated plants. The best results were obtained with SA and proline.

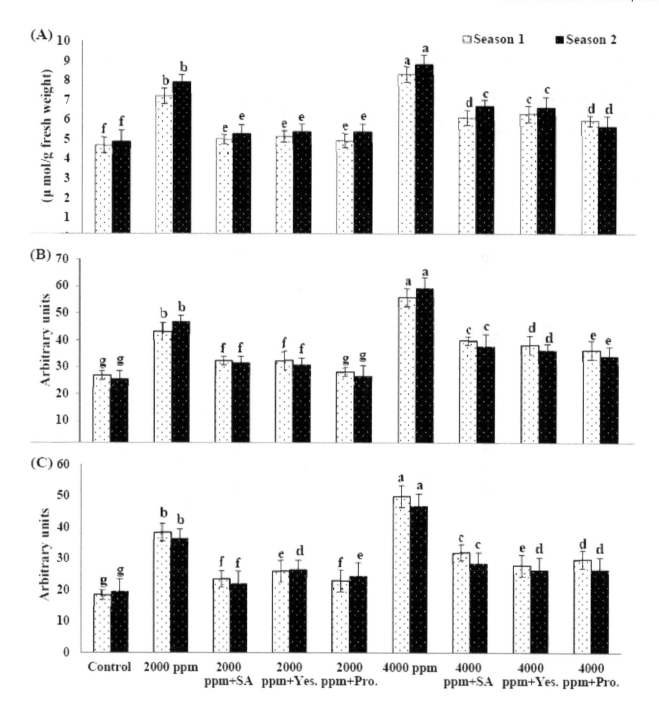

Figure 3. Effect of salinity stress (2000 and 4000 ppm NaCl) and supplementation of SA, yeast, and proline on the contents of (**A**) lipid peroxidation, (**B**) superoxide, (**C**) hydrogen peroxide in sweet pepper in the seasons of 2017 and 2018. Data is mean (±SE) of five replicates. Different letters in each Figure represent significant differences at $p < 0.05$.

3.6. Antioxidant Enzymes Activity

Antioxidant enzyme activities (CAT and POX) were assayed. The data presented in Figure 4 shows that the plants exposed to salt stress (2000 and 4000 ppm) had higher CAT and POX activity compared with control plants in both seasons. On the other hand, applications of salicylic acid (1 mM), yeast extract (6 g L^{-1}), and proline (10 mM) led to reductions in the activities of CAT and POX in the salt-stressed plants.

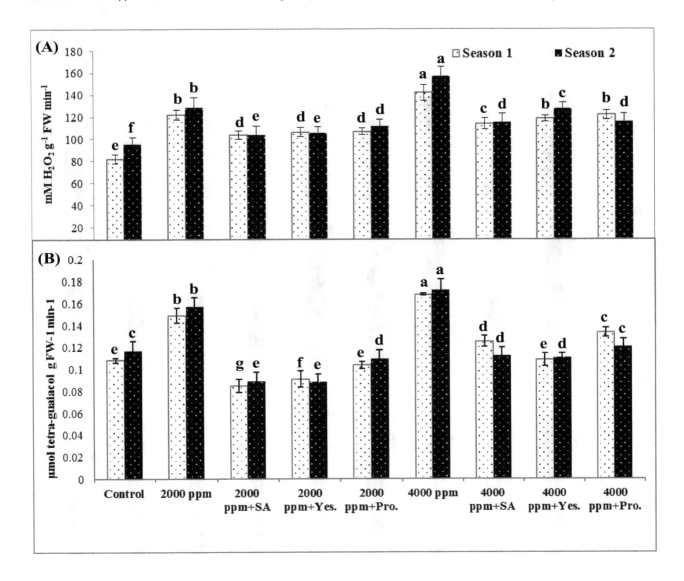

Figure 4. Effect of salinity stress (2000 and 4000 ppm NaCl) and supplementation of SA, yeast, and proline on the activity of (**A**) Catalase (CAT), (**B**) peroxides (POX) in sweet pepper in the seasons of 2017 and 2018. Data is mean (±SE) of five replicates. Different letters in each Figure represent significant differences at $p < 0.05$.

3.7. Number of Fruits per Plant, Fruit Fresh Weight, and Total Fruit Yield (Ton Hectare^{-1}).

According to our findings in Figure 5, salt stress at 2000 and 4000 ppm caused significant decreases in fruit number per plant, fruit fresh weight, and the total fruit yield (ton hectare^{-1}) in both seasons. The lowest values of these traits were recorded with salt stressed plants at 4000 ppm concentration, followed by 2000 ppm. Nevertheless, the exogenous application of salicylic acid, yeast, and proline significantly improved the number of fruits per plant, fruit fresh weight, and total fruit yield (ton hectare^{-1}) in the stressed treated plants compared with stressed untreated plants.

Interestingly, SA and proline treatments gave the maximum values of the three studied characteristics at a salinity concentration of 2000 ppm in the two seasons (Figure 5). Under salinity stress of 4000 ppm, the best results of fruit number per plant, fruit fresh weight, and total fruit yield (ton hectare^{-1}) were recorded with proline, followed by SA and yeast extract in both seasons.

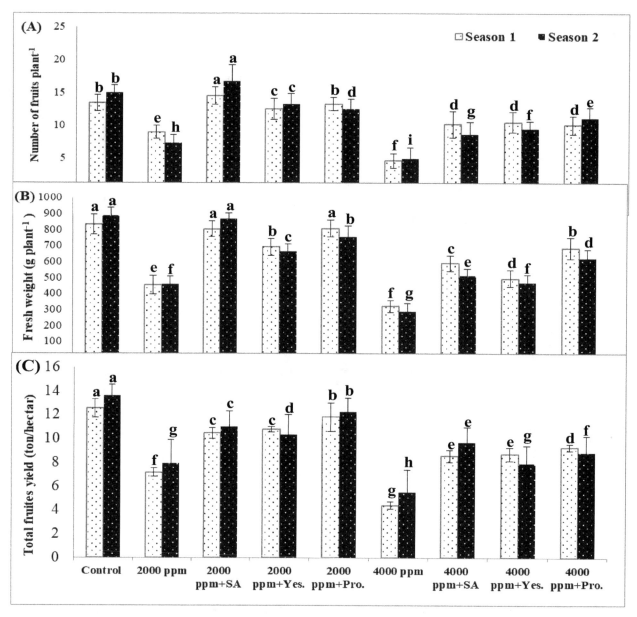

Figure 5. Effect of salinity stress (2000 and 4000 ppm NaCl) and supplementation of SA, yeast, and proline on (**A**) number of fruits plant^{-1}, (**B**) fresh weight plant^{-1}, (**C**) total fruit yield (ton hectare^{-1}) in sweet pepper in the seasons of 2017 and 2018. Data is mean (±SE) of five replicates. Different letters in each Figure represent significant differences at $p < 0.05$.

4. Discussion

The exogenous application of salicylic acid, proline, and yeast extract previously exhibited variable effectiveness levels on plant performance and tolerance to the harmful impact of salinity stress; therefore, the present study assessed and compared the effectiveness levels of these stress tolerance inducers on the growth characteristics, antioxidant levels, physiological and biochemical parameters, and yield of sweet pepper plants (*Capsicum annuum* L.) grown under the same saline conditions in order to determine which stress tolerance inducer should be recommended for the enhancement of crop performance and tolerance. In the present study, salt stress significantly decreased the aforementioned physiological parameters of sweet pepper plants. This reduction in chlorophyll a and b concentrations could be due to the effect of salinity on chlorophyll-degrading enzyme (chlorophyllase) activity, which reduces the chlorophyll synthesis level or negatively affects the structure and number of chloroplasts [40–42]. The chloroplast is one of the most vital organelles for photosynthesis and plant production, and is dramatically affected by abiotic stresses [43,44]. The obtained results indicated that foliar applications of SA (1 mM), yeast extract (6 g L^{-1}), and proline (10 mM) led to increased chlorophyll a and b contents in salt-stressed plants. These findings are in harmony with those obtained by Saleh et al. [25], Abdelaal et al. [21], and Soliman et al. [2]. These results might be due to the antioxidant scavenging influence of SA on chlorophyll degradation under saline conditions [45,46]. Correspondingly, the role of yeast extract in chlorophyll concentration enhancement might be due to the fact that yeast is rich in many essential elements, vitamins, and amino acids, which improve chlorophyll concentrations under stress conditions [21]. Moreover, our results showed that proline application minimized the harmful effects of all salinity levels on chlorophyll a and b concentrations due to its ability to function as a scavenger for ROS. Thus, proline plays a pivotal role in enzyme activation and protects chlorophyll from degradation under salt-stress conditions [47,48].

Additionally, relative water content (RWC) was decreased under salt stress. This decrease may be due to the reduction in water uptake [49] and/or its harmful effect on cell wall structure [50,51]. In contrast, RWC was significantly increased in stressed plants treated with SA, yeast, and proline. The ameliorative effects of these treatments on RWC could be due to the increase in osmoregulators, as well as to osmotic adjustment in plant cells [23,52,53].

Salt stress causes adverse effects on sweet pepper plants, including increased electrolyte leakage percentage. This increase may be due to the damaging effects on plasma membrane and selective permeability resulting in an increase in electrolyte leakage. This result is similar to that obtained in [54,55]. Conversely, the foliar application of SA, yeast, and proline led to decreased electrolyte leakage levels in all treatments. This beneficial effect could be due to the protective role of SA, yeast, and proline in plasma membrane stability and increasing soluble metabolite accumulation. A similar result was indicated by Ishikawa and Evans [56] and Huang et al. [57], who reported that osmoregulators improve plant growth and yield under various stress conditions.

Proline concentration was significantly increased in response to salt-stress conditions. This increment represents an important mechanism to minimize the deleterious impact of salinity stress and enhance plant growth [58]. The foliar application of SA, yeast, and proline under salt conditions may minimize the destructive effect of salinity on plant growth and improve proline accumulation. Similarly, SA application led to improved plant growth characteristics in maize plants under salt conditions [59]. Our results are in agreement with those of Huang et al. [60], Li et al. [61], and Gharsallah et al. [62]. Lipid peroxidation as MDA is an important factors indicating oxidative damage induced by salt stress. Lipid peroxidation was significantly boosted in salt-stressed (2000 and 4000 ppm) sweet pepper plants. Nonetheless, lipid peroxidation content was significantly decreased upon the foliar application of SA, yeast, and proline. These results may be attributed to the pivotal role of these treatments in decreasing oxidative stress damage, and consequently, in causing MDA reduction [25,48,53].

In the current study, superoxide and hydrogen peroxide, which are indicators of oxidative stress, were significantly produced in sweet pepper plants treated with NaCl at 2000 and 4000 ppm. This increase in superoxide and hydrogen peroxide production may be due to the fact that reactive oxygen species have a critical role under stress conditions in adjusting development, differentiation, redox levels, and stress signaling in the chloroplasts, mitochondria, and peroxisomes of plant cells [63,64]. Moreover, the high levels of hydrogen peroxide and superoxide are the main reasons for oxidative stress in the plant cells exposed to various stresses. Our results are supported by the findings of previous studies [65–67]. The application of SA, yeast, and proline on salt-stressed sweet pepper plants led to reductions in the formation of superoxide and hydrogen peroxide. This effect may be due to the role of these treatments in stabilizing protein structures and maintaining the redox states of plant cells, as well as stimulating antioxidant enzymes system [6,21,48]. Under salt stress (2000 and 4000 ppm), antioxidant enzyme activities were significantly increased in sweet pepper plants in order to combat the harmful impact of salt by adjusting osmotic balance. In agreement with our findings, similar results were noted in various plants under saline and drought conditions [68,69].

The activation of CAT and POX enzymes under salt conditions plays a key role in the improvement of plant defense systems. In the current study, the exogenous foliar application of SA, yeast, and proline led to improved antioxidant enzymes activity, as well as guarding the plant cells against oxidative stress and dehydration of the plasma membrane under salt-stress conditions. These results were supported by the findings reported in various plants [70–72].

The reductions of fruit numbers per plant, fruit fresh weight, and total fruit yield (ton hectare^{-1}) under salt conditions are possibly due to the adverse impacts of salinity on the growth characteristics and physiological processes such as water uptake, photosynthesis, flowering, and fruit formation, which led to diminished yields. Accordingly, the highest level of salt (4000 ppm) was adversely more effective than the lowest one (2000 ppm). The same trends of salt stress were previously described in faba bean [73] and strawberry plants [74]. Our results indicate that proline treatment was the best, followed by SA and yeast treatments. This useful effect of proline may be due to its pivotal role in osmotic regulation, enzyme activation, and protein synthesis, which consequently enhances the growth and yield characteristics of stressed plants [47,75,76]. Also, SA plays an essential role as a stress tolerance inducer via reducing the oxidative damage and enhancing plant productivity under salt stress. These results are in harmony with previous findings of Gupta and Huang [77], Ahanger et al. [78], and Husen et al. [79].

5. Conclusions

According to our findings, salt stress caused significant decreases in chlorophyll concentrations, relative water content, and fruit yields. However, lipid peroxidation, proline, electrolyte leakage, and reactive oxygen species were increased. Based on the results, the foliar application of salicylic acid (1 mM), yeast extract (6 g L^{-1}), and proline (10 mM) was an effective method by which to overcome the injurious effects of salt stress on sweet pepper plants. It may be concluded that relative water content and chlorophyll concentration, as well as antioxidant enzyme activity, were significantly modulated in the stressed treated sweet pepper plants. In contrast, electrolyte leakage and lipid peroxidation were decreased in treated sweet pepper plants under salt conditions. Thus, the application of salicylic acid, yeast, and proline led to a decrease in the harmful effects of salt stress by regulating osmolytes and antioxidants, which ultimately enhances the growth characteristics and fruit yields of sweet pepper plants. Interestingly, proline proved to be the best treatment for the further enhancement of plant performance and tolerance to salinity stress.

Author Contributions: K.A.A., L.M.E.-M., H.E., Y.M.H., E.I.I., M.E.-B., M.E.-E. and A.E. performed the experiments, analyzed the data, and wrote the manuscript. All authors have read and agreed to the published version of the manuscript.

Acknowledgments: The authors extend their appreciation to king Saud University funding through Researchers Supporting Project number (RSP-2019/118) and also to all members of PPBL and EPCRS excellence center, Fac. of Agric., Kafrelsheikh University.

References

1.   Majeed, A.; Muhammad, Z. Salinity: A Major Agricultural Problem—Causes, Impacts on Crop Productivity and Management Strategies. In *Plant Abiotic Stress Tolerance*; Hasanuzzaman, M., Hakeem, K.R., Nahar, K., Alharby, H.F., Eds.; Springer International Publishing: Cham, Switzerland, 2019; pp. 83–99, ISBN 978-3-030-06117-3.

2.   Soliman, M.H.; Alayafi, A.A.M.; El Kelish, A.A.; Abu-Elsaoud, A.M. Acetylsalicylic acid enhance tolerance of Phaseolus vulgaris L. to chilling stress, improving photosynthesis, antioxidants and expression of cold stress responsive genes. *Bot. Stud.* **2018**, *59*, 6. [CrossRef] [PubMed]

3.   Elkeilsh, A.; Awad, Y.M.; Soliman, M.H.; Abu-Elsaoud, A.; Abdelhamid, M.T.; El-Metwally, I.M. Exogenous application of β-sitosterol mediated growth and yield improvement in water-stressed wheat (Triticum aestivum) involves up-regulated antioxidant system. *J. Plant Res.* **2019**, *132*, 881–901. [CrossRef] [PubMed]

4.   Hasan, M.K.; El Sabagh, A.; Sikdar, M.S.; Alam, M.J.; Ratnasekera, D.; Barutcular, C.; Abdelaal, K.A.; Islam, M.S. Comparative adaptable agronomic traits of blackgram and mungbean for saline lands. *Plant Arch.* **2017**, *17*, 589–593.

5.   El-Esawi, M.A.; Alayafi, A.A. Overexpression of Rice *Rab7* Gene Improves Drought and Heat Tolerance and Increases Grain Yield in Rice (*Oryza sativa* L.). *Genes (Basel)* **2019**, *10*, 56. [CrossRef]

6.   Abdelaal, K.A. Effect of salicylic acid and abscisic acid on morpho-physiological and anatomical characters of faba bean plants (*Vicia faba* L.) under drought stress. *J. Plant Prod.* **2015**, *6*, 1771–1788. [CrossRef]

7.   Elkelish, A.A.; Alnusaire, T.S.; Soliman, M.H.; Gowayed, S.; Senousy, H.H.; Fahad, S. Calcium availability regulates antioxidant system, physio-biochemical activities and alleviates salinity stress mediated oxidative damage in soybean seedlings. *J. Appl. Bot. Food Qual.* **2019**, *92*, 258–266.

8.   Al Hassan, M.; Chaura, J.; Donat-Torres, M.P.; Boscaiu, M.; Vicente, O. Antioxidant responses under salinity and drought in three closely related wild monocots with different ecological optima. *AoB Plants* **2017**, *9*. [CrossRef]

9.   Al Mahmud, J.; Bhuyan, M.H.M.B.; Anee, T.I.; Nahar, K.; Fujita, M.; Hasanuzzaman, M. Reactive Oxygen Species Metabolism and Antioxidant Defense in Plants Under Metal/Metalloid Stress. In *Plant Abiotic Stress Tolerance*; Hasanuzzaman, M., Hakeem, K.R., Nahar, K., Alharby, H.F., Eds.; Springer International Publishing: Cham, Switzerland, 2019; pp. 221–257, ISBN 978-3-030-06117-3.

10. Elansary, H.O.; Szopa, A.; Kubica, P.; Ekiert, H.; Ali, H.M.; Elshikh, M.S.; Abdel-Salam, E.M.; El-Esawi, M.; El-Ansary, D.O. Bioactivities of traditional medicinal plants in Alexandria. *Evid.-Based. Complement. Altern. Med.* **2018**, *2018*. [CrossRef]

11. El-Hifny, I.M.; El-Sayed, M.A. Response of Sweet Pepper plant Growth and Productivity to Application of Ascorbic Acid and Biofertilizers under Saline Conditions. *Aust. J. Basic Appl. Sci.* **2011**, *5*, 1273–1283.

12. Hernández, J.A. Salinity Tolerance in Plants: Trends and Perspectives. *Int. J. Mol. Sci.* **2019**, *20*, 2408. [CrossRef]

13. Nguyen, H.M.; Sako, K.; Matsui, A.; Suzuki, Y.; Mostofa, M.G.; Ha, C.V.; Tanaka, M.; Tran, L.-S.P.; Habu, Y.; Seki, M. Ethanol Enhances High-Salinity Stress Tolerance by Detoxifying Reactive Oxygen Species in Arabidopsis thaliana and Rice. *Front. Plant Sci.* **2017**, *8*. [CrossRef] [PubMed]

14. Yoon, J.Y.; Hamayun, M.; Lee, S.-K.; Lee, I.-J. Methyl jasmonate alleviated salinity stress in soybean. *J. Crop. Sci. Biotechnol.* **2009**, *12*, 63–68. [CrossRef]

15. Savvides, A.; Ali, S.; Tester, M.; Fotopoulos, V. Chemical Priming of Plants Against Multiple Abiotic Stresses: Mission Possible? *Trends Plant Sci.* **2016**, *21*, 329–340. [CrossRef]

16. An, C.; Mou, Z. Salicylic Acid and its Function in Plant ImmunityF. *J. Integr. Plant Biol.* **2011**, *53*, 412–428. [CrossRef] [PubMed]

17. Rao, S.; Du, C.; Li, A.; Xia, X.; Yin, W.; Chen, J. Salicylic Acid Alleviated Salt Damage of Populus euphratica: A Physiological and Transcriptomic Analysis. *Forests* **2019**, *10*, 423. [CrossRef]

18. Brito, C.; Dinis, L.-T.; Moutinho-Pereira, J.; Correia, C.M. Drought Stress Effects and Olive Tree Acclimation under a Changing Climate. *Plants* **2019**, *8*, 232. [CrossRef]

19. Hernández-Ruiz, J.; Arnao, M. Relationship of Melatonin and Salicylic Acid in Biotic/Abiotic Plant Stress Responses. *Agronomy* **2018**, *8*, 33. [CrossRef]

20. Barnett, J.A.; Yarrow, D.; Payne, R.W.; Barnett, L. *Yeasts: Characteristics and Identification*, 3rd ed.; Cambridge University Press: Cambridge, UK; New York, NY, USA, 2000; ISBN 978-0-521-57396-2.

21. Abdelaal, K.A.; Hafez, Y.M.; El Sabagh, A.; Saneoka, H. Ameliorative effects of Abscisic acid and yeast on morpho-physiological and yield characteristics of maize plant (*Zea mays* L.) under water deficit conditions. *Fresenius Environ. Bull.* **2017**, *26*, 7372–7383.

22. Xi, Q.; Lai, W.; Cui, Y.; Wu, H.; Zhao, T. Effect of Yeast Extract on Seedling Growth Promotion and Soil Improvement in Afforestation in a Semiarid Chestnut Soil Area. *Forests* **2019**, *10*, 76. [CrossRef]

23. Kasim, W.; AboKassem, E.; Ragab, G. Ameliorative effect of Yeast Extract, IAA and Green-synthesized Nano Zinc Oxide on the Growth of Cu-stressed *Vicia faba* Seedlings. *Egypt. J. Bot.* **2017**, *57*, 1–16. [CrossRef]

24. Rouphael, Y.; De Micco, V.; Arena, C.; Raimondi, G.; Colla, G.; Pascale, S. Effect of *Ecklonia maxima* seaweed extract on yield, mineral composition, gas exchange, and leaf anatomy of zucchini squash grown under saline conditions. *J. Appl. Phycol.* **2017**, *29*, 459–470. [CrossRef]

25. Saleh, A.A.H.; Abu-Elsaoud, A.M.; Elkelish, A.A.; Sahadad, M.A.; Abdelrazek, E.M. Role of External Proline on Enhancing Defence Mechanisms of Vicia Faba L. Against Ultraviolet Radiation. *Am.-Eurasian J. Sustain. Agric.* **2015**, *9*, 13.

26. Ali, Q.; Anwar, F.; Ashraf, M.; Saari, N.; Perveen, R. Ameliorating effects of exogenously applied proline on seed composition, seed oil quality and oil antioxidant activity of maize (*Zea mays* L.) under drought stress. *Int. J. Mol. Sci.* **2013**, *14*, 818–835. [CrossRef] [PubMed]

27. Hasanuzzaman, M.; Alam, M.M.; Rahman, A.; Hasanuzzaman, M.; Nahar, K.; Fujita, M. Exogenous Proline and Glycine Betaine Mediated Upregulation of Antioxidant Defense and Glyoxalase Systems Provides Better Protection against Salt-Induced Oxidative Stress in Two Rice (*Oryza sativa* L.) Varieties. *BioMed Res. Int.* **2014**, *2014*. [CrossRef]

28. El-Amier, Y.; Elhindi, K.; El-Hendawy, S.; Al-Rashed, S.; Abd-ElGawad, A. Antioxidant System and Biomolecules Alteration in Pisum sativum under Heavy Metal Stress and Possible Alleviation by 5-Aminolevulinic Acid. *Molecules* **2019**, *24*, 4194. [CrossRef] [PubMed]

29. Qadeer, U.; Ahmed, M.; Hassan, F.; Akmal, M. Impact of Nitrogen Addition on Physiological, Crop Total Nitrogen, Efficiencies and Agronomic Traits of the Wheat Crop under Rainfed Conditions. *Sustainability* **2019**, *11*, 6486. [CrossRef]

30. Kaundun, S.S.; Jackson, L.V.; Hutchings, S.-J.; Galloway, J.; Marchegiani, E.; Howell, A.; Carlin, R.; Mcindoe, E.; Tuesca, D.; Moreno, R. Evolution of Target-Site Resistance to Glyphosate in an Amaranthus palmeri Population from Argentina and Its Expression at Different Plant Growth Temperatures. *Plants* **2019**, *8*, 512. [CrossRef]

31. Abdelaal, K.A. Pivotal Role of Bio and Mineral Fertilizer Combinations on Morphological, Anatomical and Yield Characters of Sugar Beet Plant (*Beta vulgaris* L.). *Middle East. J. Agric.* **2015**, *4*, 717–734.

32. Moran, R. Formulae for Determination of Chlorophyllous Pigments Extracted with N,N-Dimethylformamide 1. *Plant Physiol.* **1982**, *69*, 1376–1381. [CrossRef]

33. Sánchez, F.J.; de Andrés, E.F.; Tenorio, J.L.; Ayerbe, L. Growth of epicotyls, turgor maintenance and osmotic adjustment in pea plants (*Pisum sativum* L.) subjected to water stress. *Field Crop. Res.* **2004**, *86*, 81–90. [CrossRef]

34. Dionisio-Sese, M.L.; Tobita, S. Antioxidant responses of rice seedlings to salinity stress. *Plant Sci.* **1998**, *135*, 1–9. [CrossRef]

35. Bates, L.S.; Waldren, R.P.; Teare, I.D. Rapid determination of free proline for water-stress studies. *Plant Soil* **1973**, *39*, 205–207. [CrossRef]

36. Heath, R.L.; Packer, L. Photoperoxidation in isolated chloroplasts. I. Kinetics and stoichiometry of fatty acid peroxidation. *Arch. Biochem. Biophys.* **1968**, *125*, 189–198. [CrossRef]

37. Badiani, M.; De Biasi, M.G.; Colognola, M.; Artemi, F. Catalase, peroxidase and superoxide dismutase activities in seedlings submitted to increasing water deficit. *Agrochimica* **1990**, *34*, 90–102.

38. Aebi, H. Catalase in vitro. In *Methods in Enzymology*; Oxygen Radicals in Biological Systems; Academic Press: New York, NY, USA, 1984; Volume 105, pp. 121–126.

39. Hammerschmidt, R.; Nuckles, E.M.; Kuć, J. Association of enhanced peroxidase activity with induced systemic resistance of cucumber to Colletotrichum lagenarium. *Physiol. Plant Pathol.* **1982**, *20*, 73–82. [CrossRef]

40. El-Esawi, M.A.; Alaraidh, I.A.; Alsahli, A.A.; Alamri, S.A.; Ali, H.M.; Alayafi, A.A. *Bacillus firmus* (SW5) augments salt tolerance in soybean (*Glycine max* L.) by modulating root system architecture, antioxidant defense systems and stress-responsive genes expression. *Plant Physiol. Biochem.* **2018**, *132*, 375–384. [CrossRef]

41. El-Esawi, M.A.; Al-Ghamdi, A.A.; Ali, H.M.; Alayafi, A.A. *Azospirillum lipoferum* FK1 confers improved salt tolerance in chickpea (*Cicer arietinum* L.) by modulating osmolytes, antioxidant machinery and stress-related genes expression. *Environ. Exp. Bot.* **2019**, *159*, 55–65. [CrossRef]

42. El-Esawi, M.A.; Al-Ghamdi, A.A.; Ali, H.M.; Alayafi, A.A.; Witczak, J.; Ahmad, M. Analysis of genetic variation and enhancement of salt tolerance in French pea. *Int. J. Mol. Sci.* **2018**, *19*, 2433. [CrossRef]

43. Suo, J.; Zhao, Q.; David, L.; Chen, S.; Dai, S. Salinity Response in Chloroplasts: Insights from Gene Characterization. *IJMS* **2017**, *18*, 1011. [CrossRef]

44. Yang, X.; Li, Y.; Qi, M.; Liu, Y.; Li, T. Targeted Control of Chloroplast Quality to Improve Plant Acclimation: From Protein Import to Degradation. *Front. Plant Sci.* **2019**, *10*, 958. [CrossRef] [PubMed]

45. Shah, S.; Houborg, R.; McCabe, M. Response of Chlorophyll, Carotenoid and SPAD-502 Measurement to Salinity and Nutrient Stress in Wheat (*Triticum aestivum* L.). *Agronomy* **2017**, *7*, 61.

46. Bulgari, R.; Franzoni, G.; Ferrante, A. Biostimulants Application in Horticultural Crops under Abiotic Stress Conditions. *Agronomy* **2019**, *9*, 306. [CrossRef]

47. Hayat, S.; Hayat, Q.; Alyemeni, M.N.; Wani, A.S.; Pichtel, J.; Ahmad, A. Role of proline under changing environments. *Plant Signal. Behav.* **2012**, *7*, 1456–1466. [CrossRef] [PubMed]

48. Dawood, M.G.; Taie, H.A.A.; Nassar, R.M.A.; Abdelhamid, M.T.; Schmidhalter, U. The changes induced in the physiological, biochemical and anatomical characteristics of *Vicia faba* by the exogenous application of proline under seawater stress. *S. Afr. J. Bot.* **2014**, *93*, 54–63. [CrossRef]

49. Parvin, K.; Hasanuzzaman, M.; Bhuyan, M.H.M.B.; Nahar, K.; Mohsin, S.M.; Fujita, M. Comparative Physiological and Biochemical Changes in Tomato (*Solanum lycopersicum* L.) under Salt Stress and Recovery: Role of Antioxidant Defense and Glyoxalase Systems. *Antioxidants* **2019**, *8*, 350. [CrossRef]

50. Acosta-Motos, J.; Ortu?o, M.; Bernal-Vicente, A.; Diaz-Vivancos, P.; Sanchez-Blanco, M.; Hernandez, J. Plant Responses to Salt Stress: Adaptive Mechanisms. *Agronomy* **2017**, *7*, 18. [CrossRef]

51. Abdelaal, K.A.A.; Hafez, Y.M.; El-Afry, M.M.; Tantawy, D.S.; Alshaal, T. Effect of some osmoregulators on photosynthesis, lipid peroxidation, antioxidative capacity, and productivity of barley (Hordeum vulgare L.) under water deficit stress. *Environ. Sci. Pollut. Res.* **2018**, *25*, 30199–30211. [CrossRef]

52. Gholami Zali, A.; Ehsanzadeh, P. Exogenous proline improves osmoregulation, physiological functions, essential oil, and seed yield of fennel. *Ind. Crop. Prod.* **2018**, *111*, 133–140. [CrossRef]

53. Hafez, E.; Omara, A.E.D.; Ahmed, A. The Coupling Effects of Plant Growth Promoting Rhizobacteria and Salicylic Acid on Physiological Modifications, Yield Traits, and Productivity of Wheat under Water Deficient Conditions. *Agronomy* **2019**, *9*, 524. [CrossRef]

54. El-Esawi, M.A.; Alaraidh, I.A.; Alsahli, A.A.; Ali, H.M.; Alayafi, A.A.; Witczak, J.; Ahmad, M. Genetic Variation and Alleviation of Salinity Stress in Barley (*Hordeum vulgare* L.). *Molecules* **2018**, *23*, 2488. [CrossRef] [PubMed]

55. El-Esawi, M.A.; Alaraidh, I.A.; Alsahli, A.A.; Alzahrani, S.M.; Ali, H.M.; Alayafi, A.A.; Ahmad, M. *Serratia liquefaciens* KM4 Improves Salt Stress Tolerance in Maize by Regulating Redox Potential, Ion Homeostasis, Leaf Gas Exchange and Stress-Related Gene Expression. *Int. J. Mol. Sci.* **2018**, *19*, 3310. [CrossRef] [PubMed]

56. Ishikawa, H.; Evans, M.L. Electrotropism of Maize Roots: Role of the Root Cap and Relationship to Gravitropism. *Plant Physiol.* **1990**, *94*, 913–918. [CrossRef] [PubMed]

57. Huang, D.; Sun, Y.; Ma, Z.; Ke, M.; Cui, Y.; Chen, Z.; Chen, C.; Ji, C.; Tran, T.M.; Yang, L.; et al. Salicylic acid-mediated plasmodesmal closure via Remorin-dependent lipid organization. *Proc. Natl. Acad. Sci. USA* **2019**, *116*, 21274–21284. [CrossRef]

58. Verbruggen, N.; Hermans, C. Proline accumulation in plants: A review. *Amino Acids* **2008**, *35*, 753–759. [CrossRef]

59. El-Katony, T.M.; El-Bastawisy, Z.M.; El-Ghareeb, S.S. Timing of salicylic acid application affects the response of maize (*Zea mays* L.) hybrids to salinity stress. *Heliyon* **2019**, *5*, e01547. [CrossRef]

60. Huang, Z.; Zhao, L.; Chen, D.; Liang, M.; Liu, Z.; Shao, H.; Long, X. Salt Stress Encourages Proline Accumulation by Regulating Proline Biosynthesis and Degradation in Jerusalem Artichoke Plantlets. *PLoS ONE* **2013**, *8*, e62085. [CrossRef]

61. Li, T.; Hu, Y.; Du, X.; Tang, H.; Shen, C.; Wu, J. Salicylic acid alleviates the adverse effects of salt stress in Torreya grandis cv. Merrillii seedlings by activating photosynthesis and enhancing antioxidant systems. *PLoS ONE* **2014**, *9*, e109492. [CrossRef]

62. Gharsallah, C.; Fakhfakh, H.; Grubb, D.; Gorsane, F. Effect of salt stress on ion concentration, proline content, antioxidant enzyme activities and gene expression in tomato cultivars. *AoB Plants* **2016**, *8*, plw055. [CrossRef]

63. Wang, Y.; Li, X.; Li, J.; Bao, Q.; Zhang, F.; Tulaxi, G.; Wang, Z. Salt-induced hydrogen peroxide is involved in modulation of antioxidant enzymes in cotton. *Crop J.* **2016**, *4*, 490–498. [CrossRef]

64. El-Esawi, M.A.; Elkelish, A.; Elansary, H.O.; Ali, H.M.; Elshikh, M.; Witczak, J.; Ahmad, M. Genetic Transformation and Hairy Root Induction Enhance the Antioxidant Potential of *Lactuca serriola* L. *Oxid. Med. Cell. Longev.* **2017**, *2017*. [CrossRef] [PubMed]

65. Lin, C.C.; Kao, C.H. Effect of NaCl stress on H_2O_2 metabolism in rice leaves. *Plant Growth Regul.* **2000**, *30*, 151–155. [CrossRef]

66. Hernandez, M.; Fernandez-Garcia, N.; Diaz-Vivancos, P.; Olmos, E. A different role for hydrogen peroxide and the antioxidative system under short and long salt stress in *Brassica oleracea* roots. *J. Exp. Bot.* **2010**, *61*, 521–535. [CrossRef] [PubMed]

67. Li, Q.; Lv, L.R.; Teng, Y.J.; Si, L.B.; Ma, T.; Yang, Y.L. Apoplastic hydrogen peroxide and superoxide anion exhibited different regulatory functions in salt-induced oxidative stress in wheat leaves. *Biol. Plant.* **2018**, *62*, 750–762. [CrossRef]

68. Vighi, I.L.; Benitez, L.C.; Amaral, M.N.; Moraes, G.P.; Auler, P.A.; Rodrigues, G.S.; Deuner, S.; Maia, L.C.; Braga, E.J.B. Functional characterization of the antioxidant enzymes in rice plants exposed to salinity stress. *Biol. Plant.* **2017**, *61*, 540–550. [CrossRef]

69. Pérez-Labrada, F.; López-Vargas, E.R.; Ortega-Ortiz, H.; Cadenas-Pliego, G.; Benavides-Mendoza, A.; Juárez-Maldonado, A. Responses of Tomato Plants under Saline Stress to Foliar Application of Copper Nanoparticles. *Plants* **2019**, *8*, 151. [CrossRef]

70. El-Esawi, M.A.; Elansary, H.O.; El-Shanhorey, N.A.; Abdel-Hamid, A.M.E.; Ali, H.M.; Elshikh, M.S. Salicylic Acid-Regulated Antioxidant Mechanisms and Gene Expression Enhance Rosemary Performance under Saline Conditions. *Front. Physiol.* **2017**, *8*, 716. [CrossRef]

71. Đorđević, N.O.; Todorović, N.; Novaković, I.T.; Pezo, L.L.; Pejin, B.; Maraš, V.; Tešević, V.V.; Pajović, S.B. Antioxidant Activity of Selected Polyphenolics in Yeast Cells: The Case Study of Montenegrin Merlot Wine. *Molecules* **2018**, *23*, 1971. [CrossRef]

72. Mohammadrezakhani, S.; Hajilou, J.; Rezanejad, F.; Zaare-Nahandi, F. Assessment of exogenous application of proline on antioxidant compounds in three Citrus species under low temperature stress. *J. Plant Inter.* **2019**, *14*, 347–358. [CrossRef]

73. Abdul Qados, A.M.S. Effect of salt stress on plant growth and metabolism of bean plant (*Vicia faba* L.). *J. Saudi Soc. Agric. Sci.* **2011**, *10*, 7–15. [CrossRef]

74. Yildirim, E.; Karlidag, H.; Turan, M. Mitigation of salt stress in strawberry by foliar K, Ca and Mg nutrient supply. *Plant Soil Environ.* **2009**, *55*, 213–221. [CrossRef]

75. Huang, Y.; Bie, Z.; Liu, Z.; Zhen, A.; Wang, W. Protective role of proline against salt stress is partially related to the improvement of water status and peroxidase enzyme activity in cucumber. *Soil Sci. Plant Nutr.* **2009**, *55*, 698–704. [CrossRef]

76. Sharma, A.; Shahzad, B.; Kumar, V.; Kohli, S.K.; Sidhu, G.P.S.; Bali, A.S.; Handa, N.; Kapoor, D.; Bhardwaj, R.; Zheng, B. Phytohormones Regulate Accumulation of Osmolytes Under Abiotic Stress. *Biomolecules* **2019**, *9*, 285. [CrossRef] [PubMed]

77. Gupta, B.; Huang, B. Mechanism of Salinity Tolerance in Plants: Physiological, Biochemical, and Molecular Characterization. *Int. J. Genomics* **2014**, *2014*. [CrossRef] [PubMed]

78. Ahanger, M.A.; Tomar, N.S.; Tittal, M.; Argal, S.; Agarwal, R.M. Plant growth under water/salt stress: ROS production; antioxidants and significance of added potassium under such conditions. *Physiol. Mol. Biol. Plant* **2017**, *23*, 731–744. [CrossRef]

79. Husen, A.; Iqbal, M.; Sohrab, S.S.; Ansari, M.K.A. Salicylic acid alleviates salinity-caused damage to foliar functions, plant growth and antioxidant system in Ethiopian mustard (*Brassica carinata* A. Br.). *Agric. Food Secur.* **2018**, *7*, 44. [CrossRef]

13

Changes in Root Anatomy of Peanut (*Arachis hypogaea* L.) under Different Durations of Early Season Drought

Nuengsap Thangthong [1,2], **Sanun Jogloy** [1,2,*], **Tasanai Punjansing** [3], **Craig K. Kvien** [4], **Thawan Kesmala** [1,2] **and Nimitr Vorasoot** [1,2]

[1]   Department of Agronomy, Faculty of Agriculture, Khon Kaen University, Khon Kaen 40002, Thailand; nuengsap.th@gmail.com (N.T.); thkesmala@gmail.com (T.K.); nvorasoot@gmail.com (N.V.)

[2]   Peanut and Jerusalem Artichoke Improvement for Functional Food Research Group, Khon Kaen University, Khon Kaen 40002, Thailand

[3]   Department of Biology, Faculty of Science, Udonthani Rajaphat University, Udonthani 41000, Thailand; tasanaipun@gmail.com

[4]   Crop & Soil Sciences, National Environmentally Sound Production Agriculture Laboratory (NESPAL), The University of Georgia, Tifton, GA 31793, USA; ckvien@uga.edu

*   Correspondence: sjogloy@gmail.com

Abstract: Changes in the anatomical structure of peanut roots due to early season drought will likely affect the water acquiring capacity of the root system. Yet, as important as these changes are likely to be in conferring drought resistance, they have not been thoroughly investigated. The objective of this study was to investigate the effects of different durations of drought on the root anatomy of peanut in response to early season drought. Plants of peanut genotype ICGV 98305 were grown in rhizoboxes with an internal dimension of 50 cm in width, 10 cm in thickness and 120 cm in height. Fourteen days after emergence, water was withheld for periods of 0, 7, 14 or 21 days. After these drought periods, the first and second order roots from 0–20 cm below soil surface were sampled for anatomical observation. The mean xylem vessel diameter of first- order lateral roots was higher than that of second- order lateral roots. Under early season drought stress root anatomy changes were more pronounced in the longer drought period treatments. Twenty-one days after imposing water stress, the drought treatment and irrigated treatment were clearly different in diameter, number and area of xylem vessels of first- and second-order lateral roots. Plants under drought conditions had a smaller diameter and area of xylem vessels than did the plants under irrigated control. The ability of plants to change root anatomy likely improves water uptake and transport and this may be an important mechanism for drought tolerance. The information will be useful for the selection of drought durations for evaluation of root anatomy related to drought resistance and the selection of key traits for drought resistance.

Keywords: xylem vessel; water stress; root anatomy

1. Introduction

In many areas of the tropics, peanut production is mostly in rain-fed and semi-arid areas with low and unpredictable rainfall and rain distribution. In these areas, drought stress can occur at any growth stage, resulting in yield loss of 22–53% [1]. Drought stress also increases *Aspergillus flavus* infection and aflatoxin contamination by 2–17% [2]. However, drought stress at a pre-flowering growth stage sometimes actually increases yield [3]. Irrigation, planting date selection and drought resistant varieties can improve yield and reduce aflatoxin contamination during periods of drought. However, management of irrigation requires an available water source and investment in additional equipment.

Planting date selection, while less expensive than irrigation is not as effective because rainfall and rain distribution are often unpredictable. The use of drought resistant varieties is a promising and sustainable choice in need of further development. When selecting for drought resistance in peanut, yield and biomass during drought are often used as selection criteria. Yet this selection method is complicated by high genotype by environment interaction. Many physiological and morphological traits have been suggested as surrogate traits for drought resistance to increase selection efficiency, yet measurement for these traits are often quite variable.

Root traits are known to improve drought resistance [4] and are therefore important for plant breeding programs. Improving the water acquiring capacity of crops to extract water from the soil profile during drought is one example. Root traits such as large root systems (root dry weight), root length density and the percentage of root length density that respond to drought have been investigated in peanut [1,3,5–7].

Anatomical parameters, such as xylem vessel number and diameter, have been positively correlated with dry matter production under stress in chili (*Capsicum annum* L.) [8]. Drought resistant varieties of several plant species have been reported to have a higher number of vessel cells and a larger xylem cross-section than susceptible varieties of chili (*Capsicum annum* L.) [8], tomato (*Lycopersicon esculentum*) [9] and grape (*Vitis vinifera* L.) [10]. As in the above studies of other plants, it is likely that studies on the fine root structure of peanut, especially under drought stress, will lead to a better understanding of why some peanut genotypes yield better during a drought than others.

Cell-wall ingrowths or phi-thickening have been reported in loquat (*Eriobotrya japonica* Lindl.) root [11], apple (*Pyrus malus*) [12], geranium (*Pelargonium hortorum*) roots [12] Sibipiruna (*Caesalpinia peltophoroides*) [13] with solute movement (salt stress) [12], water logging [13], and drought stress [11]. Although the effect of early season drought on ingrowths and phi-thickenings has not been investigated in peanut and further investigations are necessary to understand phi-thickenings. The response of phi-thickening might be related to the transport processes in the peanut root.

Root anatomy is interesting, and it might play an important role in plant response to drought. The types of lateral roots during root growth were recognized in peanut [14]. The different types and different structures may be related to different functions. The structure of the first order lateral roots helps determine the efficacy of the axial water transport system, yet the structure within the second order lateral roots helps determine the efficacy of the water uptake process. Unfortunately, this useful information has not been thoroughly investigated in peanut. The objective of this study was to investigate the effects of different durations of drought imposition on the root anatomy of peanut in response to early season drought. The information will be useful for selection of drought durations for evaluation of root anatomy related to drought resistance and selection of key traits for drought resistance.

2. Materials and Methods

2.1. Experimental Design and Plant Material

The experiment was conducted under a rainout shelter at the Field Crop Research Station of Khon Kaen University, Khon Kaen, Thailand (16°28′ N, 102°48′ E, 200 m. above sea level). The peanut genotype—ICGV 98305—was subjected to four water treatments (0, 7, 14 or 21 days without irrigation), each beginning 14 days after plant emergence (DAE). The experiment used a completely randomized design with three replications and was conducted for two seasons during July–September 2013 and March–May 2014.

ICGV 98305 is a drought resistant line from ICRISAT known for high root length density in the deep sub-soil during periods of drought [1,15].

2.2. Preparation and Irrigation of Rhizobox Experiment

The plants were grown in rhizoboxes with internal dimensions of 50 cm in width, 10 cm in thickness and 120 cm in height (Figure 1a). The rhizoboxes were filled with dry soil to obtain bulk density of 1.57 Mg m^{-3} and height of 115 cm, and water was added to achieve field capacity. Peanut seeds were planted in the center of rhizobox, 5 cm below the soil surface. At 3 days after emergence (DAE), the seedlings were thinned to obtain 1 plant per rhizobox. The front side of the rhizoboxes was transparent and covered with black sheet, and all sides of rhizoboxes were then covered with aluminum foil to reduce light absorption and temperature increase (Figure 1b).

The root needle-board method [16] was used for the observation of root growth and distribution with a minor modification for size and spacing of needles. The root system of the plant in the box was held in place by needles attached to back board of rhizoboxes and projecting out to the transparent front. The needle spacing was 5 × 5 cm^2. The needle columns started 2.5 cm from left and right margins and the needle rows were started at 12.5 cm from the top of rhizobox and continued at 5 cm intervals to the bottom of the box (Figure 1c).

Soil moisture contents for field capacity and permanent wilting point were determined to be 11.13% and 3.40%, respectively. Water was supplied to the rhizoboxes through horizontal tubes which were installed at 5, 15, 35, 55, 75, 95 and 115 cm below the soil surface. For each rhizobox, water was first supplied at field capacity and all three drought treatments (7, 14 and 21 days without added water) began 14 DAE. The fourth treatment was kept at field capacity for the entire experimental period. The field capacity was maintained uniformly throughout the soil profile by using the six watering tubes. Drainage holes, 1.5 cm in diameter, were placed at the bottom of the rhizobox. Drained water was replenished at the same amount. Crop evapotranspiration was calculated as the sum of water lost through plant transpiration and soil evaporation, as described by Reference [17];

$$ETcrop = ETo \times Kc \tag{1}$$

where ETcrop is crop water requirement (mm/day), ETo is evapotranspiration of a reference under specified conditions calculated using the pan evaporation method, and Kc is the peanut water requirement coefficient.

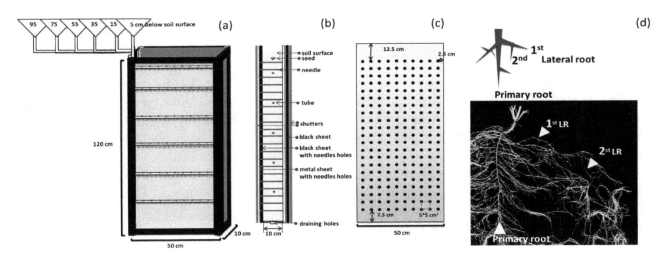

Figure 1. Diagrammatic representation and dimension of rhizobox with six tubes of irrigation (**a**), cross section showing the different elements of the system (**b**), spacing of needle at backside of rhizoboxes (**c**) and taproot system of a rhizobox-grown peanut (**d**).

2.3. Crop Management

Phosphorus as triple superphosphate (Ca(H$_2$PO$_4$)$_2$H$_2$O) (Chia tai company limited, Phranakhonsiayutthaya, Thailand) at the rate of 122.3 kg ha^{-1} and potassium as potassium chloride (KCl;

60% K_2O) (Chia tai company limited, Phranakhonsiayutthaya, Thailand) at the rate of 62.5 kg ha^{-1} were applied to the soil before planting. A water-diluted commercial peat-based inoculum of *Bradyrhizobium* (mixture of strains THA 201 and THA 205; Department of Agriculture, Ministry of Agriculture and Cooperatives, Bangkok, Thailand) was applied 5 cm below the soil surface through the irrigation tubes. Seeds were treated with captan (3a,4,7,7a-tetrahydro-2-[(trichloromethyl) thio]-1H-isoindole-1, 3(2H)-dione, Erawan Agricultural Chemical Co., Ltd., Bangkok, Thailand.) at the rate of 5 g kg^{-1} seeds before planting. Carbosulfan [2-3-dihydro-2,2-dimethylbenzofuran-7-yl (dibutylaminothio) methylcar-bamate 20% (*w/v*) water soluble concentrate] (FMC AG Ltd., Bangkok, Thailand) at 2.5 L ha^{-1} was applied weekly to control thrips, and methomyl [S-methyl-N-((methylcarbamoyl)oxy) thioacetimidateand methomyl [(E,Z)-methyl N-{[(methylamino) carbonyl]oxy}ethanimidothioate] 40% soluble powder (Du Pont Co., Ltd., Bangkok, Thailand) at 1.0 kg ha^{-1} was used to control mites. Weeds were controlled by hand weeding.

2.4. Data Collection

Rainfall, relative humidity, pan evaporation, maximum and minimum temperature and solar radiation were recorded daily from planting to 35 DAE at a weather station located 50 m from the experiment. Soil physical and chemical properties were analyzed before planting. Soil samples for analysis were taken from the mixed pile of soil used for this experiment. The soil's physical properties in the experiment were analyzed for percentage sand, silt and clay. The soil chemical properties were analyzed for pH, organic matter, total N, available P, exchangeable K and exchangeable Ca.

2.5. Soil Moisture Content

Soil moisture content was determined gravimetrically using a micro auger method at 10, 25, 65, and 85 cm soil depths at 14, 21, 28 and 35 DAE. Soil moisture content for each rhizobox was calculated as;

$$\text{Soil moisture content (\%)} = ((\text{wet weight} - \text{dry weight})/\text{dry weight}) \times 100 \qquad (2)$$

2.6. Observation of Root Anatomy

Roots were collected at 7, 14 and 21 days after water withholding began. At the sampling date, the shoot in each box was cut at the soil surface and the roots were carefully washed with a fine spray of tap water to remove soil. Rhizobox needles helped roots maintained the approximate position they were in the soil profile.

Root samples for anatomical observation were taken from 0–20 cm below soil surface. The first-, and second-order lateral roots (Figure 1d) were taken at approximately 5 cm from the root tips from each treatment. The root sampling strategy (5 cm from the tip, and 20 cm deep) was as suggested from a previous rhizotron study [18] in which peanut root growth rates of drought and well-watered treatments were 12.6 and 21.9 cm per week, respectively Therefore, we took the root samples for anatomical study at 5 cm from the root tips, as roots at this position would be expected to be significantly affected by drought. The samples were fixed in a formaldehyde (Sigma-Aldrich; Bangkok, Thailand, 36.5–38% in H_2O)-glacial acetic acid (Fisher Chemical)-40% ethanol-solution (FAA$_{40}$). Dehydration of the samples was accomplished by adding a series of alcohol concentrations at 10% intervals from 40% to 70%. Free-hand cross sections were stained with Safranin O (Dye content ≥ 85%; Sigma-Aldrich). Anatomical characteristics of the root samples were observed using a Nikon eclipse 50i optical microscope with ocular and stage micrometers. The microscope's digital camera (Nikon DS-Fi1, Shingawa-ku, Tokyo, Japan) was used for photographs. All transverse sections of roots were measured and recorded for diameter and area of the xylem vessels of first-order and second-order lateral roots. Xylem vessel elements consisted of protoxylem and metaxylem. Although the identification of these xylem tissues was difficult, we were able to classify them into two groups by diameter. Smaller xylem vessels were equal to or smaller than the overall mean diameter of xylem vessels and bigger xylem vessels were larger than the mean diameter of xylem vessels. The cell-wall ingrowths were compared in both the drought and well-watered treatments using the cortical layers of both first-order and second-order lateral roots.

2.7. Data Analysis

The statistical analysis was performed using the statistix-8 program as a completely randomized design. An analysis of variance and least significance difference (LSD) tests were used to compare differences at $p \leq 0.05$.

3. Results

3.1. Meteorological Data and Soil Data

The meteorological details for the two years were collected (data not shown) and are described in Field Crops Research (2016) [19]. Daily air temperatures ranged from 22.7 to 36.8 °C in 2013 and 20.2 to 40.5 °C in 2014. Relative humidity (RH) values ranged between 63–88% in 2013 and 47–87% in 2014. The means of evaporation (E0) were 4.5 mm in 2013 and 5.7 mm in 2014. While rain did not directly fall on the experimental plants, as it was conducted in a rainout shelter; it did affect relative humidity and evapotranspiration.

Differences between years were observed for maximum temperature (T-max) and minimum temperature (T-min) as the trial in 2013 was conducted during the cooler rainy season (May–July) than the 2014 trial conducted from March–May.

3.2. Soil Moisture Content and Relative Water Content

Soil moisture content and relative water content are described in Reference [19]. Soil moisture content measured at field capacity was 11.13% and permanent wilting point was 3.40%. Soil moisture content for non-stress conditions was similar to those at field capacity. However, soil moisture content at field capacity (FC) in the lower soil layers was slightly higher than 11.13% at the initiation of drought stress. Drought and well-irrigated treatments were clearly different at all sampling dates, especially at top soil layers of 10 and 25 cm. The differences between drought and well-irrigated treatments were small in lower soil layers and the treatments became similar at 65 and 85 cm except at 28 and 35 DAE in 2014.

3.3. Observation of Root Anatomy

Peanut has a dicotyledonous root system with a single taproot and branched first-, second-, and higher order lateral roots (Figure 1d). In this study, the anatomy of first- and second- order lateral roots was observed.

3.3.1. First order Lateral Root

Combined analysis of variance for total vessel numbers, bigger vessel numbers, smaller vessel number, total vessel diameter (μm), bigger vessel diameter (μm), smaller vessel diameter (μm), total vessel area (μm²), bigger vessel area (μm²), smaller vessel area (μm²) of the first order lateral root in 2013 and 2014 are shown in Table 1. Significant differences in total vessel numbers, bigger vessel numbers, total vessel area and bigger vessel area were observed in different durations and seasons. The interactions between duration and treatment (D × T) were also significant for total vessel numbers and smaller vessel area traits.

Central cylinders of first order lateral roots had an almost triarch arrangement of the vascular bundles (Figures 2 and 3). Within these bundles, the xylem vessels showed a wide range in size. For ease of discussion, we classified the vessels into two groups (large and small) based on their diameter. Large vessels had a diameter greater than the mean (16.06 μm) of all vessels, and small vessels had a diameter less than the mean.

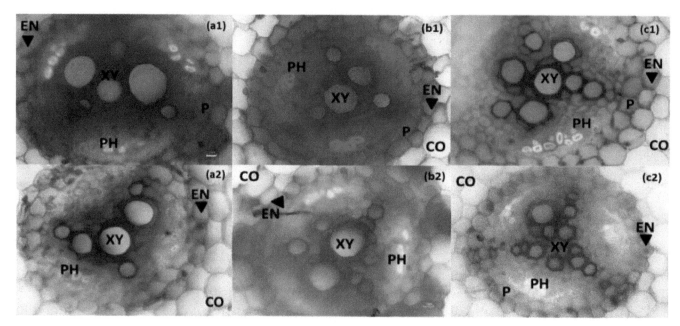

Figure 2. Freehand cross sections of first order lateral roots of peanut under well-irrigated conditions (**a1, b1** and **c1**) and drought stress conditions (**a2, b2** and **c2**) at 21, 28 and 35 DAE, respectively. CO, cortex; EN, endodermis; P, pericycle; PH, phloem; XY, xylem; Scale bar = 10 μm; 40×.

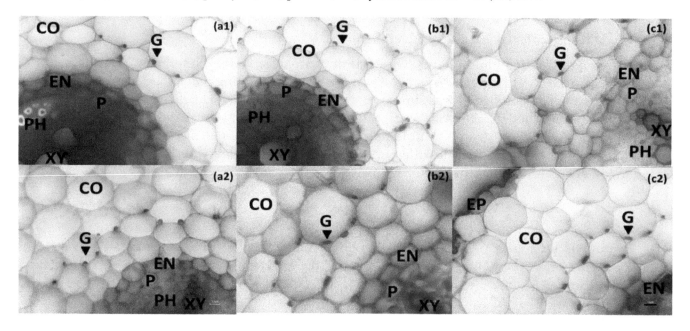

Figure 3. Freehand cross sections of first-order lateral roots under well-irrigated conditions (**a1, b1** and **c1**) and drought stress conditions (**a2, b2** and **c2**) at 21, 28 and 35 days after plant emergence (DAE). CO, cortex; EN, endodermis; G, phi-thickening or cell wall ingrowth; P, pericycle; PH, phloem; XY, xylem; Scale bar = 10 μm; 40×.

Total xylem numbers per cross-section of first order lateral roots (Figure 4) in the first and second seasons were not significantly different between drought and well-irrigated treatments at 21, 28 and 35 DAE with one exception at 35 DAE in 2014. At 35 DAE in 2014, the drought treatments had higher vessel numbers, in the small diameter vessels, than did well-irrigated treatments. At 35 DAE in 2013, stress and non-stress treatments were not significantly different for the total number of vessels, yet, like in 2014, stress tended to reduce the number of bigger vessels and increase the number of smaller vessel.

Table 1. Mean square from the combined analysis of variance for total vessel numbers, bigger vessel numbers, smaller vessel numbers, total vessel diameter (μm), bigger vessel diameter (μm), smaller vessel diameter (μm), total vessel area (μm²), bigger vessel area (μm²), smaller vessel area (μm²) of the first order lateral root in 2013 and 2014.

Source	DF	Total Vessel Numbers	Bigger Vessel Numbers	Smaller Vessel Numbers	Total Vessel Diameter (μm)	Bigger Vessel Diameter (μm)	Smaller Vessel Diameter (μm)	Total Vessel Area (μm²)	Bigger Vessel Area (μm²)	Smaller Vessel Area (μm²)
Duration (D)	2	32.028 **	6.19 *	6.91 ns	14.87 ns	8.81 ns	2.04 ns	6,586,518 **	5185627 **	39262 ns
Season (S)	1	42.25 **	11.11 **	1.01 ns	0.79 ns	5.52 ns	0.29 ns	2,160,885 ns	1,053,634 ns	214,114 *
Treatment (T)	1	0.03 ns	4.00 ns	0.01 ns	3.16 ns	30.24 ns	0.07 ns	2,012,582 ns	2,117,714 ns	110,969 ns
D × S	2	3.25 ns	4.30 *	33.47 ns	6.97 ns	43.05 ns	1.32 ns	1,576,123 ns	1,723,383 ns	33,482 ns
D × T	2	8.36 *	0.75 ns	2.40 **	42.04 *	1.39 ns	5.29 ns	1,801,293 ns	3,447,407 *	358,589 **
S × T	1	0.30 ns	1.78 ns	7.65 ns	0.65 ns	2.24 ns	0.97 ns	56,394 ns	11,259 ns	28,413 ns
D × S × T	2	1.36 ns	1.03 ns	5.21 ns	6.77 ns	6.52 ns	2.90 ns	66,096 ns	106,452 ns	91,367 ns
Pooled error	24	1.80	1.13	5.21	10.98	13.75	2.10	566,481	792,286	40,728
Total	35									

ns, *, ** = non-significant and significant at $p < 0.05$ and $p < 0.01$ probability levels, respectively, durations (7, 14 and 21 days without added water), treatments (well-watered and water stress) and seasons (2013 and 2014).

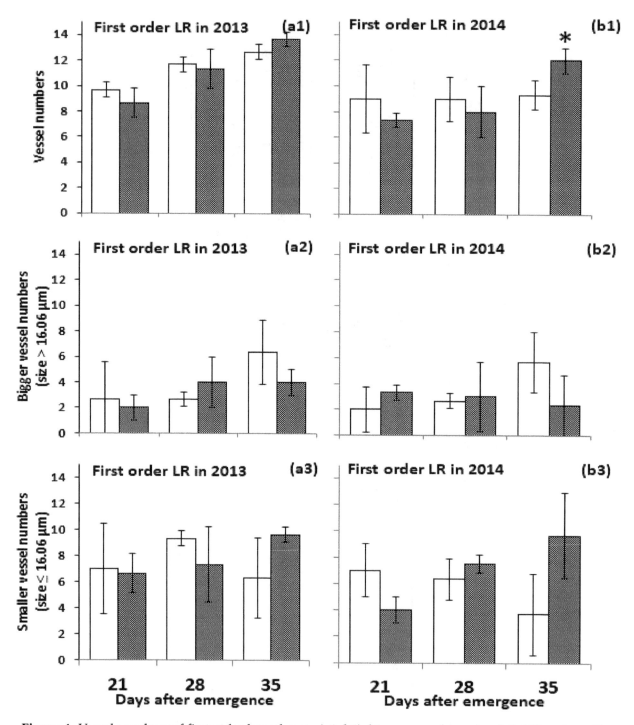

Figure 4. Vessel numbers of first order lateral roots (**a1, b1**), bigger vessel number (**a2, b2**) and smaller vessel number (**a3, b3**) of peanut at 21, 28 and 35 DAE in 2013 (**a**) and in 2014 (**b**); Significant at * $p \leq 0.05$, non-stress treatments (□) and stress treatments (■).

Vessel diameters in first order lateral roots (Figure 5) under non-stress and drought stress treatments varied between 4.03 to 41.09 μm (data not shown, unpublished data). Yet, the total vessel area in smaller vessels increased in both 2013 and 2014 and the total vessel area in the large vessels decreased in 2013 and slightly reduced in 2014 when the stress treatments, were compared to the well-watered control (Figure 6) in both 2013 and 2014. Stress and non- stress treatments were not significantly different for vessel diameter at all durations of drought stress. However, the average vessel diameter of long duration stress at 35 DAE and 21 days after irrigation withholding in each season tended to reduce. Figure 5 showed that the diameter of bigger xylem vessels in each season and the diameter of smaller xylem vessels were not significantly different except for the diameter of smaller xylem vessels at 35 DAE in 2014. The diameters of smaller xylem vessels were smaller in size under long duration stress at 35 DAE and 21 days after irrigation withholding compared to under well-watered treatment in 2014.

A significant reduction was observed in the diameter of the smaller xylem vessels and the diameter of the bigger xylem vessels tended to reduce, ultimately reducing total xylem area per root cross section.

The area of total xylem vessel elements in roots grown under stress conditions was significantly lower than those grown under non-stress conditions and these differences in area increased as the length of stress increased. Non-stress and stress treatments were significantly different for the area of total xylem vessels and the area of bigger vessels at 35 DAE. Stress treatment reduced the area of total vessels in 2013 and to a smaller exert the area tended to reduce in 2014.

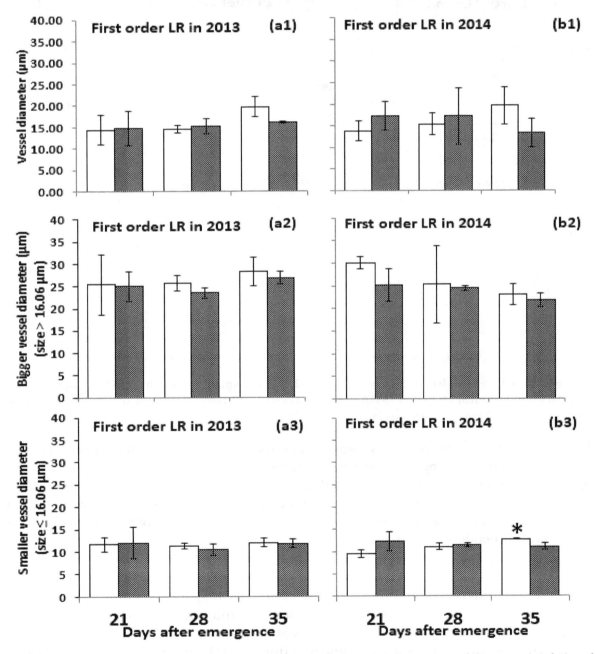

Figure 5. Average vessel diameter of first order lateral roots (**a1, b1**), bigger vessel diameter (**a2, b2**) and smaller vessel diameter (**a3, b3**) of peanut at 21, 28 and 35 DAE under well-irrigate and drought stress in 2013 (**a**) and in 2014 (**b**); Significant at * $p \leq 0.05$, non-stress treatments (□) and stress treatments (■).

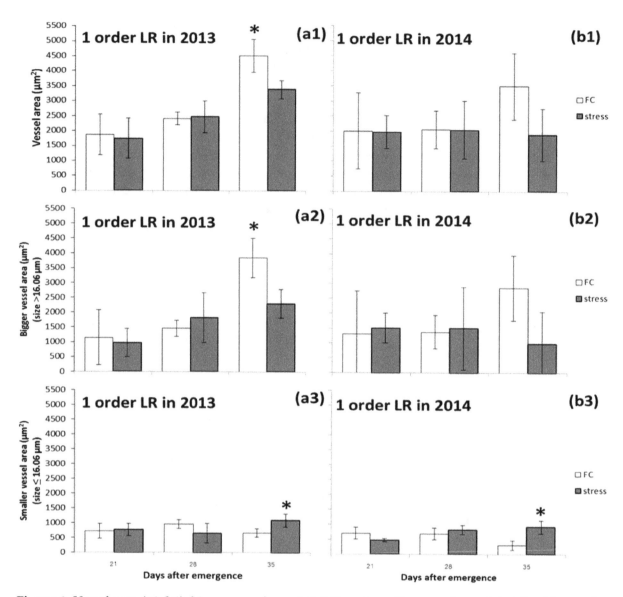

Figure 6. Vessel area (**a1, b1**), bigger vessel area (**a2, b2**) and smaller vessel area (**a3, b3**) of first order lateral roots of peanut at 21, 28 and 35 DAE in 2013 (**a**) and in 2014 (**b**); Significant at * $p \leq 0.05$, non-stress treatments (□) and stress treatments (■).

The cell-wall ingrowths in the first order lateral roots were detected in the cortical cells under both well-watered and drought stress treatments (Figure 3). The cell-wall ingrowths were localized at the opposite side of the intercellular spaces adjacent to the endodermis except in under drought at 28 DAE (Figure 3b2). The cell-wall ingrowths were found in two positions which were on the opposite side of the intercellular spaces and cell-cell conjunction. The 1–2 layers of this cell were found and indicated as the peri-endodermal layer.

3.3.2. Second Order Lateral Root

Combined analysis of variance for total vessel numbers, bigger vessel numbers, smaller vessel number, total vessel diameter (μm), bigger vessel diameter (μm), smaller vessel diameter (μm), total vessel area (μm²), bigger vessel area (μm²), smaller vessel area (μm²) of the second order lateral root in 2013 and 2014 are shown in Table 2. Differences in duration (D) and treatment (T) were significant ($p \leq 0.01$ and $p \leq 0.05$) for most traits. Season (S) was significant for total vessel numbers and bigger vessel numbers. The interactions between duration × treatment (D × T) and duration × season (D × S) were also significant for some traits.

The structure of second order lateral roots differed from that of the first order lateral roots. First order lateral roots are thicker, and the stele and vascular bundle tissues are more extensive than in the second order lateral roots. Second order lateral roots had an almost diarch and triarch organization of vascular bundles (Figures 7 and 8). Average value of vessel diameter was 14.21 μm (data not shown, unpublished data).

Drought and well-irrigated treatments at all durations were not significantly different for number of total xylem per cross-section of second order lateral roots (Figure 9) in 2013 and 2014. Drought and well-watered treatments were also not significantly different in the number of bigger vessels but the number of bigger vessels tended to reduce at 35 DAE, whereas the number of smaller vessels increased at 35 DAE (21 days after water withholding began).

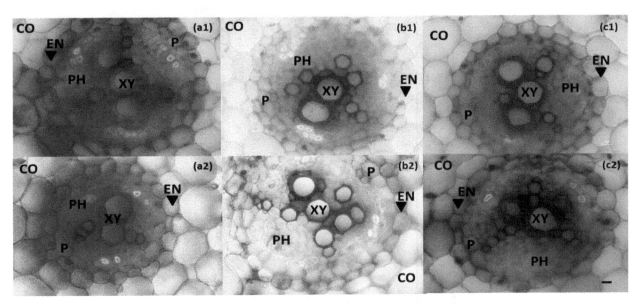

Figure 7. Freehand cross sections of second order lateral roots of peanut under well-irrigated conditions (**a1**, **b1** and **c1**) and drought stress conditions (**a2**, **b2** and **c2**) at 21, 28 and 35 DAE. CO, cortex; EN, endodermis; P, pericycle; PH, phloem; XY, xylem; Scale bar = 10 μm; 40×.

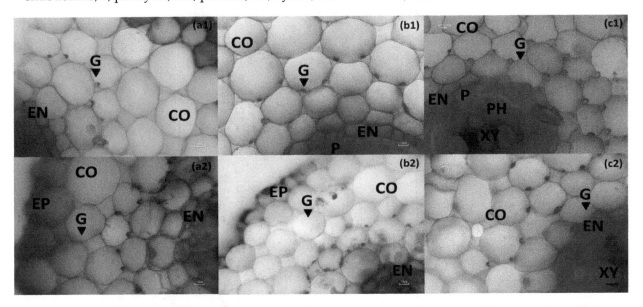

Figure 8. Freehand cross sections of second-order lateral roots under well-irrigated conditions (**a1**, **b1** and **c1**) and drought stress conditions (**a2**, **b2** and **c2**) at 21, 28 and 35 DAE. CO, cortex; EN, endodermis; G, phi-thickening or cell wall ingrowth; P, pericycle; PH, phloem; XY, xylem; Scale bar = 10 μm; 40×.

Table 2. Mean square from the combined analysis of variance for total vessel numbers, bigger vessel numbers, smaller vessel number, total vessel diameter (μm), bigger vessel diameter (μm), smaller vessel diameter (μm), total vessel area (μm²), bigger vessel area (μm²), smaller vessel area (μm²) of the second order lateral root in 2013 and 2014.

Source	DF	Total Vessel Numbers	Bigger Vessel Numbers	Smaller Vessel Numbers	Total Vessel Diameter (μm)	Bigger Vessel Diameter (μm)	Smaller Vessel Diameter (μm)	Total Vessel Area (μm²)	Bigger Vessel Area (μm²)	Smaller Vessel Area (μm²)
Duration (D)	2	11.44 **	6.19 *	3.03 ns	26.52 **	4.63 ns	9.02 **	2,437,286 **	1,689,456 **	59,535 *
Season (S)	1	18.78 **	1.11 **	1.00 ns	1.41 ns	13.96 ns	0.23 ns	1,155,729 *	860,956 ns	24,033 ns
Treatment (T)	1	7.11 *	4.00 ns	31.78 **	34.54 **	35.64 *	0.19 ns	1,747,821 **	2,743,513 **	95,334 *
D × S	2	1.44 ns	4.36 *	2.08 ns	12.83 **	11.36 ns	2.66 *	1,174,800 **	1,167,283 **	1126 ns
D × T	2	3.11 ns	0.75 ns	5.86 *	3.55 *	31.43 *	0.00 ns	413,242 ns	669,743 ns	22,934 ns
S × T	1	1.00 ns	1.78 ns	0.11 ns	3.08 ns	0.23 ns	0.04 ns	158,148 ns	138,356 ns	1298 ns
D × S × T	2	1.00 ns	1.02 ns	0.36 ns	3.95 ns	3.39 ns	1.85 ns	285,065 ns	199,055 ns	15,256 ns
Pooled error	24	1.17	1.14	1.33	1.84	7.91	0.08	185,616	202,569	13,384
Total	35									

ns, *, ** = non-significant and significant at $p < 0.05$ and $p < 0.01$ probability levels, respectively, durations (7, 14 and 21 days without added water), treatments (well-watered and water stress) and seasons (2013 and 2014).

Means for the vessel diameter of second-order lateral roots (Figure 10) of all treatments varied between 4.29 to 38.48 μm (data not shown). Stress treatment significantly reduced the vessel diameter of second-order lateral roots at 35 DAE with drought imposition for 21 days in 2013 and slightly reduced the vessel diameter of second-order lateral roots at 35 DAE with drought imposition for 21 days in 2014. Stress treatment significantly reduced the diameter of bigger xylem vessels in 2014 at 35 DAE with drought imposition for 21 days and stress treatment also reduced the diameter of bigger xylem vessels in 2013, although the reduction was not significant. Stress treatment did not significantly affect the diameter of smaller xylem vessel diameter in 2013 and 2014.

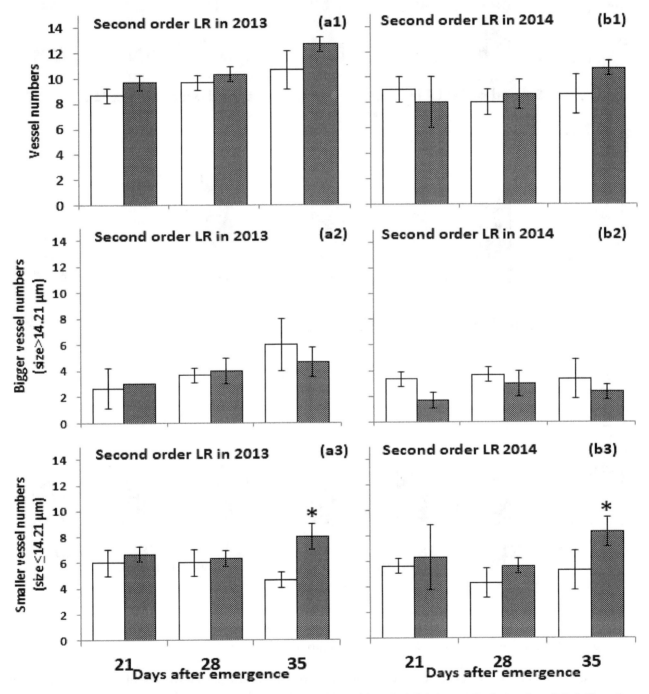

Figure 9. Vessel numbers of second order lateral roots (**a1, b1**), bigger vessel number (**a2, b2**) and smaller vessel number (**a3, b3**) of peanut at 21, 28 and 35 DAE in 2013 (**a**) and in 2014 (**b**); Significant at * $p \leq 0.05$, non-stress treatments (□) and stress treatments (■).

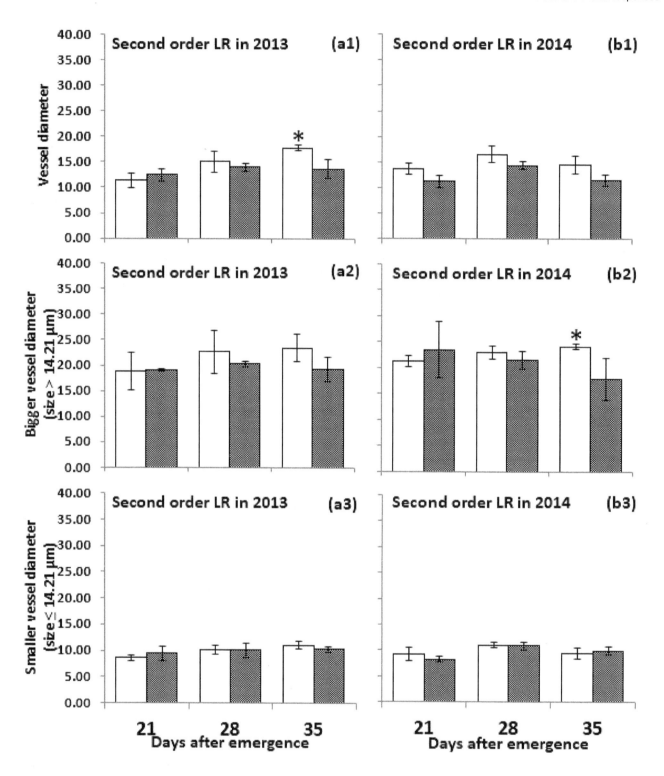

Figure 10. Vessel diameter of second order lateral roots (**a1, b1**), bigger vessel diameter (**a2, b2**) and smaller vessel diameter (**a3, b3**) of peanut at 21, 28 and 35 DAE in 2013 (**a**) and in 2014 (**b**); Significant at * $p \leq 0.05$, non-stress treatments (□) and stress treatments (■).

Because stress treatment reduced the diameters of the average xylem vessels and bigger xylem vessels, the area of vessels per cross section of each season and the area of bigger vessels in 2014 was reduced at 35 DAE, although the reduction was not significant and the area of bigger vessels area was significantly reduced at 35 DAE in 2013 (Figure 11). The area of smaller xylem vessels per cross section under stress treatment was increased.

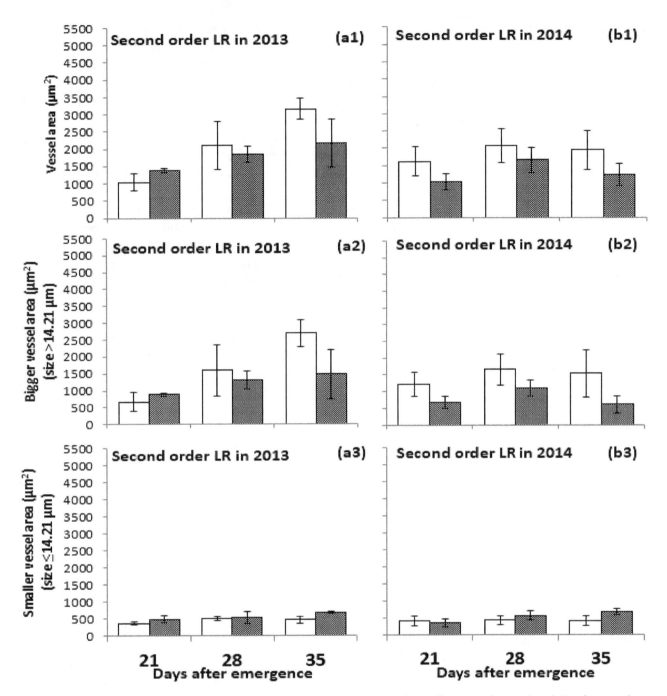

Figure 11. Vessel area (**a1, b1**), bigger vessel area (**a2, b2**) and smaller vessel area (**a3, b3**) of second order lateral roots of peanut at 21, 28 and 35 DAE in 2013 (**a**) and in 2014 (**b**); Significant at * $p \leq 0.05$, non-stress treatments (□) and stress treatments (■).

Cell-wall ingrowths appeared in the cortical cells of the second order lateral root under both conditions (Figure 8). The 1–2 layers of cell-wall ingrowths were found in the peri-endodermal layer.

4. Discussion

Weather conditions may be a key factor affecting the root anatomy of peanut. The experiment was conducted for two years. In the rainy season, air temperature and humidity were low, but in the summer to the early rainy season, air temperature and humidity were rather high. Soil moisture in the drought and well-watered treatments were clearly different in the upper soil layers. Soil moisture content for drought stress treatment at 28 and 35 DAE at the 10 cm of soil layer was less than 3.4%

(the permanent wilting point). However, soil moisture content for drought stress treatment at 65 cm and 85 cm of soil levels was higher than the permanent wilting point. The rate of water loss in 2013 was slower than in 2014, and soil moisture content at 21 days after irrigation withholding in 2013 was similar to those at 14 days after irrigation withholding in 2014.

The responses of plants to water stress depend on many things including timing and the intensity and duration of the drought. Root anatomy and root growth, like other plant parts, are sensitive to drought [20]. In this study, the long duration of the early season drought changed the root anatomical traits of peanut. Long periods of stress caused a significant increase in the number of xylem vessels in first and second order lateral roots but a significant decrease in the vessel diameter and the area of these first and second order lateral roots.

In both seasons, the mean xylem vessel diameters of first order lateral roots was higher than that of the second order lateral roots. The reduction in vessel diameter of first order lateral roots was higher than that of the second order lateral roots and these results may explain the differential root functions. The reduction in vessel diameter of first order lateral roots will better support the transport system's hydraulic conductivity according to Poiseuille's law [21]. In hot pepper, drought stress significantly reduced the diameters of xylem vessels in all of cultivars [8]. Vessel diameter is closely and positively correlated with volume of water flow and therefore it is correlated with the 'safety' of the conductive system [22,23]. The large vessel size under water deficit resulted in xylem cavitation [24]. The narrower diameter of metaxylem vessels maintain the water column, lowers the risk of cavitation, increases water flow resistance and saves water columns in narrower capillaries from damage [25]. Formation of narrower vessels occurring in drought-tolerant dicotyledons (including short-lived perennials and annuals with secondary structure) will likewise be advantageous when the plants are grown under drought [26,27].

Morphometric measurements on xylem vessels showed that the vessels of water-stressed plants had lower sectional areas. These results suggested that the reduction in vessel sectional area due to a diminished growth in response to water stress was the main factor affecting conductivity. Under a water deficit environment, roots develop to help extract soil moisture which being held at greater surface tension [28]. Deep root growth and large xylem diameter in deep roots may also increase the ability of roots to mine more water in deep soil when water in deep soil is abundant [29]. However, small and fine roots with greater specific root length enable plants to efficiently increase water uptake and maintain plant productivity under drought by increasing surface area and root length in contact with soil water, especially at deeper soil with available water [19,29].

The ability of plant to take up water is highly influenced by the number and size of the water conductive elements [25]. The change in number and size of the vessel xylem could help maintain water uptake under water stress [8].

In Ferna'ndez-Garcı'a, Lo'pez-Berenguer, and Olmos book chapter on the role of phi Cells under abiotic stress the authors noted that phi thickening is not the exception in the root anatomy [30]. They noted that the literature has described 16 different families, covering more than 100 species, which present the phi thickening in the roots. The phi thickening is classified into three types based on their root cell location: Type I, the most frequently found phi cell layer, is located in contact with the endodermis. Type II phi cell layer is located in contact with the epidermis and Type III phi cell layers are located in the inner cortical cells but not in contact with either the epidermis or the endodermis. In this study, cell-wall ingrowths were detected in the cortical cells of all first and second order lateral roots under well-watered and drought stress treatments. The 1–2 layers of these cells were localized at the opposite side of the intercellular spaces adjacent to the endodermis. The cell-wall ingrowths layers were indicated as the peri-endodermal layer and also called phi-thickening [31]. In previous studies, phi-thickening was induced under salt stress [30,31] and drought stress [11]. Phi-thickening of loquat roots grown under drought stress developed dramatically compared to normal conditions and the formation of phi-thickening was thought to be a defense mechanism against water stress. As the functions of these cells are difficult to determine precisely, phi thickening would play a role in controlling the water and solute rate of transportation through cell walls [32]. In peanut, cell wall ingrowth development in cortical cells might be a drought resistance mechanism for peanut roots as well. In this study, the 1–2 layers of cell-wall ingrowths were detected in both well-watered and drought stress treatments which were not significantly different for number of cell-wall ingrowths layers. However, the cells could be seen at higher magnification and using an electron microscope.

5. Conclusions

Under early season drought stress, root anatomy changes were more pronounced in the longer drought period treatments. At 21 days after imposing water stress, the drought treatment and irrigated treatment were clearly different in diameter, number and area of xylem vessels of first- and second-order lateral roots. Plants under drought conditions had smaller diameter and area of xylem vessels than did the plants under irrigated control. The ability of plant to change root anatomy likely improves water uptake and transport, and this may be an important mechanism for drought avoidance.

Author Contributions: Conceptualization, N.T., S.J., T.P. and N.V.; methodology, N.T., S.J. and N.V.; validation, N.T., S.J. and N.V.; formal analysis, N.T.; investigation, N.T.; resources, S.J.; data curation, N.T.; writing—original draft preparation, N.T.; writing—review and editing, T.K., C.K.K.; supervision, S.J.; funding acquisition, S.J.

Acknowledgments: This study was funded by the Royal Golden Jubilee Ph.D. Program (6.A.KK/ 53/ E.1). Assistance was also received from Peanut and Jerusalem artichoke Improvement Project for the Functional Food Research Group, Plant Breeding Research Center for Sustainable Agriculture and the Thailand Research Fund for providing financial support through the Senior Research Scholar Project of Sanun Jogloy (Project no. RTA6180002). Thailand Research Fund (TRF) (IRG 578003), Khon Kaen University (KKU) and Faculty of Agriculture, KKU are acknowledged for providing financial support for training on manuscript preparation. The manuscript was critical reviewed by Ian Charles Dodd.

References

1. Songsri, P.; Jogloy, S.; Vorasoot, N.; Akkasaeng, C.; Patanothai, A.; Holbrook, C.C. Root distribution of drought-resistant peanut genotypes in response to drought. *J. Agron. Crop Sci.* **2008**, *194*, 92–103. [CrossRef]

2. Girdthai, T.; Jogloy, S.; Vorasoot, N.; Akkasaeng, C.; Wongkaew, S.; Holbrook, C.C.; Patanothai, A. Associations between physiological traits for drought tolerance and aflatoxin contamination in peanut genotypes under terminal drought. *Plant Breed.* **2010**, *129*, 693–699. [CrossRef]

3. Jongrungklang, N.; Toomsan, B.; Vorasoot, N.; Jogloy, S.; Boote, K.; Hoogenboom, G.; Patanothai, A. Rooting traits of peanut genotype with different yield response to pre-flowering drought stress. *Field Crops Res.* **2011**, *120*, 262–270. [CrossRef]

4. Russell, R.S. *Plant Root System: Their Function and Interaction with the Soil*; McGRAW-HILL Book Company (UK) Limited: Oxford, UK, 1982.

5. Jongrungklang, N.; Toomsan, B.; Vorasoot, N.; Jogloy, S.; Boote, K.; Hoogenboom, G.; Patanothai, A. Classification of root distribution patterns and their contributions to yield in peanut genotypes under mid-season drought stress. *Field Crops Res.* **2012**, *127*, 181–190. [CrossRef]

6. Koolachart, R.; Jogloy, S.; Vorasoot, N.; Wongkaew, S.; Holbrook, C.; Jongrungklang, N.; Kesmala, T.; Patanothai, A. Rooting traits of peanut genotypes with different yield responses to terminal drought. *Field Crops Res.* **2013**, *149*, 366–378. [CrossRef]

7. Rucker, K.S.; Kvien, C.K.; Holbrook, C.C.; Hook, J.E. Identification of peanut genotypes with improved drought avoidance traits. *Peanut Sci.* **1995**, *22*, 14–18. [CrossRef]

8. Kulkarni, M.; Phalke, S. Evaluating variability of root size system and its constitutive traits in hot pepper (*Capsicum annum* L.) under water stress. *Scr. Hortic.* **2009**, *120*, 159–166. [CrossRef]

9. Kulkarni, M.; Deshpande, U. Comparative studies in stem anatomy and morphology in relation to drought tolerance in tomato (*Lycopersicon esculentum*). *Am. J. Plant Physiol.* **2006** *1*, 82–88. [CrossRef]

10. Kulkarni, M.; Borse, T.; Chaphalkar, S. Anatomical variability in grape (*Vitis venifera*) genotypes in relation to water use efficiency (WUE). *Am. J. Plant Physiol.* **2007**, *2*, 36–43. [CrossRef]

11. Pan, C.X.; Nakao, Y.; Nii, N. Anatomical development of Phi thickening and the Casparian strip in loquat roots. *J. Jpn. Soc. Hortic. Sci.* **2006**, *75*, 445–449. [CrossRef]

12. Peterson, C.A.; Emanuel, M.E.; Weerdenburg, C.A. The permeability of phi thickenings in apple (*Pyrus malus*) and geranium (*Pelargonium hortorum*) roots to an apoplastic fluorescent dye tracer. *Can. J. Bot.* **1981**, *59*, 1107–1110. [CrossRef]

13. Henrique, P.D.; Alves, J.D.; Goulart, P.D.P.; Deuner, S.; Silveira, N.M.; Zanandrea, I.; de Castro, E.M. Physiological and anatomical characteristics of sibipiruna plants under hypoxia. *Ciencia Rural* **2010**, *40*, 70–76. [CrossRef]

14. Tajima, R.; Abe, J.; Lee, O.N.; Morita, S.; Lux, A. Developmental changes in peanut root structure during root growth and root-structure modification by nodulation. *Ann. Bot.* **2008**, *101*, 491–499. [CrossRef] [PubMed]

15. Jongrungklang, N.; Toomsan, B.; Vorasoot, N.; Jogloy, S.; Boote, K.; Hoogenboom, G.; Patanothai, A. Drought tolerance mechanisms for yield responses to pre-flowering drought stress of peanut genotypes with different drought tolerant levels. *Field Crops Res.* **2013**, *144*, 34–42. [CrossRef]

16. Kano-Nakata, M.; Inukai, Y.; Wade, L.J.; Siopongco, J.D.; Yamauchi, A. Root development, water uptake, and shoot dry matter production under water deficit conditions in two CSSLs of rice: Functional roles of root plasticity. *Plant Prod. Sci.* **2011**, *14*, 307–317. [CrossRef]

17. Doorenbos, J.; Pruitt, W.O. Calculation of crop water requirement. In *Crop Water Requirements*; FAO of The United Nation: Rome, Italy, 1992; pp. 1–65.

18. Meisner, C.A.; Karnok, K.J. Peanut root response to drought stress. *Agron. J.* **1992**, *84*, 159–165. [CrossRef]

19. Thangthong, N.; Jogloy, S.; Pensuk, V.; Kesmala, T.; Vorasoot, N. Distribution patterns of peanut roots under different durations of early season drought stress. *Field Crops Res.* **2016**, *198*, 40–49. [CrossRef]

20. Boyer, J.S. Leaf enlargement and metabolic rates in corn, soybean, and sunflower at various leaf water potentials. *Plant Physiol.* **1970**, *46*, 233–235. [CrossRef] [PubMed]

21. Steudle, E.; Carol, A.P. How does water get through roots. *J. Exp. Bot.* **1998**, *49*, 775–788. [CrossRef]

22. Carlquist, S. Further concepts in ecological wood anatomy, with comments on recent work in wood anatomy and evolution. *Aliso* **1980**, *9*, 459–553. [CrossRef]

23. Salleo, S.; Lo Gullo, M.A. Xylem cavitation in nodes and internodes of whole *Chorisia insignis* H.B. et K. plants subjected to water stress: Relations between xylem conduit size and cavitation. *Ann. Bot.* **1986**, *58*, 431–441. [CrossRef]

24. Willson, J.C.; Jackson, R.B. Xylem cavitation caused by drought and freezing stress in four co-occurring *Juniperus* species. *Physiol. Plant.* **2006**, *127*, 374–382. [CrossRef]

25. Vasellati, V.; Oesterheld, M.; Medan, D.; Loreti, J. Effects of flooding and drought on anatomy of *Paspalum dialatatum*. *Ann. Bot.* **2001**, *88*, 355–360. [CrossRef]

26. Carlquist, S. Wood anatomy of Gentianaceae, tribe Helieae, in relation to ecology, habit, systematics, and sample diameter. *Bull. Torrey Bot. Club.* **1985**, *112*, 59–69. [CrossRef]

27. Arnold, D.H.; Mauseth, J.D. Effects of environmental factors on development of wood. *Am. J. Bot.* **1999**, *86*, 367–371. [CrossRef] [PubMed]

28. Comas, L.H.; Mueller, K.E.; Taylor, L.L.; Midford, P.E.; Callahan, H.S.; Beerling, D.J. Evolutionary patterns and biogeochemical significance of angiosperm root traits. *Int. J. Plant Sci.* **2012**, *173*, 584–595. [CrossRef]

29. Comas, L.; Becker, S.; Cruz, V.; Byrne, P.; Dierig, D. Root traits contributing to plant productivity under drought. *Front. Plant Sci.* **2013**, *4*, 442. [CrossRef] [PubMed]

30. Fernandez-Garcia, N.; Lopez-Perez, L.; Hernandez, M.; Olmos, E. Role of phi cells and the endodermis under salt stress in *Brassica oleracea*. *New Phytol.* **2009**, *181*, 347–360. [CrossRef] [PubMed]

31. López-Pérez, L.; Fernández-García, N.; Olmos, E.; Carvajal, M. The Phi thickening in roots of broccoli plants: An acclimation mechanism to salinity. *Int. J. Plant Sci.* **2007**, *168*, 1141–1149. [CrossRef]

32. Mackenzie, K. The development of the endodermis and phi layer of apple roots. *Protoplasma* **1976**, *100*, 21–32. [CrossRef]

Permissions

All chapters in this book were first published in MDPI; hereby published with permission under the Creative Commons Attribution License or equivalent. Every chapter published in this book has been scrutinized by our experts. Their significance has been extensively debated. The topics covered herein carry significant findings which will fuel the growth of the discipline. They may even be implemented as practical applications or may be referred to as a beginning point for another development.

The contributors of this book come from diverse backgrounds, making this book a truly international effort. This book will bring forth new frontiers with its revolutionizing research information and detailed analysis of the nascent developments around the world.

We would like to thank all the contributing authors for lending their expertise to make the book truly unique. They have played a crucial role in the development of this book. Without their invaluable contributions this book wouldn't have been possible. They have made vital efforts to compile up to date information on the varied aspects of this subject to make this book a valuable addition to the collection of many professionals and students.

This book was conceptualized with the vision of imparting up-to-date information and advanced data in this field. To ensure the same, a matchless editorial board was set up. Every individual on the board went through rigorous rounds of assessment to prove their worth. After which they invested a large part of their time researching and compiling the most relevant data for our readers.

The editorial board has been involved in producing this book since its inception. They have spent rigorous hours researching and exploring the diverse topics which have resulted in the successful publishing of this book. They have passed on their knowledge of decades through this book. To expedite this challenging task, the publisher supported the team at every step. A small team of assistant editors was also appointed to further simplify the editing procedure and attain best results for the readers.

Apart from the editorial board, the designing team has also invested a significant amount of their time in understanding the subject and creating the most relevant covers. They scrutinized every image to scout for the most suitable representation of the subject and create an appropriate cover for the book.

The publishing team has been an ardent support to the editorial, designing and production team. Their endless efforts to recruit the best for this project, has resulted in the accomplishment of this book. They are a veteran in the field of academics and their pool of knowledge is as vast as their experience in printing. Their expertise and guidance has proved useful at every step. Their uncompromising quality standards have made this book an exceptional effort. Their encouragement from time to time has been an inspiration for everyone.

The publisher and the editorial board hope that this book will prove to be a valuable piece of knowledge for researchers, students, practitioners and scholars across the globe.

List of Contributors

Weixing Li, Siyu Pang, Zhaogeng Lu and Biao Jin
College of Horticulture and Plant Protection, Yangzhou University, Yangzhou 225009, China

Elham Mehri Eshkiki
Department of Agricultural Biotechnology, Payame Noor University (PNU), Tehran, Iran

Zahra Hajiahmadi and Amin Abedi
Department of Biotechnology, Faculty of Agricultural Sciences, University of Guilan, Rasht, Iran

Mojtaba Kordrostami
Nuclear Agriculture Research School, Nuclear Science and Technology Research Institute (NSTRI), Karaj, Iran

Cédric Jacquard
Resistance Induction and Bioprotection of Plants Unit (RIBP) — EA4707, SFR Condorcet FR CNRS 3417, University of Reims Champagne-Ardenne, Moulin de la Housse, CEDEX 2, BP 1039, 51687 Reims, France

Leandro Pereira-Dias, Daniel Gil-Villar, Adrián Rodríguez-Burruezo and Ana Fita
Instituto de Conservación y Mejora de la Agrodiversidad Valenciana (COMAV), Universitat Politècnica de València, 46022 Valencia, Spain

Vincente Castell-Zeising
Departamento de Producción Vegetal, Universitat Politècnica de València, 46022 Valencia, Spain

Ana Quiñones and Ángeles Calatayud
Instituto Valenciano de Investigaciones Agrarias (IVIA), 46113 Moncada, Valencia, Spain

Ricardo Gil-Ortiz
Institute for Plant Molecular and Cell Biology (UPV-CSIC), Universitat Politècnica de València, 46022 Valencia, Spain

Miguel Ángel Naranjo and Marcos Caballero-Molada
Institute for Plant Molecular and Cell Biology (UPV-CSIC), Universitat Politècnica de València, 46022 Valencia, Spain
Fertinagro Biotech S.L., Polígono de la Paz, C/ Berlín s/n, 44195 Teruel, Spain

Antonio Ruiz-Navarro and Carlos García
Centre for Soil and Applied Biology Science of Segura (CEBAS-CSIC), Espinardo University Campus, 30100 Murcia, Spain

Sergio Atares
Fertinagro Biotech S.L., Polígono de la Paz, C/ Berlín s/n, 44195 Teruel, Spain

Oscar Vicente
Institute for the Preservation and Improvement of Valencian Agrodiversity (COMAV), Universitat Politècnica de València, 46022 Valencia, Spain

Toi Ketehouli, Kue Foka Idrice Carther, Muhammad Noman, Fa-Wei Wang, Xiao-Wei Li and Hai-Yan Li
College of Life Sciences, Engineering Research Center of the Chinese Ministry of Education for Bioreactor and Pharmaceutical Development, Jilin Agricultural University, Changchun 130118, China

Emuejevoke D. Vwioko and Marcus E. Imoni
Department of Plant Biotechnology, Faculty of Life Sciences, University of Benin, Benin City, Nigeria

Mohamed A. El-Esawi
Botany Department, Faculty of Science, Tanta University, Tanta 31527, Egypt
Sainsbury Laboratory, University of Cambridge, Cambridge CB2 1LR, UK

Abdullah A. Al-Ghamdi, Hayssam M. Ali and Monerah A. Al-Dosary
Botany and Microbiology Department, College of Science, King Saud University, Riyadh 11451, Saudi Arabia

Mostafa M. El-Sheekh
Botany Department, Faculty of Science, Tanta University, Tanta 31527, Egypt

Emad A. Abdeldaym
Vegetable Crops Department, Faculty of Agriculture, Cairo University, Giza, Egypt

Ali Anwar
Institute of Vegetables and Flowers, Chinese Academy of Agricultural Sciences, Beijing 100081, China
Graduate School of International Agricultural Technology and Crop Biotechnology Institute/Green Bio Science & Technology, Seoul National University, Pyeongchang 25354, Korea

Jun Wang, Xianchang Yu, Chaoxing He and Yansu Li
Institute of Vegetables and Flowers, Chinese Academy of Agricultural Sciences, Beijing 100081, China

Yingying Li, Qiuqiu Zhang, Lina Ou, Dezhong Ji, Tao Liu, Rongmeng Lan, Xiangyang Li and Linhong Jin
State Key Laboratory Breeding Base of Green Pesticide and Agricultural Bioengineering, Key Laboratory of Green Pesticide and Agricultural Bioengineering, Ministry of Education, Guizhou University, Huaxi District, Guiyang 550025, China

Shuangjie Jia, Hongwei Li, Yanping Jiang, Yulou Tang, Guoqiang Zhao, Yinglei Zhang, Yongchao Wang, Jiameng Guo, Qinghua Yang and Ruixin Shao
The Collaborative Center Innovation of Henan Food Crops, National Key Laboratory of Wheat and Maize Crop Science, Henan Agricultural University, Zhengzhou 450046, China

Shenjiao Yang and Husen Qiu
Farmland Irrigation Research Institute, CAAS/ National Agro-ecological System Observation and Research Station of Shangqiu, Xinxiang 453002, China

Narges Moradtalab and Roghieh Hajiboland
Department of Plant Science, University of Tabriz, Tabriz 51666-16471, Iran

Nasser Aliasgharzad
Department of Soil Science, University of Tabriz, Tabriz 51666-16471, Iran

Tobias E. Hartmann and Günter Neumann
Institute of Crop Science, University of Hohenheim, 70593 Stuttgart, Germany

Marta Muñoz
Instituto Agroforestal Mediterráneo (IAM), Universitat Politècnica de València, Camino de Vera s/n, 46022 Valencia, Spain
SEIPASA S.A. C/Ciudad Darío, Polígono Industrial La Creu naves 1-3-5, L'Alcudia, 46250 Valencia, Spain

Natalia Torres-Pagán and Mercedes Verdeguer
Instituto Agroforestal Mediterráneo (IAM), Universitat Politècnica de València, Camino de Vera s/n, 46022 Valencia, Spain

Rosa Peiró
Centro de Conservación y Mejora de la Agrodiversidad Valenciana (COMAV), Universitat Politècnica de València, Camino de Vera s/n, 46022 Valencia, Spain

Rubén Guijarro
SEIPASA S.A. C/Ciudad Darío, Polígono Industrial La Creu naves 1-3-5, L'Alcudia, 46250 Valencia, Spain

Adela M. Sánchez-Moreiras
Department of Plant Biology and Soil Science, Faculty of Biology, University of Vigo, Campus Lagoas-Marcosende s/n, 36310 Vigo, Spain

CITACA. Agri-Food Research and Transfer Cluster, Campus da Auga. University of Vigo, 32004 Ourense, Spain

Khaled A. Abdelaal and Yaser M. Hafez
Plant Pathology and Biotechnology Lab., Excellence Center (EPCRS), Faculty of Agriculture, Kafrelsheikh University, Kafrelsheikh 33516, Egypt

Lamiaa M. EL-Maghraby
Agricultural Biochemistry Department, Faculty of Agriculture, Zagazig University, Zagazig 44511, Egypt

Hosam Elansary
Plant Production Department, College of Food and Agricultural Sciences, King Saud University, Riyadh 11451, Saudi Arabia
Floriculture, Ornamental Horticulture and Garden Design Department, Faculty of Agriculture, Alexandria University, Alexandria 21526, Egypt

Eid I. Ibrahim
Rice Biotechnology Lab., Rice Research Dep., Field Crops Research Institute, Sakha, Kafr El-Sheikh 33717, ARC, Egypt

Mostafa El-Banna
Agricultural Botany Department, Faculty of Agriculture, Mansoura University, Mansoura 35516, Egypt

Mohamed El-Esawi
Botany Department, Faculty of Science, Tanta University, Tanta 31527, Egypt
Sainsbury Laboratory, University of Cambridge, Cambridge CB2 1LR, UK

Amr Elkelish
Botany Department, Faculty of Science, Suez Canal University, Ismailia 41522, Egypt

Nuengsap Thangthong, Sanun Jogloy, Thawan Kesmala and Nimitr Vorasoot
Department of Agronomy, Faculty of Agriculture, Khon Kaen University, Khon Kaen 40002, Thailand
Peanut and Jerusalem Artichoke Improvement for Functional Food Research Group, Khon Kaen University, Khon Kaen 40002, Thailand

Tasanai Punjansing
Department of Biology, Faculty of Science, Udonthani Rajaphat University, Udonthani 41000, Thailand

Craig K. Kvien
Crop & Soil Sciences, National Environmentally Sound Production Agriculture Laboratory (NESPAL), The University of Georgia, Tifton, GA 31793, USA

Index

Printed in the USA
CPSIA information can be obtained
at www.ICGtesting.com
JSHW051405091023
49903JS00006B/279